GIVEN TO CHR

BY RICK HENDRICKS

# Lecithins

Edited by

Bernard F. Szuhaj
Gary R. List

American Oil Chemists' Society

Mention of firm names or trade products does not imply endorsement or recommendation by the editors or contributors over other firms or similar products not mentioned.



**Library of Congress Cataloging in Publication Data**
Main entry under title:

Lecithins

(AOCS monograph ; 12)
Includes bibliographies and index.
1. Lecithin. I. Szuhaj, Bernard F. II. List, Gary R. III. Series.
TP453.L43L43 1985 661.8 85-7470

LCN

Printed in the United States of America.

# Preface

For the first time since the publication of Witcoff's monograph, *The Phosphatides,* information regarding the manufacture and use of lecithin has been brought together in one volume. This volume has attempted to cover the sources, manufacture, chemistry and use of the lecithin in a manner as thorough and technically correct as possible. The book begins by covering the structure, nomenclature and chemistry of lecithin; then proceeds through a discussion of the commercial sources including soybean, cottonseed, corn and sunflower and rapeseed. Chapters then follow on the manufacture of modified and natural lecithin, food uses and nonfood uses. Although the book is intended to cover primarily plant lecithin, a chapter on animal phosphatides is included. Several areas in which lecithin has assumed considerable importance are in baking, health, nutrition and flavor chemistry, and chapters on these topics are included. The editors and contributing authors feel confident that the material assembled here will be useful to those working in the field of fats, oils, food technology, health and nutrition.

# Foreword

Lecithin continues to play an important role in the manufacture of numerous foods, and it continues to find new outlets in new food uses.

It is estimated that worldwide production of lecithin amounts to more than 100,000 tons annually. Although the soybean yields most of the lecithin produced industrially, other oilseeds such as cottonseed, corn, rapeseed and sunflower represent new commercial sources of this valuable material.

Much of the information relating to manufacture and use of lecithin is dispersed widely throughout the open and patent literature, so there is a definite need for a monograph detailing the sources, manufacture and uses of lecithin in both food and nonfood applications. This work was undertaken to fulfill the need for an up-to-date, comprehensive volume on the chemistry and uses of lecithin.

# Table of Contents

# Chapter One
## Occurrence, Structure, Composition and Nomenclature

**C. R. Scholfield**
Northern Regional Research Center
Agricultural Research Service
US Department of Agriculture
1815 North University Street
Peoria, IL 61604

### Introduction

In modern usage, lecithin generally refers to a complex, naturally occurring mixture of phosphatides obtained by water-washing crude vegetable oil and separating and drying the hydrated gums. In addition to the phosphatides, such products contain triglycerides and other substances that are removed in an emulsion with the gums. Soybean lecithin, the most common commercial product, has been reported to contain 25–35% triglycerides and smaller amounts of other nonphosphatide materials (1–4). In commercial practice, the term also applies to plastic and fluidized lecithin in unbleached, single-bleached and double-bleached forms (3–5) and to other refined and modified products such as the acetone-insoluble phosphatides remaining after removal of the triglycerides with acetone (2,3,6,7).

In older literature, and to some extent today, lecithin also refers to the choline ester of phosphatidic acids. In 1967, the International Union of Pure and Applied Chemistry (IUPAC) and the International Union of Biochemistry (IUB) Committee on Biochemical Nomenclature published a document for discussion on the Nomenclature of Lipids (8). It was followed in 1976 by revised Recommendations of the IUPAC–IUB Commission on Biochemical Nomenclature (9). These documents have been published in a number of journals, including the references cited. The document for discussion (8) suggests "the term 'lecithin' is permitted but not recommended to designate a 1,2-diacyl-*sn*-glycero-3-phosphorylcholine. The recommended generic term for such compounds is 3-*sn*-phosphatidylcholine." Phosphatidylcholine and analogous terms such as phosphatidylethanolamine and phosphatidylinositol are commonly used in current literature. The recommendations (9) also state a preference for the term phospholipid

(PL), although phosphatide is still often used in oil and fat literature.

The commission found that its most complex problem concerned the distinguishing of stereoisomers of glycerol derivatives, for which at least four different systems had been proposed. It suggests a stereospecific numbering system, in which for glycerol "the carbon atom that appears on top in that Fischer projection that shows a vertical carbon chain with the secondary hydroxyl group to the left is designated as C-1." To indicate such numbering, the prefix *sn* is used.

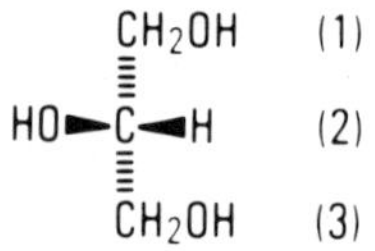


Klyashchitskii et al. (10) have proposed a stereospecific numbering system for substituted myoinositols in PL.

## Occurrence and Structure of Phospholipids in Lecithin

Phospholipids seem to be of universal occurrence in living organisms. As constituents of cell walls and active participants in metabolic processes they appear to be essential to life. Although they have been studied for more than 150 years, PL with their similar physical and chemical properties, their variety of fatty acids, their tendency toward hygroscopicity and oxidation have been extremely difficult to isolate and study in pure form. In addition to the reviews on lecithin already cited (1–7), useful reviews on structure and composition include those of Witcoff (11) and of Strickland (12). Phospholipids from innumerable animal tissues, microorganisms, and lower as well as higher plants, have been studied. In higher plants there are large differences between lipids from photosynthetic tissue (13) and those from seeds and other nonphotosynthetic tissues.

This chapter will consider principally seed PL from which the mixtures commonly considered as lecithin are obtained. Soybean oil is the largest source of commercial lecithin, but PL from other vegetable oils including corn, cottonseed, linseed, peanut, rapeseed, safflower and sunflower have been studied or used. Structures of some PL that occur in lecithins are shown in Figure 1.

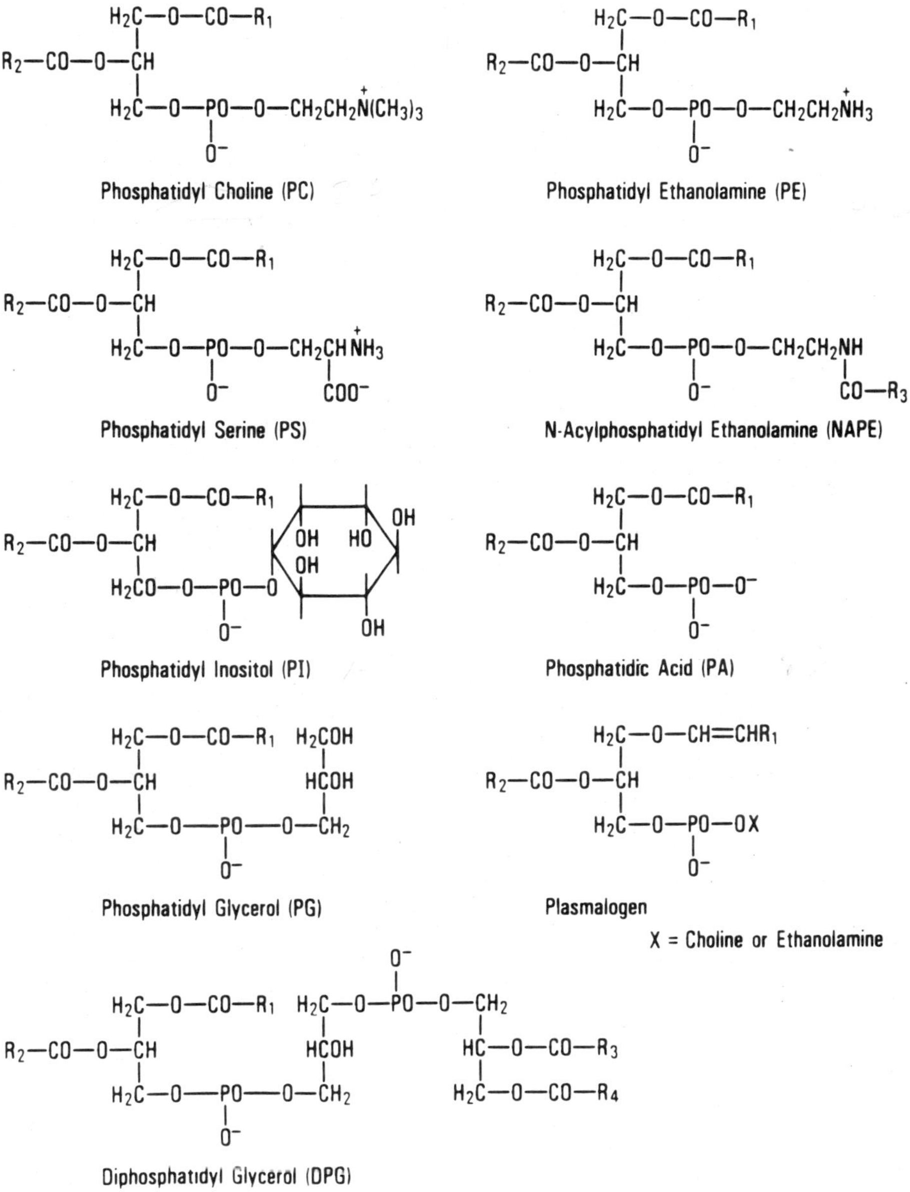


Fig 1-1. Some phospholipids reported in plant seed lecithins.

**Phosphatidylcholine and Phosphatidylethanolamine**

Phosphatidylcholine (PC) and phosphatidylethanolamine (PE) are the two best known PL. Their general structures were deduced from hydrolysis products by early workers (11). The structures shown in Figure 1 represent the -3-*sn* isomers. Hydrolysis of PC and PE results in both $\alpha$- and $\beta$-glycerophosphoric acid, and this fact was formerly considered evidence for both $\alpha$- and $\beta$-PL. However, Baer et al. have shown that hydrolysis of the synthetic compounds L-$\alpha$-glycerylphosphorylcholine (14), L-$\alpha$-glycerylphosphorylethanolamine (15), L-$\alpha$-PC (16) or L-$\alpha$-PE (15)—all corresponding to the 3-*sn* compounds in the stereospecific numbering nomenclature—yield mixtures of $\alpha$- and $\beta$-glycerophosphoric acids closely resembling those reported from natural PC and PE. In reviewing this evidence, Baer stated that the concept of the occurrence of $\beta$-PC and $\beta$-PE in nature is no longer tenable (17). Although $\beta$-PL have occasionally been mentioned in recent literature, there seems to be no instance of their isolation or evidence for their existence in natural materials.

Phosphatidylethanolamine formerly was known as cephalin and was considered to be insoluble in alcohol. It is now known that, in addition to PE, the alcohol-insoluble cephalin fraction contains phosphatidylserine (PS), inositol PL and other materials. Preparation of pure PE from the alcohol-insoluble fraction is difficult, and possibly the first pure PE was prepared by Rudy and Page (18) from the alcohol-soluble portion of human brain PL. A similar product was isolated by Folch (19), who proposed the name phosphatidylethanolamine. Soybean PE has been found in both the alcohol-soluble and alcohol-insoluble fractions (20–22), and a 30-transfer countercurrent distribution of the alcohol-soluble fraction between hexane and 90% methanol resulted in only partial separation of PE from the completely alcohol-soluble PC.

In PC, some neutralization occurs between the strong phosphoric acid group and the strong quaternary base. The zwitterion structure shown in Figure 1 was proposed by Jukes (23) on the basis of titrations of mixtures of PC and PE, and was confirmed by similar results of Fischgold and Chain (24). Although PE, like PC, can exist as a zwitterion, its weakly basic amino group allows it to exist also as a neutral molecule, as an anion or as a cation. PC is a monomer in polar solvents, but in nonpolar solvents association produces apparent molecular weights reported from 3,000 to 16,000 (25,26).

More recent work shows that the $^1$H–NMR spectrum of PE in $CDCL_3$ solution has a single line for ionic hydrogen, indicating rapid proton interchange between $PO_4H$ and $NH_3^+$ and showing that the two groups interact inter- or intramolecularly by forming hydrogen bonds in $CDCl_3$ solution

(27). The line moved downfield when PC was added. The interaction of PC with PE or with phosphatidic acid in soybean oil was also shown by changes in critical micelle concentration and by chemical shifts in $^{31}P$–NMR spectra (28).

**TABLE 1-I**
**Fatty Acid Composition of Individual Soybean Phospholipids, Percentage**

| Phospholipid | Reference | 16:0 | 18:0 | 18:1 | 18:2 | 18:3 | |
|---|---|---|---|---|---|---|---|
| PC | 29 | 15.8 | 6.3 | 13.0 | 62.9 | 2.0 | |
| PC | 30 | 20.5 | 5.5 | 10.5 | 58.8 | 4.6 | |
| PC | 31 | 14.9 | 3.2 | 13.7 | 62.7 | 5.2 | |
| PC | 32 | 16.3 | 4.0 | 7.5 | 64.8 | 6.0 | (16:1, 0.5%, other 0.9%) |
| 1-position PC | 32 | 34 | 8 | 8 | 43 | 5 | |
| 2-position PC | 32 | 1 | 2 | 5 | 85 | 7 | |
| 1-position PC | 33 | 34.4 | 7.6 | 30.2 | 23.8 | 4.0 | |
| 2-position PC | 33 | — | — | 15.8 | 71.8 | 13.1 | |
| PE | 30 | 31.6 | 3.2 | 8.7 | 53.2 | 3.2 | |
| PE | 32 | 15.8 | 4.4 | 8.4 | 64.7 | 5.9 | |
| 1-position PE | 32 | 29 | 5 | 5 | 56 | 6 | |
| 2-position PE | 32 | 1 | 3 | 10 | 75 | 6 | |
| NAPE O-acyl | 34 | 34.6 | 4.2 | 9.8 | 47.7 | 2.8 | (18:4, 0.9%) |
| NAPE N-acyl | 34 | 64.8 | 8.8 | 16.9 | 9.3 | 0.2 | |
| PA | 30 | 34.0 | 8.1 | 11.9 | 44.7 | 1.3 | |
| PI | 30 | 47.7 | 8.2 | 4.9 | 36.2 | 2.8 | |
| PI | 35 | 25.6 | 11.7 | 9.1 | 44.4 | 6.4 | (others, 2.7%) |
| 1-position PI | 35 | 71.9 | 17.3 | 4.6 | 6.7 | 0 | (others, 0.5%) |
| 2-position PI | 35 | 1.3 | 0.5 | 9.7 | 74.3 | 12.5 | (others, 1.4%) |

Fatty acid compositions for some samples of soybean PC and PE are listed in Table I. Privett and Blank (36) and Mangold (37) by ozonolysis of PC and thin layer chromatography of the resulting aldehyde "core" have shown four fatty acid groups in PC: $\alpha$-saturated-$\beta$-unsaturated, $\alpha$-unsaturated-$\beta$-unsaturated, $\alpha$-unsaturated-$\beta$-saturated, and a small amount of $\alpha$-$\beta$-saturated. Enzymatic hydrolysis of soybean PC by Debuch (33) and of soybean PC and PE by Harwood (32) shows that only very small amounts of saturated acids are in the 2-position. Harwood (32) also found classes of both PC and PE with from one to five double bonds per molecule, and Crawford et al. (38) isolated fractions from soybean PC ranging in unsaturation from dilinolenyl to stearyl-linolyl PC. Kimura et al. (39–42) isolated fractions containing arachidonic, eicosatrienoic and eicosadienoic acids from soybean PC and PE.

### Phosphatidylserine

Folch first reported phosphatidylserine (PS) in PL from ox brain (43) and determined its general structure (44). Baer and Maurukas (45) by comparison with a synthetic preparation established the structure as 3-*sn*-phosphatidyl-L-serine. It was reported in peanut PL by Hutt et al. (46) and in soybean and peanut by Van Handel (47), and has since been found in other seed PL.

### N-Acyl Phosphatidylethanolamine

N-acyl phosphatidylethanolamine (NAPE) was found in wheat flour PL and established as a monosubstituted fatty acid amide of PE by Bomstein (48). Dawson et al. (49) studied the structure of peaseed NAPE and detected NAPE in soybean lipids. Aneja et al. (34) isolated NAPE from commercial soybean lecithin and determined its structure to be N-acyl-O-(1,2-diacyl-*sn*-glycero-3-phosphoryl) ethanolamine. As shown in Table I, the N-acyl fatty acids are more saturated than the O-acyl. Wilson and Rinne (50) found NAPE to be more abundant in developing soybean seed and to decrease at maturity.

### Phosphatidic Acid

Phosphatidic acid (PA), the moiety common to all the glycerophospholipids, was first found in cabbage leaves by Chibnall and Channon (51). Hanahan and Chaikoff (52,53) have shown that in cabbage and carrots PA results from enzymatic action on the glycerophosphatides. In soybeans (50,54), PA has been found in greater amounts in developing than in mature seeds; in immature seeds it is increased by freezing and thawing prior to analysis (55) but decreased by extraction procedures that inactivate enzymes (56–58). Chapman and Robertson (59) and Nakayama et al. (58) find that PA also increases in PL from soybeans stored at high temperature and humidity. The amount of PA seems, therefore, to be dependent on enzymatic reactions during storage and processing of seed. In degumming soybean oil, much PA remains in the degummed oil as nonhydratable PL (28,60,61).

### Inositol Phospholipids

Inositol was first found in soybean PL by Klenk and Sakai (62), and early work was done by Woolley (63) and by Folch (64). By countercurrent distribution (CCD) between hexane and aqueous methanol, the alcohol-insoluble materials from soybean (20), corn (65) and linseed PL (66) were separated into two inositol-containing fractions. In soybeans, the fraction of lower partition coefficient—more soluble in methanol—was found to contain PE and an inositol-containing phosphatidic acid (21). This inositol compound corresponds to phosphatidylinositol (PI). The glycerophosphoinositide structure in material from commercial soybean lecithin was determined by Okuhara and Nakayama (67). Materials to which they assigned the same structure had been isolated from wheat germ (68) and ox heart lipids (69) by Faure and Morelec-Coulon. Ballou and Pizer (70) showed that inositol was present as L-myo-inositol-l-phosphate. Subsequent studies have established that PI together with PC and PE are among the most abundant PL in plant seed lecithins.

The fraction of higher partition coefficient—more soluble in hexane—from CCD of alcohol-insoluble soybean PL contained, in addition to inositol, the sugars galactose, mannose and arabinose (71). Carter et al. (72) found that CCD of a purified inositol lipid with a hexane/methanol system gave results essentially identical with those above, and Hörhammer et al. (73) by CCD of the alcohol-insoluble fraction of commercial soybean lecithin found lyso-PI in addition to PI and the sugar-containing phosphoinositide.

Van Handel (47,74) found evidence for a long-chain base in soybean PL, and Carter et al. (75,76) identified phytosphingosine (4D-hydroxysphinganine [8] or 1,3,4 trihydroxy-2-amino-octadecane) and dehydrophytosphingosine ([8E]-4D-hydroxysphingenine or 1,3,4-trihydroxy-2-amino-8-*trans* octadecene) in PL from a number of plant seeds. Ratios of saturated to unsaturated base were 15:85 in flax, 20:80 in soybean, 50:50 in peanut and 90:10 in corn.

After mild alkali saponification to destroy glycerol-containing lipids, Carter et al. (72) isolated a new complex lipid called phytoglycolipid (PGL) from soybean, corn, flaxseed, peanut, sunflower, cottonseed and wheat PL. From PGL, a tetrasaccharide was prepared containing phytosphingosine or its unsaturated analog, fatty acids, inositol, glucosamine, glucuronic acid and mannose, as shown in Figure 2. In the intact PGL, galactose, arabinose and fucose are attached through glucosamine to form penta-, hexa-, hepta- and octasaccharides (72,77–81). The possibility remained that the alkaline hydrolysis employed in preparation of PGL might have cleaved a more polar moiety (72). Purification by CCD was unsuccessful because impurities were bound as mixed chelated calcium and magnesium salts. However, after conversion to the sodium form (82), PGL was isolated by CCD without alkaline treatment from corn, soybean, flax and safflower PL, and thereby confirmed as the original material in the PL (83).

CCD of the sodium form PL also separated a second component, ceramide-phosphate-polysaccharide. It was devoid of hexosamine, but contained inositol, hexuronic acid, galactose, arabinose, fucose and mannose (78,83). Carter

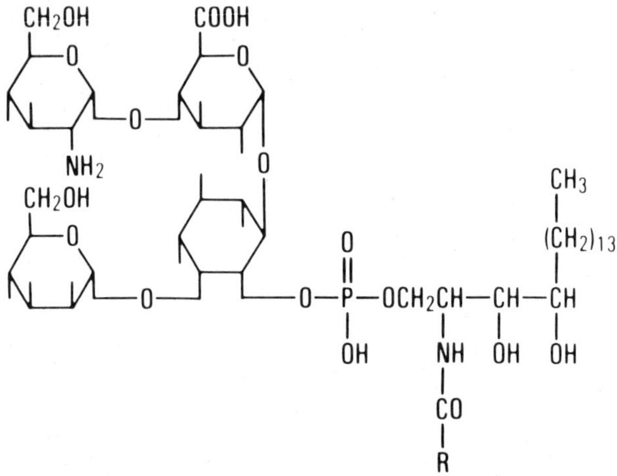

Fig. 2. Phytoglycolipid tetrasaccharide from phospholipids.

and Kisic (83) also found a material called lipophytin, present more in corn and soybean than in linseed or safflower, that appeared at the interface of their hexane/butanol/methanol/water CCD solvent system. It consisted of a complex mixture of inositol polyphosphates, fatty acids and amino acids probably bound as Ca and Mg salts.

The fatty acids of the isolated PGL from soybeans consisted of 5% cerebronic acid and 95% palmitic plus stearic acids. From corn PGL there was obtained 65% mixed $C_{24}$ and $C_{26}$ $\alpha$-hydroxy acids, 11% lignoceric acid and 24% mixed palmitic and stearic acids (72). From soybean sphingophospholipids, Sen Gupta (84) isolated 20% nonhydroxylated fatty acids and 80% $\alpha$-hydroxy acids. All were saturated. The nonhydroxylated acids ranged from 16:0 to 26:0, mainly 16, 18 and 22–25. The $\alpha$-hydroxy acids consisted of 53% nervonic acid with others ranging from 16 to 28 carbons.

**Other Glycerophospholipids**

Phosphatidylglycerol (PG) was found in PL of *Scenedesmus* and of higher plants by Benson and Maruo (85), and its structure has been established as 1,2 diacyl-*sn*-glycero-3-phosphoryl-1′-*sn*-glycerol (86–88). It is an important PL of chloroplasts and photosynthetic tissue. Although it has been reported in soybean (89,90) and sunflower PL (91), it is found in much smaller amounts in seed lecithins. *Bis*-phosphatidic acid, the fully acylated analog of PG, has been reported in developing soybeans by Stearns and

Morton (92), but was not found in the mature seed.

Diphosphatidylglycerol (DPG) was first isolated from beef heart by Panghorn (93) as cardiolipin, one of the essential components in a serological test for syphilis. Cardiolipin was shown to be identical with DPG by DeHaas et al. (94). It was found in plant materials by Benson and Strickland (95) and by Coulon-Morelec and Douce (96), and has been reported in soybean PL by Szuhaj et al. (97), Harwood (90) and Privett et al. (54).

Roughan et al. (56) found with developing soybean cotyledons that, unless enzymes are inactivated before extraction with methanol-containing solvents, variable amounts of phosphatidylmethanol (PM) are formed by phospholipase D-catalyzed transphosphatidylation of PC and PE. They point out the possibility that in chromatographic separations artefact PM may be mistaken for NAPE and *bis*-phosphatidic acid. In view of the variable degree of phospholipase D inhibition found in mature beans by Nakayama et al. (58), this possibility would appear to be present with mature beans also.

**Plasmalogens**

Plasmalogens or alk-l-enyl PL were first reported by Feulgen and Voit (98) as materials stained by fuchsin-sulfurous acid, a reagent for aldehydes, in histological studies. Their chemistry and that of the related alkyl ether lipids has been reviewed by Snyder (99). They are apparently more abundant in animal materials, and earlier reports of their occurrence in plant seeds were based only on color tests for aldehydes. Feulgen et al. (100) reported their presence in soybeans; Lovern (101) reported aldehydes equivalent to ca. 7% palmitaldehyde in soybean and peanut lecithin; and Wagenknecht (102) found them in green peas. Mangold (99, chap. XVI) pointed out the limited value of the plasmal color reaction for the alk-l-enyl ethers, since autoxidation products can interfere. However, Kaufmann et al. (103) have isolated 1-alk-1$^1$-enyl-2-acyl ethanolamine- and choline-phosphoglycerides from green peas. Mangold states that "In the opinion of the author there is now no doubt that alk-1-enylacylglycerophosphorylcholine and alk-1-enylacylglycerophosphorylethanolamine do indeed occur in plant tissues, but the presence of alkylglycerols or their derivatives has yet to be proved."

**Lysophospholipids**

Lyso- or monoacyl forms of all the common glycerophospholipids have been reported in plant lecithins. These may be formed by enzymatic hydrolysis of the diacyl forms. The lipolytic enzymes have been reviewed by

Brockerhoff and Jensen (104). They report that phospholipase 2 from snake venom or pancreas removes fatty acids at the 2-position. Phospholipase 1, which removes fatty acids at the 1-position, is less well known but has been prepared from animal tissues and bacteria. Fatty acids in the 2-position tend to migrate to the more stable 1-position (105). There seems to have been little study of the corresponding plant phospholipases.

## NonPhospholipid Materials in Lecithin

### Triglycerides

In the degumming process, various non-PL substances are removed in an emulsion with the PL. In commercial lecithin, triglycerides are the largest non-PL component (1–4). Additional oil or fatty acids are sometimes added to increase fluidity.

### Pigments

From soybean oil, xanthophylls are removed with the gums but carotene remains in the oil. In one sample, lutein made up 75% of the carotenoid pigments in the gums (106). Peroxide bleaching destroys the carotenoids and leaves a variable amount of brown color. The amount of brown-colored fluorescent substances is increased by heating (106–110).

### Carbohydrates

Triglycerides and some other non-PL materials can be removed from the PL by extraction with acetone. Free carbohydrates, as well as some sugars bound to lipid constituents, remain with the PL. One sample of oil-free soybean lecithin contained 7% free sugar made up of 45% sucrose, 9% raffinose and 47% stachyose (71).

### Sterols

Sterols and their derivatives are also present. Lepage (111) found esterified sterylglucoside in soybean lecithin. Kiribuchi et al. (112) reported free sterols 0.5%, sterylglucoside (SG) 2.1%, and esterified sterylglucoside (ESG) 2.6%, and Popov et al. (113) found free sterols 0.74% and combined SG and ESG 2.68%. Steryl esters were present in neglible amounts. Lepage (111) found in ESG 59% $\beta$-sitosterol, 19% stigmasterol and 22% sterol A. In the light of later work (112,114), the compound designated as sterol A may be campesterol. Kiribuchi et al. (112) found free sterols, SG and ESG all to contain ca. 50% β-sitosterol, 30% stigmasterol and 20% campesterol. Milkova et al. (114) found in the free sterols 33% sitosterol, 37% stigmasterol with other $C_{29}$ dienols, 5% campesterol, 5% dehydrocampesterol and 20% of a substance with molecular weight 428. In SG and ESG they found sitosterol 80%, campesterol 18% and stigmasterol 2%. As shown in Table II, all found fatty acids in ESG more saturated than typical soybean triglycerides.

**TABLE 1-II**
**Fatty Acid Percentage Composition of Esterified Steryl Glucosides**

| Reference | 16:0 | 16:1 | 18:0 | 18:1 | 18:2 | 18:3 |
|---|---|---|---|---|---|---|
| Lepage (111) | 33.7 | 0.9 | 7.0 | 8.8 | 47.4 | 2.2 |
| Kiribuchi et al. (112) | 42.7 | | 9.6 | 13.7 | 31.2 | 2.8 |
| Milkova et al. (114) | 52.2 | | 7.5 | 17.9 | 22.4 | |
| Singh and Privett (115) | 33.9 | | 6.9 | 9.1 | 43.2 | 7.6 |

## Galactosyl Glycerides

Monogalactosyldiglyceride (MGDG) and digalactosyldiglyceride (DGDG) were first reported in wheat flour lipids by Carter et al. (116,117). They also found DGDG in wheat germ oil and found both in corn PL. These compounds were shown to be 2,3-diacyl-1-β-D-galactopyranosyl-D-glycerol and 2,3-diacyl-1-(α-D-galactopyranosyl-1,6-D-galactopyranosyl)-D-glycerol as shown in Figure 3 (118). Although they are major components of photosynthetic tissue (119), they appear to be present in much smaller amounts in most seed lecithins. An exception is *Briza humilis,* whose seed lipids were reported to contain 49% DGDG and 29% MGDG (120,121). Singh and Privett (115) reported the fatty acids of soybean DGDG to be highly unsaturated with 46.6% linolenic acid.

## Cerebrosides

Cerebrosides in plant lipids were also first found by Carter et al. (117,122,123) in wheat flour. They contained one of four long-chain bases: phytosphingosine, dehydrophytosphingosine, dihydrosphingosine, and an incompletely identified isomer of sphingosine. The sugar was glucose and the major fatty acid was α-hydroxystearic acid. In soybeans Singh and Privett (115) found

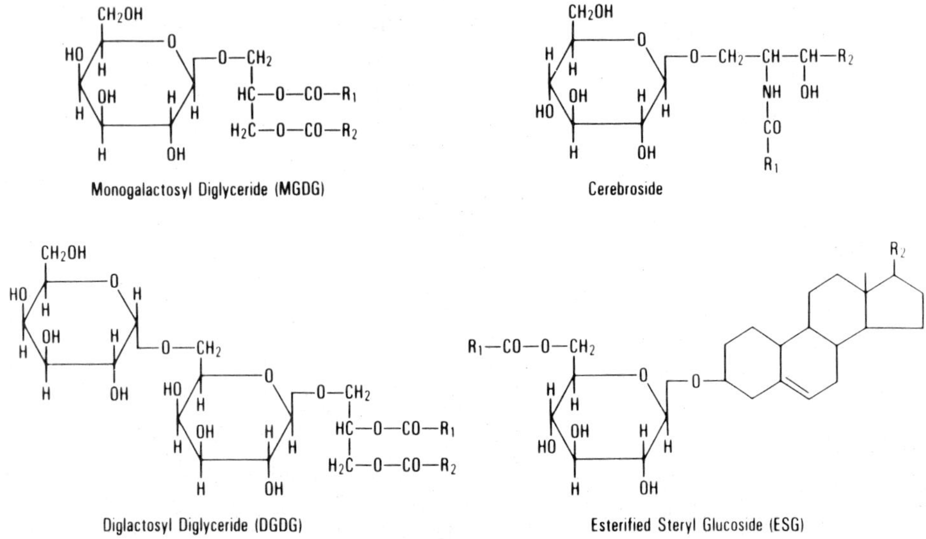


Fig. 1-3. Some nonphospholipid compounds reported in plant seed lecithins.

dehydrophytosphingosine and two unidentified long-chain bases apparently containing additional unsaturation.

## Composition of Lecithins

### Phospholipid Composition

Values for the composition of soybean lecithin are found in most of the review articles cited (1–5). Some other values are summarized in Table III. Major components of many lecithins are PC, PE and PI. Mukhamedova and Akramov (126) report them to be the main PL of seeds of 22 plant species. These PL are well separated by thin layer chromatography and in some publications are the only ones for which quantitative data are given.

In addition to data for commercial lecithin, data have been published for PL obtained by other methods including chloroform/methanol extraction. Some of these data have been included in Table III, although they are not strictly comparable because of the difference in isolation methods. Nielson (127) investigated the difficultly extractable material that was extracted from soybeans by 80% hexane/20% absolute alcohol after hexane extraction. He obtained 0.5% difficulty extractable material, a little less than the amount extracted by

**TABLE 1-III**
**Components of Soybean Lecithin, Percentage**

| Reference | PC | PE | PI | PA | PS |
|---|---|---|---|---|---|
| Wagner and Wolff[a] (124) | 12–15 | 8–9.5 | 7.0 | 0.9 | 6.3 |
| Negishi et al.[b] (125) | 29.5 | 22.3 | 13.0 | | 5.9 |
| Privett et al.[c] (54) | 45.0 | 26.3 | 14.1 | 5.0 | |
| Erdahl et al.[d] (89) | 29.0 | 23.5 | 15.1 | 7.0 | |
| Wilson and Rinne[e] (50) | 46.0 | 25.0 | 17.4 | 0 | |
| Harwood[f] (90) | 43.6 | 21.4 | 17.5 | 1.5 | |
| Chapman and Robertson[g] (59) | 35.2 | 26.3 | 13.5 | 0 | |
| Chapman[h] (30) | 39.0 | 23.3 | 20.3 | 4.8 | |
| Rydhag and Wilton[i] (31) | 41 | 34 | 19 | 9 | |
| Rydhag and Wilton[j] (31) | 33 | 32 | 21 | 14 | |

[a]Commercial soybean lecithin. Also lyso PC 0.50, lyso PE 0.40, lyso PI 0.75, PGL 6–7, unknown 4–5.
[b]Commercial soybean lecithin.
[c]$CHCL_3/CH_3OH$ extraction. Also PG + DPG 3.3, unknown 6.3.
[d]Commercial oil-free lecithin. Also 7.9% including NAPE, lyso PC, lyso PE and other unknown.
[e]$CH_2CL_3/CH_3OH:H_2O$ extraction. Also PG 3.6, DPG 3.4, NAPE 4.8.
[f]$CHCl_3/CH_3OH$ extraction. Also DPG 3.7, PGL 6.1, NAPE 1.1, unidentified 4.3.
[g]$CHCl_3/CH_3OH$ extraction. Composition of major PL.
[h]$CHCl_3/CH_3OH$ extraction. Also unknown 12.5.
[i]Acetone-precipitated soybean PL.
[j]Granulated soybean PL.

hexane. This difficultly extractable material, in addition to PC, PE, PS and probably PI, contained d-inositol as the methyl ether, pinitol, a high-molecular-weight phosphatidic acid, and a number of nitrogen-containing substances.

Szuhaj et al. (97), by silicic acid column and thin layer chromatography, found in soybean lecithin PC, PE, PI, triglycerides, diglycerides, monoglycerides, free fatty acids, free sterols, steryl esters, DPG, lyso-PC, ESG, SG, DGDG, cerebroside and two unidentified glycolipids. Erdahl et al. (89), in one of the most comprehensive quantitative studies, separated azolectin—a commercial oil-free lecithin—into 2.5% neutral lipid, 15.0% glycolipid and 82.5% PL fractions. The neutral lipids included 2.0% triglycerides, 0.12% fatty acids, 0.23% sterols and 0.15% unidentified including mono- and diglycerides, sterol esters, pigments and other unknowns. Glycolipids included 6.2% ESG, 3.7% SG + cerebroside, 1.7% DGDG and 3.4% unidentified components. The PL composition is included in Table III.

### Fatty Acid Composition

Fatty acid composition of the total seed PL from different plants resembles that of the triglycerides. Rao et al. (128) and Vijayalakshmi and Rao (129)

**TABLE 1-IV**
**Percentage Composition Fatty Acids of Soybean Lecithin**

| Reference | 14:0 | 14:1 | 16:0 | 16:1 | 18:0 | 18:1 | 18:2 | 18:3 | 20:0 | 20–22 Unsaturated |
|---|---|---|---|---|---|---|---|---|---|---|
| Hilditch and Zaky (130) | | | 11.7 | 8.6 | 4.0 | 9.8 | 55.0 | 4.0 | 1.4 | 5.5 |
| Rzhekhin et al.[a] (131) | | | 18.1 | | 3.7 | 22.4 | 40.0 | 5.0 | 2.3 | 6.2 |
| Vijayalakshmi and Rao[b] (129) | | | 42.7 | 7.0 | 11.7 | 17.0 | 20.0 | 1.6 | | |
| Daga (132) | 1.9 | tr | 26.7 | | 9.3 | 25.1 | 37.0 | | | |
| Daga (133) | 0.3 | 1.2 | 25.5 | | 10.3 | 39.4 | 17.1 | 6.2 | | |
| Rhyhag and Wilton[c] (31) | | | 21.5 | | 4.3 | 7.2 | 60.9 | 6.1 | | |
| Rhyhag and Wilton[d] (31) | | | 18.9 | | 4.1 | 6.8 | 60.8 | 9.2 | | |

[a]Also 2.3% unidentified.
[b]$CHCl_3/CH_3OH$ extraction.
[c]Acetone-precipitated.
[d]Granulated.

note the absence of unusual fatty acids in PL from a number of sources and suggest "PL appear to pick up predominantly acids of the $C_{16}$ and $C_{18}$ series in various proportions, despite the fact that other fatty acids are available in the fatty acid pool." An exception is the $\alpha$-hydroxy acids noted in amide linkage with long-chain bases.

Fatty acid compositions reported for a number of soybean PL are listed in Table IV. The values fall into two groups: one with total unsaturation like or slightly lower than typical soybean triglycerides, and one with an average of less than one double bond per fatty acid molecule. No reason is apparent, but comparison with fatty acid compositions of PC, PE and PI in Table I suggest the higher unsaturation is more common. The data show a higher palmitic acid content and generally a lower oleic acid content than triglycerides. Linoleic acid is generally slightly higher than would be expected for the same level of unsaturation in triglycerides. Comparison with PL from other sources suggests the slightly lower unsaturation and higher palmitic acid is common, but the lower oleic acid was noted only in soybeans.

## References


1. Iveson, H. T., Soybean Dig. 21:18 (1961).
2. Van Nieuwenhuyzen, W., JAOCS 53:425 (1976).
3. Wolf, W. J., and D. J. Sessa, in Encyclopedia of Food Science, edited by M. S. Peterson and A. H. Johnson, The Avi Publishing Company, Inc., Westport, CT, 1978, pp. 461–467.
4. Eichberg, J., in Kirk-Othmer Encyclopedia of Chemical Technology, 3rd edn., Vol. 14, John Wiley & Sons, New York, 1981, pp. 250–269.
5. Stanley, J., in Soybeans and Soybean Products, edited by K. S. Mackley, Vol. II, Interscience Publishers, Inc., New York, 1951, pp. 593–647.
6. Sullivan, D. R., and B. F. Szuhaj, JAOCS 52: Program Abstract 39, 66th annual AOCS meeting (February 1975).
7. Van Nieuwenhuyzen, W., Ibid. 58:886 (1981).
8. IUPAC–IUB Commission on Biochemical Nomenclature, Ibid. 44:548A (1967).
9. IUPAC–IUB Commission on Biochemical Nomenclature, Lipids 12:455 (1977).
10. Klyashchitskii, B. A., V. I. Shvets and N. A. Preobrazhenskii, Ibid. 3:393 (1969).
11. Witcoff, H., in The Phosphatides, Reinhold Publishers Corp., New York, 1951.
12. Strickland, K. P., in Form and Function of Phospholipids, edited by G.

B. Ansell, J. N. Hawthorne and R. M. C. Dawson, Elsevier Scientific Publishing Company, Amsterdam, 1973, pp. 9–42.
13. Kates, M., in Advances in Lipid Research, Vol. 8, edited by R. Paoletti and P. Kritchevsky, Academic Press, New York, 1970, pp. 225–265.
14. Baer, E., and M. Kates, J. Biol. Chem. 175:79 (1948).
15. Baer, E., H. C. Stancer and I. A. Korman, Ibid. 200:251 (1953).
16. Baer, E., and M. Kates, Ibid. 185:615 (1950).
17. Baer, E., JAOCS 42:257 (1965).
18. Rudy, H., and I. H. Page, Hoppe-Seyler's Z. Physiol. Chem. 193:251 (1930); Chem. Abstr. 25:718 (1931).
19. Folch, J., J. Biol. Chem. 146:35 (1942).
20. Scholfield, C. R., H. J. Dutton, F. W. Tanner Jr. and J. C. Cowan, JAOCS 25:368 (1948).
21. Scholfield, C. R., and H. J. Dutton, J. Biol. Chem. 208:461 (1954).
22. Scholfield, C. R., and H. J. Dutton, Ibid. 214:633 (1955).
23. Jukes, T. H., Ibid. 107:783 (1934).
24. Fischgold, H., and E. Chain., Proc. Royal Soc. (London) B117:239 (1935); Chem. Abstr. 29:5465 (1935).
25. Price, H. I., and W. C. McC. Lewis, Biochem. J. 23:1030 (1929); Chem. Abstr. 24:870 (1930).
26. Faure, M., and J. Legault-Demare, Bull. Soc. Chim. Biol. 32:509 (1950); Chem. Abstr. 45:679 (1951).
27. Seimiya, T., N. Ashida, M. Hayashi, T. Muramatsu and I. Hara, Chem. Phys. Lipids 21:69 (1978).
28. Kanamoto, R., Y. Wada, G. Miyajima and M. Kito, JAOCS 58:1050 (1981).
29. Thornton, M. H., C. S. Johnson and M. A. Ewan, Oil Soap 21:85 (1944).
30. Chapman, G. W., Jr., JAOCS 57:299 (1980).
31. Rydhag, L., and I. Wilton, Ibid. 58:830 (1981).
32. Harwood, J. L., Phytochemistry 15:1459 (1976).
33. Debuch, H., Hoppe-Seyler's Z. Physiol. Chem. 306:279 (1957).
34. Aneja, R., J. S. Chadha and J. A. Knaggs, Biochem. Biophys. Res. Commun. 36:401 (1969).
35. Summa, S., and J. B. Mudd, Plant Physiol. 45:712 (1970).
36. Privett, O. S., and M. L. Blank, JAOCS 40:70 (1963).
37. Mangold, H. K., Ibid. 38:708 (1961).
38. Crawford, C. G., R. D. Plattner, D. J. Sessa and J. J. Rackis, Lipids 15:91 (1980).
39. Kimura, S., M. Motoki and K. Shibasaki, Nippon Shokuhin Kogyo Gakkai-Shi 16:425 (1969); Chem. Abstr. 73:129716z (1970).

40. Kimura, S., and K. Shibasaki, Ibid., 16:369 (1969); Chem. Abstr. 73:129715y (1970).
41. Kimura, S., and K. Shibasaki, Ibid. 17:22 (1970); Chem. Abstr. 73:117186p (1970).
42. Kimura, S., Y. Sano and K. Shibasaki, Ibid. 17:123 (1970); Chem. Abstr. 73:117187q (1970).
43. Folch, J., J. Biol. Chem. 139:973 (1941).
44. Folch, J., Ibid. 174:439 (1948).
45. Baer, E., and J. Maurukas, Ibid. 212:25 (1955).
46. Hutt, H. H., T. Malkin, A. G. Poole and P. R. Watt, Nature 165:314 (1950).
47. Van Handel, E., The Chemistry of Phosphoaminolipids, D. B. Centen's Uitgever-Maatschappij N. V., Amsterdam, 1954.
48. Bomstein, R. A., Biochem. Biophys. Res. Commun. 21:49 (1965).
49. Dawson, R. M. C., N. Clarke and R. H. Quarles, Biochem. J. 114:265 (1969).
50. Wilson, R. F., and R. W. Rinne, Plant Physiol. 54:744 (1974).
51. Chibnall, A. C., and H. J. Channon, Biochem. J. 21:233 (1927).
52. Hanahan, D. J., and I. L. Chaikoff, J. Biol. Chem. 169:699 (1947).
53. Hanahan, D. J., and I. L. Chaikoff, Ibid. 172:191 (1948).
54. Privett, O. S., K. A. Dougherty, W. L. Erdahl and A. Stolyhwo, JAOCS 50:516 (1973).
55. Wilson, R. F., and R. W. Rinne, Plant Physiol. 57:270 (1976).
56. Roughan, P. G., C. R. Slack and R. Holland, Lipids 13:497 (1978).
57. Phillips, F. C., and O. S. Privett, Ibid. 14:949 (1979).
58. Nakayama, Y., K. Saio and M. Kito, Cereal Chem. 58:260 (1981).
59. Chapman, G. W., Jr., and J. A. Robertson, JAOCS 54:195 (1977).
60. Hvolby, A., Ibid. 48:503 (1971).
61. Mounts, T. L., G. R. List and A. J. Heakin, Ibid. 56:883 (1979).
62. Klenk, E., and R. Sakai, Hoppe-Seyler's Z. Physiol. Chem. 258:33 (1939).
63. Woolley, D. W., J. Biol. Chem. 147:581 (1943).
64. Folch, J., Fed. Proc. 6:252 (1947).
65. Scholfield, C. R., T. A. McGuire and H. J. Dutton, JAOCS 27:352 (1950).
66. McGuire, T. A., and F. R. Earle, Ibid. 28:328 (1951).
67. Okuhara, E., and T Nakayama, J. Biol. Chem. 215:295 (1955).
68. Faure, M., and M. J. Morelec-Coulon, Compt. Rend. 236:1104 (1953); Chem. Abstr. 47:6675i (1953).
69. Faure, M., and M. J. Morelec-Coulon, Ibid. 238:411 (1954); Chem. Abstr. 48:6481h (1954).

70. Ballou, C. B., and L. I. Pizer, J. Am. Chem. Soc. 82:3333 (1960).
71. Scholfield, C. R., H. J. Dutton and R. J. Dimler, JAOCS 29:293 (1952).
72. Carter, H. E., W. D. Celmer, D. S. Galanos, R. H. Gigg, W. M. E. Lands, J. H. Law, K. L. Mueller, T. Nakayama, H. H. Tomizawa and E. Weber, Ibid. 35:335 (1958).
73. Hörhammer, L., H. Wagner and J. Hölzl, Biochem Z. 330:591 (1958).
74. Van Handel, E., Recl. Trav. Chim. 72:763 (1953).
75. Carter, H. E., W. D. Celmer, W. E. M. Lands, K. L. Mueller and H. H. Tomizawa, J. Biol. Chem. 206:613 (1954).
76. Carter, H. E., and H. S. Hendrickson, Biochemistry 2:389 (1963).
77. Carter, H. E., R. H. Gigg, J. H. Law, T. Nakayama and E. Weber, J. Biol. Chem. 233:1309 (1958).
78. Carter, H. E., D. S. Galanos, H. S. Hendrickson, B. Jann, T. Nakayama, Y. Nakazawa and B. Nichols, JAOCS 39:107 (1962).
79. Carter, H. E., S. Brooks, R. H. Gigg, D. R. Stroback and T. Suami, J. Biol. Chem. 239:743 (1964).
80. Carter, H. E., B. E. Betts and D. R. Strobach, Biochemistry 3:1103 (1964).
81. Carter, H. E., D. R. Strobach and J. N. Hawthorne, Ibid. 8:383 (1969).
82. Carter, H. E., and E. J. Weber, Lipids 1:16 (1966).
83. Carter, H. E., and A. Kisic, J. Lipid Res. 10:356 (1969).
84. Sen Gupta, A. K., Fette Seifen Anstrichm. 76:440 (1974).
85. Benson, A. A., and B. Maruo, Biochim. Biophys. Acta 27:189 (1958).
86. Benson, A. A., and M. Miyano, Biochem. J. 81:31P (1961).
87. Allen, G. F., P. Good, H. F. Davis and S. D. Fowler, Biochim. Biophys. Res. Commun. 15:424 (1964).
88. Haverkate, F., and L. L. M. VanDeenen, Biochim. Biophys. Acta 106:78 (1965).
89. Erdahl, W. L., A. Stolyhwo and O. S. Privett, JAOCS 50:513 (1973).
90. Harwood, J. L., Phytochemistry 14:1985 (1975).
91. Grewal, S. S., P. S. Sukhija and I. S. Bhatia, Qual. Plant Plant Foods Hum. Nutr. 28:211 (1978); Chem. Abstr. 90:135105s (1979).
92. Stearns, E. M., Jr., and W. T. Morton, Lipids 12:451 (1977).
93. Pangborn, M. C., J. Biol. Chem. 143:247 (1942).
94. DeHaas, G. H., P. P. M. Bonsen and L. L. M. Van Deenen, Biochim. Biophys. Acta 116:114 (1966).
95. Benson, A. A., and E. H. Strickland, Ibid. 41:328 (1960).
96. Coulon-Morelec, M. J., and R. Douce, Bull. Soc. Chim. Biol. 50:1547 (1968); Chem. Abstr. 70:103041p (1969).
97. Szuhaj, B. F., I. L. Balazs and L. D. Williams, JAOCS 47:Program Abstract 175, 44th Fall meeting, AOCS (July 1970).

98. Feulgen, R., and K. Voit, Arch. Ges. Physiol. (Pflügens) 206:389 (1924); Chem. Abstr. 19:1155 (1925).
99. Snyder, F. (ed.), Ether Lipids Chemistry and Biology, Academic Press, New York, 1972.
100. Feulgen, R., P. Feller and G. Andresen, Ber. Oberhess. Ges. Nat. Heilkd. Giessen, Naturwiss. Abt. 24:270 (1949) C.A. 45:2044 (1951).
101. Lovern, J. A., Nature 169:969 (1952).
102. Wagenknecht, A. C., Science 126:1288 (1957).
103. Kaufmann, H. P., Y. Hamza and H. K. Mangold, Chem. Phys. Lipids 6:325 (1971).
104. Brockerhoff, H., and R. G. Jensen, Lipolytic Enzymes, Academic Press, New York, 1974.
105. Slotboom, A. J., and P. P. M. Bonsen, Chem. Phys. Lipids 5:301 (1970).
106. Scholfield, C. R., and H. J. Dutton, JAOCS 31:258 (1954).
107. Scholfield, C. R., and H. J. Dutton, Ibid. 32:169 (1955).
108. Zuev, E. I., V. V. Klyuchkin and V. P. Rzhekhim, Tr. Vses. Nauchno. Issled. Inst. Zhirov. 27:117 (1970); Chem. Abstr. 76:5780Ik (1972).
109. Klyuchkin, V. V., E. I. Zuev and V. L. Loseva, Ibid. 27:127 (1970); Chem. Abstr. 76:57818w (1972).
110. Zuev, E. I., and V. V. Klyuchikin, Ibid. 27:136 (1970); Chem. Abstr. 76:57819x (1972).
111. Lepage, M., J. Lipid Res. 5:587 (1964).
112. Kiribuchi, T., T. Mizunaga and S. Funahashi, Agric. Biol. Chem. 30:770 (1966).
113. Popov, A., Ts. Milkova and N. Marekov, Nahrung 19:547 (1975).
114. Milkova, Ts., A. Popov, A. Selva and V. Vettori, Ibid. 21:7 (1977).
115. Singh, H., and O. S. Privett, Lipids 5:692 (1970).
116. Carter, H. E., R. H. McCluer and E. D. Slifer, J. Am. Chem. Soc. 78:3735 (1956).
117. Carter, H. E., K. Ohno, S. Nojima, C. L. Tipton and N. Z. Stanacev, J. Lipid Res. 2:215 (1961).
118. Carter, H. E., R. A. Hendry and N. Z. Stanacev, Ibid. 2:223 (1961).
119. Wintermans, J. F. G. M., Biochim. Biophys. Acta 44:49 (1960).
120. Smith, C. R., Jr., and I. A. Wolff, Lipids 1:123 (1966).
121. White, G. A., B. C. Willingham, W. Calhoun and R. W. Miller, Econ. Bot. 30:193 (1976).
122. Carter, H. E., R. A. Hendry, S. Nojima and N. Z. Stanacev, Biochim. Biophys. Acta 45:402 (1960).
123. Carter, H. E., R. A. Hendry, S. Nojima, N. Z. Stanacev and K. Ohno, J. Biol. Chem. 236:1912 (1961).
124. Wagner, H., and P. Wolff, Fette Seifen Anstrichm. 66:425 (1964).

125. Negishi, T., H. Hayashi, S. Ito and Y. Fujino, Obihiro Chikusan Daigaku Gakujutsu Kenkyu Hokuku Dai-Bu 5:97 (1967); Chem. Abstr. 68:56722h (1968).
126. Mukhamedova, Kh. S., and S. T. Akramov, Khim. Prir. Soedin. 505 (1977); Chem. Abstr. 87:180700f (1977).
127. Nielson, K., JAOCS 47:217 (1960).
128. Rao, S. V., M. M. Paulose and B. Vijayalakshmi, Lipids 2:88 (1967).
129. Vijayalakshmi, B., and S. V. Rao, Chem. Phys. Lipids 9:82 (1972).
130. Hilditch, T. P., and Y. A. H. Zaky, Biochem. J. 36:815 (1942).
131. Rzhekhin, V. P., N. I. Pogonkina and I. A. Slov'eva, Maslob.-Zhir. Prom. 30:11 (1964); Chem. Abstr. 62:8018b (1965).
132. Daga, H. G. M., Indian Chem. J. 11:17 (1976); Chem. Abstr. 85:190977g (1976).
133. Daga, H. G. M., Paintindia 27:21 (1977); Chem. Abstr. 90:82543z (1979).

# Chapter Two

## The Manufacture of Soybean Lecithins

**Frank J. Flider**
Riceland Foods, Inc.
650 S. Shackleford Road
Little Rock, AR 72211

### Introduction

The term "lecithin" is used to describe a diverse group of commercially available phosphatide products. Available lecithin products include fractions containing one or more phosphatides, deoiled phosphatides, phosphatides with oil and/or surfactant blends, and chemically modified phosphatides. Lecithin sources include soybeans, egg yolk, cottonseed and corn. Due to relatively high concentrations of phosphatides and availability, the soybean is the primary source of commercial lecithin. As such, this chapter describes the manufacture of soybean lecithin products.

Sullivan (Table I) divided lecithin products into 3 major categories: (a) natural, (b) refined and (c) chemically modified. The manufacture of natural and refined grades will be discussed in depth; chemically modified lecithins will be discussed elsewhere.

### Seed Preparation and Extraction

All operations prior to, and during, the manufacture of lecithin can affect the quality of the finished product. The highest quality lecithin is obtained from sound and mature soybeans containing a minimum of burnt and fragmented beans. Frost-damaged beans and beans subjected to prolonged storage periods can result in reduced lecithin yields. Processing of immature beans can result in a high chlorophyll content, which is difficult to remove by traditional bleaching methods.

Impurities such as iron-containing soils, exogenous plant materials and biocides must be reduced by careful cleaning and dehulling prior to processing. Removal of these impurities during later processing stages is difficult.

**TABLE 2-I**
**Lecithin Types[a]**

I. Natural
  A. Plastic
    1. Unbleached
    2. Single-bleached
    3. Double-bleached
  B. Fluid
    1. Unbleached
    2. Single-bleached
    3. Double-bleached
II. Refined
  A. Custom-blended natural
  B. Oil-free phosphatides
    1. As is
    2. Custom-blended
  C. Fractionated oil free phosphatides
    1. Alcohol-soluble
      a. As is
      b. Custom-blended
    2. Alcohol-insoluble
      a. As is
      b. Custom-blended
III. Modified chemically

[a]Ref. 1.

In the USA, the solvent most commonly used for extraction is hexane, although other solvent systems such as a hexane/ethanol/water mixture may be employed (2). Hexane extracts only about half of the total phosphatides of the soybean, but the lecithin produced contains smaller amounts of carbohydrates, bitter principles and color (3).

Careful control of extraction conditions is necessary to assure a minimum of color generation in the lecithin. The brown color of lecithin is largely the result of heating the oil during solvent stripping and is believed to be the result of an aldehyde-amine reaction or Aldol condensation (4,5). Oxidation may also be involved. Brown color formation may be minimized by reducing the final stage stripping temperature as much as possible while still achieving acceptable crude oil flashpoints.

## Filtration

Crude oil from which lecithin is to be recovered is usually filtered prior to degumming to remove residual meal fines and seed fragments. Although more difficult, dry lecithin can also be filtered. Careful filtration results in a highly clarified lecithin with little or no residual hexane insoluble matter.

Brian (6) describes two methods of miscella filtration. The first employs a monofilament nylon or cotton muslin cloth stretched over metal frames in a liquid and vapor-tight housing. Since no filter aid is used, the lecithin appears cloudy due to residual meal fines.

The second method employs wire cloth filters. The filter leaves are covered with either a 60 mesh or 24 × 110 weave cloth to retain the filter aid which is added to the miscella at a concentration of 0.05% by weight. The typical cloth area is 15 sq ft/gal/min of crude oil. The lecithin obtained by this method is less cloudy due to retention of the meal fines.

Highly clarified lecithin products can be produced by filtering crude oil. Plate and frame filters are the most widely used although industrial wire cloth filters are also used. Dry oil is heated to 82 C and 0.1% filter aid is added with agitation. Press dressings vary widely, but one cloth found suitable is 10 oz cotton duck. An area of cloth equal to 40 sq ft/gal/min of crude oil is recommended.

## Degumming

The first actual step in the manufacture of lecithin is degumming, which is depicted in figure 1. The addition of water to crude oil hydrates the phosphatides, rendering them oil-insoluble.

Wide ranges of degumming conditions are reported in the literature as summarized in Table II. In batch degumming, which is the principal method utilized in the USA, 2% water by volume is added to oil heated to 70 C and agitated for ½–1 hr. (5,7). Continuous degumming is widely practiced in Europe. Crude oil, heated to 80 C, is mixed with water metered into a continuous in-dwell pipeline agitator, held for a short period of time, and centrifugally separated (3,5,7).

In degumming, steam, steam condensate, deionized water or city water may be used. As the proper water concentration is difficult to control when using steam, water or condensate is preferred. A portion of the degumming water can be substituted with hydrogen peroxide. In addition to its bleaching action, it will render the gums sterile.

The hardness of the degumming water is an important consideration since phosphatidylethanolamine and phosphatidic acid can be flocculated and inactivated by high levels of calcium and magnesium ions (8). Water degumming removes ca. 90% of the phosphatides present in the crude oil (9). The phosphatides remaining in the oil, known as nonhydratable

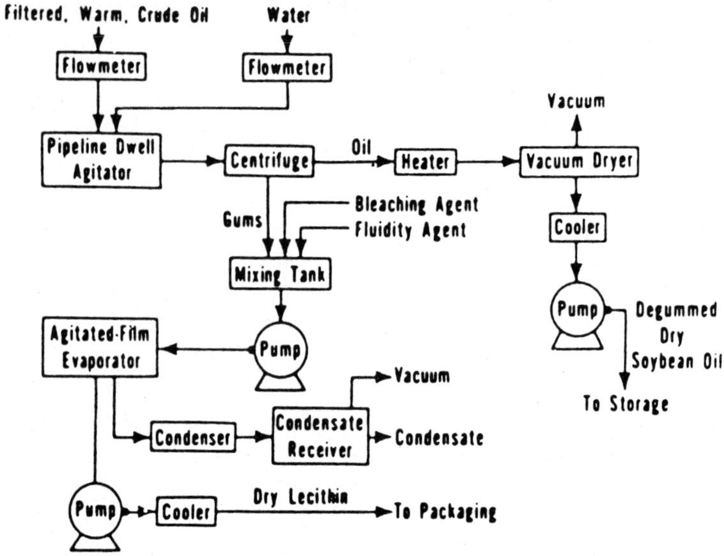


Fig 2-1. Flowsheet for degumming soybean oil and crude lecithin production (6).

**TABLE 2-II**
**Degumming Conditions from the Literature**[a]

| Parameter | Quantity | Reference |
|---|---|---|
| Water: | 75% wt of gums | Crauer (1972) |
| | 1–2.5% | Brian (1976) |
| | 2–3% | Van Niewenhuyzen (1976) |
| | 3% | Bernardini (1973) |
| | 1% | Norris (1964) |
| | 2% | Carr (1976) |
| | Equal to wt gums | Braae (1976) |
| | 2–5% | Andersen (1953) |
| Temperature: | 32–49 C | Norris (1964) |
| | 50–70 C | Van Niewenhuyzen (1976) |
| | 65–75 C | Bernardini (1973) |
| | 70 C | Carr (1976) |
| | 95 C | Andersen (1953) |
| Agitation: | Vigorous | Bernardini (1973) |
| | Mechanical agitation | Carr (1976) |
| Time: | 30–60 min | Carr (1976) |
| | 10–15 min | Braae (1976) |

[a]Ref. 5.

phosphatides, are largely calcium and magnesium salts of the phosphatides. It was thought that the nonhydratable phosphatides were predominately β-phosphatides, but their natural occurrence in soybean oil is no longer considered tenable (4).

The addition of 0.2% of 85% phosphoric acid to oil heated to 70–90 C prior to degumming precipitates the calcium and magnesium as insoluble phosphates. The resulting lecithin is darker in color than water-degummed lecithin (10).

One producer currently utilizes an acetic anhydride pretreatment prior to water addition. Acetic anhydride is added, at a concentration of 0.1%, to oil heated to 60 C (11). The reaction, which is complete within minutes, results in a degummed oil containing less than 25 ppm phosphorus.

Other degumming agents studied in the past include acetic, citric, oxalic, boric (12) and nitric (13) acids. None of these reagents is currently being used to produce commercial lecithin.

List et al. (5) recently investigated the effects of time, temperature, agitation and water concentration on finished lecithin quality. Hydration was found to proceed very rapidly and very little benefit was observed with hydration times exceeding 5–15 min. Temperature had very little effect on phosphorus removal, although 70 C appeared to be optimum. Agitation rates of 200–500 rpm had no significant effect on phosphorus removal. A degumming water concentration of 2% was determined to be optimum.

In commercial processes, the amount of degumming water required is 1.5–2.0%, which is roughly equivalent to the phosphatide content of crude oil. Too little water will result in a dark viscous gums phase and a hazy oil phase which contains unhydrated phosphatides. Too much water will result in a free water phase, a fluid yellowish-brown gums phase, and a hazy degummed oil phase (5, 14).

Various authors have presented data indicating that temperature during degumming has very little effect on phosphorus removal. However, temperature appears to have an effect on the acetone-insoluble content of crude lecithin. In one processing plant, degumming at 49 C yielded a crude lecithin with an acetone-insolubles content of 63–65%. Raising the temperature to 60 C while maintaining the same $\Delta$ P resulted in a lecithin containing 68–73% acetone-insolubles, which is desirable. Table III gives a typical analysis range for a good quality crude unbleached lecithin.

In general, higher degumming temperatures at a given centrifuge pressure differential ($\Delta$ P) result in higher lecithin AI concentrations. The relative relationship between $\Delta$ P, AI and temperature is shown in Figure 2. In actual practice, operational focus should be placed on $\Delta$ P rather than on

**TABLE 2-III**
**Analysis of Good Quality Crude Lecithin**

| | |
|---|---|
| Acetone-insolubles (%) | 68–73 |
| Acid value (mg KOH/g) | 19–24 |
| Moisture (%) | 0.3–0.7 |
| Hexane-insolubles (%) | 0.0–0.05 |
| Color (Gardner undiluted) | 15–17 |
| Peroxide value (meq/kg) | 0–5 |

backpressure; changes in flow rates have a greater effect on $\Delta$ P than backpressure, which could result in undesirable AI fluctuations.

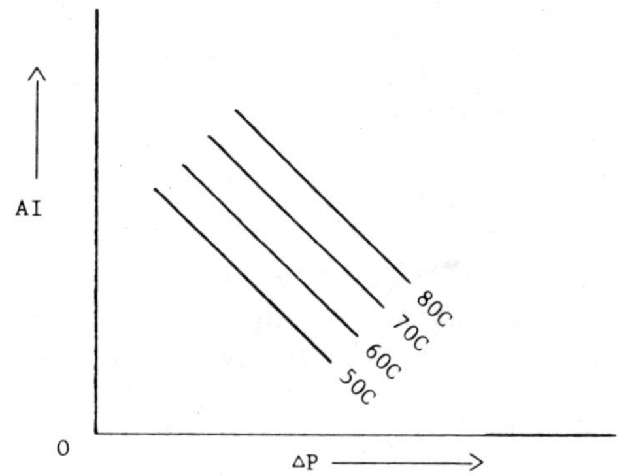


Fig 2-2. Relative Relationship Between AI and $\Delta$P at given degumming temperatures (C).

Degumming at higher temperatures (70–80 C) results in a more consistent lecithin AI on a day-to-day basis. In addition, AI problems caused by low quality soybeans can be diminished. Color degradation of lecithin can be minimized by hydrating at 60 C and increasing the temperature to the desired degumming temperature immediately prior to centrifugation.

Three types of centrifuges are used to separate the gums from the oil. These are the disc bowl, tubular bowl and concentric plate centrifugal contactor (10). Since these can be hermetically sealed, exposure of the oil and gums to air is avoided. Some newer disc-type centrifuges employ a solids-ejecting feature. Accumulated solids such as meal fines and built-up gums can be discharged on a regular timed basis, reducing shutdowns and manual cleaning frequency of the centrifuges. The capacity of this centrifuge is 35,000 lb/hr of crude soybean oil. The AI of the crude lecithin can be controlled by adjusting the oil discharge pressure control valve. High concentrations of AI (65–72%) in crude lecithins may result in a lower quality degummed oil as determined by the heat break test (7,10).

Under optimum degumming conditions, a crude lecithin AI content of 73–75% can be realized. When water comes into contact with a solution of phosphatides and glycerides, mixed double layers that contain both phosphatides and glycerides are formed. The energy of interaction is at a maximum, which corresponds to the most stable mixed layer, when phosphatides and glycerides are present in a ratio of 70:30 (5).

## Bleaching

The color of soybean lecithin is due to carotenoids, brown bodies or melanoids and porphyrins (3,4). Other factors influencing the color of soybean lecithin include: (a) age, origin and quality of soybeans, (b) pretreatment prior to crushing, (c) the depth and thickness of the flakes and temperatures during solvent extraction, (d) conditions during degumming and (e) lecithin processing conditions (5,15).

To produce a lighter-colored product, the gums are treated with hydrogen peroxide and, if necessary, benzoyl peroxide. The National Soybean Processors Association Year Book and Trading Rules specifies 3 color grades for lecithin: natural, bleached and double-bleached (Table IV). This nomenclature is technically incorrect as it is more descriptive of a process than of the product. Under optimum conditions, it is possible to obtain lecithin meeting bleached lecithin specifications without the use of peroxide. Likewise, it is possible to produce double-bleached lecithin using only hydrogen peroxide.

Depending on the quality of the gums, 0.3–1.5% hydrogen peroxide (30% concentration) is used to bleach lecithin. A reaction time of 30–60 min is usually required. List determined that bleaching is time-dependent, requiring 15–30 min for optimum results (5). The longer time required for bleaching on a commercial scale is probably due to inefficient agitation, with more time being required for sufficient mixing of the peroxide with the gums. Bleaching temperatures, which are dependent on degumming temperatures and any ensuing heat loss, typically range between 50 and 60 C.

As indicated earlier, lecithin may be bleached by replacing a portion of the degumming water with hydrogen peroxide. On a commercial scale, this method of bleaching is less efficient than mixing peroxide directly with the gums. No data could be found on whether degumming with hydrogen peroxide has an effect on the resulting degummed oil.

To manufacture double-bleached lecithin, benzoyl peroxide is usually

**TABLE 2-IV**
**Soybean Lecithin Specifications[a]**

| | Fluid natural lecithin | Fluid bleached lecithin | Fluid double-bleached lecithin |
|---|---|---|---|
| Analysis | | | |
| Acetone-insolubles, min (%) | 62 | 62 | 62 |
| Moisture, max (%) | 1 | 1 | 1 |
| Hexane-insolubles, max (%) | 0.3 | 0.3 | 0.3 |
| Acid value, max | 32 | 32 | 32 |
| Color, Gardner, max | 10 | 7 | 4 |
| Viscosity, poises, @ 77 F, max | 150 | 150 | 150 |
| | **Plastic natural lecithin** | **Plastic bleached lecithin** | **Plastic double-bleached lecithin** |
| Analysis | | | |
| Acetone-insolubles, min (%) | 65 | 65 | 65 |
| Moisture, max (%) | 1 | 1 | 1 |
| Hexane-insolubles, max (%) | 0.3 | 0.3 | 0.3 |
| Acid value, max | 30 | 30 | 30 |
| Color, Gardner, max | 10 | 7 | 4 |
| Penetration, max (mm) | 22 | 22 | 22 |

[a]Ref. 16.

used in conjunction with hydrogen peroxide. Benzoyl peroxide may be added to the gums during hydrogen peroxide bleaching or slurried in oil and added to the dried gums. A concentration of 0.3–0.5% is typically used.

Although the National Soybean Processors Association (NSPA) (Table IV) specifies diluted Gardner colors, it has become increasingly more common to express lecithin colors as undiluted Gardner colors. Diluted colors of 10, 7 and 4 approximately correspond to undiluted Gardner colors of 17, 14 and 12, respectively. The diluted and undiluted colors usually correspond very well in natural and bleached lecithin. In double-bleached lecithin, correlation can be achieved only if benzoyl peroxide is used. When only hydrogen peroxide is used for the manufacture of double-bleached lecithin, the diluted color is often 5 or 6 rather than the specified 4.

## Drying

After centrifugal separation and bleaching, the gums, which contain 25–50% moisture, are dried. It is necessary to dry the gums as soon as possible after degumming as microbial activity may become evident within just a few hours.

The gums are dried to a moisture content of less than 1%, typically 0.3–0.75%. Two types of dryers are used in lecithin processing. The first is a vacuum batch dryer equipped with rotating ball-shaped coils through which water circulates to maintain the lecithin at 60–70 C (10). Operated under an absolute pressure of 20–60 mm Hg, ca. 3–5 hr is required to dry a batch of lecithin (15). Some European processors prefer this type of dryer, maintaining that their low operating temperatures minimize charring (10).

In the USA, batch dryers have, for the most part, been replaced by continuous, thin-film evaporators. There are two types, operating on either the vertical or horizontal axis.

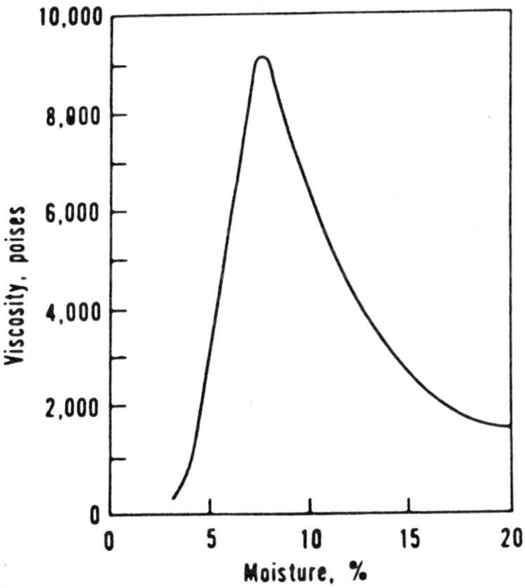


Fig. 2–3. Viscosity of lecithin sludge at 158 F (70 C) in relation to moisture content (15).

As indicated by Van Nieuwenhuyzen's data (Fig. 3), the viscosity of hydrated lecithin increases as it is dried to a maximum of over 9,000 poises at 6–7% moisture content and then falls off sharply. As such, the horizontal type is preferred as there is less tendency for the lecithin film to break.

Little data is available for sizing thin-film dryers and operating conditions vary widely. Since residence time or flow rate is relatively inflexible in most commercial operations, temperature and vacuum must be adjusted accordingly. Temperatures ranging from 80 to 105 C and absolute pressures of 25–300 mm Hg are used. Residence time may range from 2 min to less than 1 min.

In addition to drying lecithin to an acceptable moisture level, the drying operation serves to reduce the peroxide value. At dryer temperatures exceeding 98 C, residual peroxide is removed more rapidly than at lower temperatures, regardless of vacuum efficiency. In highly bleached products, the peroxide value of freshly dried lecithin should be ca. 50–75 meq/kg.

Due to lecithin's thermal sensitivity, careful control of drying conditions is critical. Immediately after drying, lecithin should be cooled to 55–60 C. If the lecithin will not be processed any further, it should be cooled to 35–50 C.

## Standardization and Fluidization

According to NSPA specifications, fluid grades of lecithin contain a minimum of 62% acetone-insolubles and plastic grades a minimum of 65%.

Standardized fluid lecithins typically contain 62–64% acetone-insolubles and have an acid value range of 28–32 mg KOH/g. To meet these specifications, soy fatty acids and soybean oil are blended with the crude lecithin. For optimum batch-to-batch consistency, an AI of 63% and an AV of 30 are targeted; diluents required are calculated according to the following formulae:

$$\%\text{ Fatty acids} = [30 - (\frac{63}{\text{Crude AI}} \times \text{Crude AV})] \times \frac{1}{2}$$

$$\%\text{ Oil} = \left(100 \times [(\frac{\text{Crude AI}}{63}) - 1]\right) - \%\text{ Fatty acids required}$$

A blend time of at least 2–4 hr at 57–60 C is recommended.

The viscosity of standardized fluid lecithin ranges between 7,500 and 10,000 centipoises at 25 C.

Plastic grades of lecithin usually have an acetone-insolubles content of 65–68%. If required, soybean oil is used for AI adjustment. Fatty acids should not be used.

Other lecithin viscosity modifying agents include monovalent ions, divalent ions, phosphoric acid and acetic anhydride. Monovalent ions such as $Na^+$ or $K^+$ are attracted to the negatively charged phosphoryl radical of the phosphatide molecule. This tends to increase the crystalline order of the phospholipid molecules, thus increasing viscosity. Divalent ions such as $Ca^{2+}$ and $Mg^{2+}$ will reduce the crystalline order and consequently reduce viscosity. This principle is used to produce fluid commercial lecithin products containing 66–70% phosphatides with no added fatty acids.

Phosphoric acid fluidization is discouraged due to the tendency of the treated lecithin to darken more rapidly on heating. In addition, unless the addition rate and temperature are carefully controlled, hydrolysis may result. The use of acetic anhydride for degumming results in a fluid lecithin. This may be due in part to acetylation of phosphatidylethanolamine by the acetic anhydride.

For nonedible grades of lecithin, acidulated and dried soapstock may be substituted for the more expensive fatty acids. Degummed oil is recommended for the oil diluent. Crude oil may be used, but AI and AV are slightly more difficult to control due to the presence of phosphatides. Refined, bleached and deodorized oils may be used but serve no advantage. The preponderance of crude oil in crude lecithin negates any organoleptic benefits which may be derived from a refined oil.

## Custom-Blended Lecithins

Lecithin is often blended with other oils, carriers or surfactants to enhance or change functionality.

Water-dispersible lecithins may be produced by introducing 5–20% of a hydrophilic surfactant such as a polysorbate or ethoxylated monoglycerides. A mixture of lecithin and 10–20% nonionic surfactant has great utility in nonedible applications of water-dispersible lecithin. Lecithin and high melting mono- and diglycerides may be mixed together and spray-cooled, resulting in a flaked or powdered product (8).

Lecithin products containing 50–54% phosphatides present special manufacturing problems. When diluted with soybean oil and fatty acids under typical blending conditions, these products have a tendency to separate into 2 phases: an oil/fatty acid phase containing less than 20% AI and a heavier lecithin phase. At 0–5 C, separation may occur in as little as a few days. Stability can be increased by substituting a portion or all of the soybean oil with other oils such as peanut oil, partially hydrogenated soybean oil (95 IV), cottonseed oil or a low melting coconut oil fraction. In addition, blend time is increased to 4–8 hr. If possible, the blending temperature should be maintained at 60 C.

## Deoiled Lecithin

Deoiled lecithin is produced by acetone fractionation of crude lecithin. Being acetone-insoluble, the phosphatides are precipitated and recovered essentially free of oil. The Food Chemicals Codex defines deoiled lecithin as containing 90% minimum phosphatides but, in actual practice, the phosphatide concentration is 95% or greater.

When contacted with acetone, the phosphatides precipitate as a fine free-flowing powder. Considering the fact that a large proportion of the phosphatide molecule consists of oleic and linoleic acids, the interaction of phosphatides with acetone, relative to other fatty solvents, is remarkable. Substantial quantities of acetone are held by the precipitated phosphatides, suggesting that acetone plays a role in inducing matrix formation of the molecules with a high degree of crystalline order. As such, acetone may be considered to be a solvent of crystallization (17).

**TABLE 2-V**
**Specifications for Crude Lecithin for Deoiling**

| | | |
|---|---|---|
| Acetone-insolubles (%) | 65 | min |
| Acid value (mg KOH/g) | 25 | max |
| Moisture (%) | 0.7 | max |
| Hexane-insolubles (%) | 0.05% | max |
| Color (Gardner undiluted) | 12–16 | max |
| Peroxide value (meq/kg) | 2 | max |

Crude lecithin quality is a major factor for optimum production of deoiled lecithin. Table V gives specifications for crude lecithin for deoiling and Table VI gives the effects of using a low quality crude lecithin. Where high quality crude lecithin is unavailable in suitable quantities, it is often the practice to blend crude lots in order to minimize these effects.

Depending on the type and efficiency of the extraction equipment, the acetone/crude lecithin ratio necessary to achieve a 95% phosphatide product is 10–20:1 (v/v). In batch extraction, the tank is charged with acetone prior to crude lecithin addition. Crude is then introduced into the vessel with agitation until an acetone/crude lecithin volume ratio of ca. 5:1 is achieved. The mixture is agitated for 20–30 min, after which time the phospholipids are allowed to settle. The acetone miscella is then removed and the vessel charged again with acetone for a second extraction. A single batch may be extracted 2–4 times to obtain the desired phosphatide concentration. In continuous extraction, crude lecithin and acetone are simultaneously metered

**TABLE 2-VI**
**Effects of Low Quality Crude Lecithin on Deoiled Lecithin**

| Crude lecithin parameter | Value | Effect |
|---|---|---|
| Dark color | > 16 Gardner | Dark granules |
| High acid value | > 25 mg KOH/g | Tacky granules, granulation problems |
| High divalent metal concentration | > 0.2% | Tacky granules, granulation problems |
| High peroxide value | > 5 meq/kg | Decreased shelf life, poor organoleptic profile |
| Low acetone-insolubles | < 65% | Extraction inefficiency |
| High moisture | > 0.7% | Sticky granules, poor flow properties, inconsistent granule rigidity |

into the vessel. Within limits, acetone consumption can be decreased by increasing residence time in the continuous extractor, increasing raw material efficiency.

After extraction, the deoiled lecithin is recovered as a cake containing 25–50% acetone. The acetone concentration of the cake is critical for optimum granulation. Too little acetone will result in the formation of a high concentration of fines and powder. Too much acetone will result in a "salt and pepper" effect caused by agglomeration of the fines and powder during granulation. The fine and powder output is 5–50% of the total deoiled material, depending on production conditions.

After granulation, the remainder of the acetone is removed by drying, preferably in a moving bed forced air dryer. Fluid bed dryers are not used to a great extent due to their tendency to break apart the granules, generating as much as 50% fines and powder. After drying, the acetone content of the product should be well below 50 ppm and preferably below 25 ppm. Mesityl oxide, the aldol condensation product of acetone, may be formed if excess acetone is present. When the lecithin has been sufficiently dried, mesityl oxide formation is not a problem. The acetone should be monitored on a regular basis for mesityl oxide and its precursor, diacetone alcohol. If either of these compounds are present at significant concentrations, the acetone should be distilled prior to use.

The dried deoiled lecithin is sized by sieving through a series of screens. Fines and powder may be reworked by reintroducing them into the extraction vessel.

The tocopherols are removed from the lecithin by the extraction process, rendering the deoiled lecithin less oxidatively stable than crude lecithin.

The high surface/volume ratio of deoiled lecithin also contributes to reduced stability. To stabilize the lecithin, mixed soya tocopherols are added back at a level of 500 ppm.

Deoiled lecithin should be packaged as soon as possible after production to prevent moisture pick up. The moisture content of a deoiled lecithin initially containing 0.5% moisture will double in as little as 15 min on exposure to atmospheric conditions of 25 C and 50% relative humidity.

The acetone-soluble fraction, known as high acid oil, is a mixture of triglycerides, fatty acids and, to a lesser degree, sterols, color bodies and odor and flavor components. High acid oil is similar to crude oil in composition and may be metered back into the crude oil stream.

## Refined Deoiled Lecithin

Deoiled lecithin can be blended with carriers such as cocoa butter, hard butters or other diluents to obtain a product with physical or functional characteristics substantially different from standard fluid and deoiled lecithins.

Relatively low concentrations of deoiled lecithin may be dissolved in certain carriers without the use of a solvent. When higher concentrations of phosphatides are desired, it is often necessary to use a solvent.

The desired end product dictates the process to be used. Deoiled lecithin is readily soluble in short- and medium-chained triglycerides and straight-chained triglycerides such as hydrogenated oils. Using a simple dry blend, a solution containing 40% phosphatides can be readily achieved. Because natural antioxidants are removed during deoiling, it is usually necessary to add antioxidant to the final products.

To obtain higher levels of phosphatides (up to 77%), water, acetone or a combination of both are used as the solvent. A kettle type steam-jacketed vacuum dryer is charged with deoiled lecithin/acetone slurry and the excess acetone is evaporated. Addition of a small percentage (1–2%) of water or hydrogen peroxide will aid in evaporation of the acetone. A portion of the desired carrier is also added to the kettle. The combination of temperature, water, acetone and carrier causes the phosphatides to "melt," which facilitates blending. Once the melt has been achieved, the remainder of the carrier is added and excess acetone and water are removed. Injection of steam aids in the removal of acetone from the product.

Another method of manufacturing refined lecithin uses water as the only solvent. The deoiled lecithin is first hydrated with water, the carrier is added and the resulting emulsion is then dried. An advantage to this system is a better organoleptic profile than can be achieved when using acetone. The major disadvantage of using water instead of solvent is the increased energy expenditure necessary to remove the water, as well as the greater possibility of thermal abuse to the product.

## Fractionated Lecithins

Alcohol fractionation redistributes the phosphatides according to their respective hydrophilic and lipophilic properties. The alcohol-soluble fraction is typified by a high concentration of phosphatidylcholine whereas the alcohol-insoluble fraction contains a high concentration of the inositol phosphatides. Phosphatidylethanolamine is partitioned equally between the 2 fractions (Table VII).

**TABLE 2-VII**
**Approximate Compositional Analysis of Commercial Lecithin Fractions[a]**

| Typical composition (%) | Oil-free lecithin | Alcohol-soluble lecithin | Alcohol-insoluble lecithin |
|---|---|---|---|
| Chemical lecithin | 29 | 60 | 4 |
| Chemical cephalin | 29 | 30 | 29 |
| Inositol and other phosphatides (including glycolipids) | 32 | 2 | 55 |
| Soybean oil | 3 | 4 | 4 |
| Miscellaneous | 7 | 4 | 8 |
| Emulsion type favored | Either oil in water or water in oil | Oil in water | Water in oil |

[a]Ref. 1.

The PC/PE ratios of fractionated lecithins are largely determined by processing variables such as alcohol polarity, concentration, lecithin/alcohol ratio, temperature and extraction time. By extracting with 90% ethanol, alcohol-soluble fractions with PC/PE ratios of 8:1 can be obtained (8). The fractions may be blended with other surfactants or carriers to obtain desired functionalities.

Alcohol-soluble crude lecithin can be further treated with alumina to yield a lecithin fraction high in phosphatidylcholine and free of phosphatidylethanolamine and inositol phosphatides (18).

## Storage and Handling

When maintained at 20–25 C in closed containers, liquid lecithin can be kept for years with no significant deterioration in quality or functionality. Storage temperatures of bleached products in particular should be maintained within this range. As residual peroxide decomposes, bleached lecithins will experience rapid color reversion at elevated temperatures. In general, the greater the degree of bleaching, the more unstable the lecithin relative to color reversion. As such, low storage temperatures are critical for highly bleached products.

Freezing temperatures should be avoided when storing liquid lecithin. Although freezing has no effect on functionality, physical separation of the oil and phosphatides may occur after prolonged frozen storage. Physical separation is more likely to occur with low AI (50–55%) products than higher AI (63–70%) products. The phosphatides may be redispersed in the oil by simply remixing at 40–60 C.

To facilitate handling and shipping, bulk lecithin storage temperatures of 35–50 C are acceptable. For prolonged storage, temperatures should be maintained below 35 C to minimize color degradation.

If not exposed to light and moisture, deoiled lecithin can be stored for at least two years at temperatures not exceeding 25 C. Deoiled lecithin may be stored below 0 C if desired, however, due to its hygroscopic nature, frozen product should be allowed to come to ambient temperature prior to exposing to the atmosphere.

## Acknowledgments

The author is grateful to Mr. Paul McElroy for his assistance in the preparation of this chapter.

## References

1. Sullivan, D. R., and B. F. Szuhaj, JAOCS Program Abstract 39, AOCS meeting, April 1975.
2. Hayes, L. P., and R. P. Simms, U.S. Patent 3,878,232 (1975).
3. Sartoretto, P., in Kirk-Othmer Encyclopedia of Chemical Technology, 2nd edn., Vol. 12, Interscience Publishers, Div. of John Wiley & Sons, Inc., New York and London, 1967, pp. 343–361.
4. Scholfield, C. R., and H. J. Dutton, JAOCS 31:258 (1954).

5. List, G. R., J. M. Avellaneda and T. L. Mounts, Ibid. 58:892 (1981).
6. Brian, R., Ibid. 53:27 (1976).
7. Klein, K., and L. S. Craver, Ibid. 51:368 (1974).
8. Van Nieuwenhuyzen, W., Ibid. 58:886 (1981).
9. Myers, N. W., Ibid 34:93 (1957).
10. Brekke, O. L., in Handbook of Soy Oil Processing and Utilization, American Soybean Association and American Oil Chemists' Society, 1980, pp. 71–83.
11. Hayes, L. P., and H. Wolff, U.S. Patent 2,782,216 (1957).
12. Andersen, A. J. C., Refining of Oils and Fats for Edible Purposes, Pergamon Press, London, 1962, pp. 28–42.
13. Guillaumin, R., Rev. Fr. Corps Gras 9:486 (1962).
14. Crauer, L. S., Presented at meeting of International Society for Fat Research, Goteborg, Sweden, The DeLaval Separator Co., Poughkeepsie, NY, 1972.
15. Van Nieuwenhuyzen, W., Ibid. 53:425 (1976).
16. NSPA, Yearbook and Trading Rules of the National Soybean Processors Association, 1978–79.
17. Aneja, R., J. S. Chadha and R. W. Yoell, Fette Seifen Anstrichm. 73:643 (1971).
18. Betzing, H., U.S. Patent 4,235,793 (1980).

# Chapter Three

## Corn Lecithin

**Evelyn J. Weber**
Agricultural Research Service
U.S. Department of Agriculture
S-320 Turner Hall
University of Illinois
1102 S. Goodwin Ave.
Urbana, IL 61801

### Introduction

Commercial corn "lecithin" was available from the mid 1930s through the 1950s. Patents were issued for commercial preparations of corn "phosphatides" (1,2) and for products containing corn lecithin, such as cosmetics (3), ointments (4), foaming agents (2) and rust inhibitors (5). In recent years, the only source of commercial lecithin in the USA has been soybean (6).

The first detailed analysis of commercial corn phosphatides was published by Scholfield et al. in 1950 (7). These researchers used countercurrent distribution to fractionate corn and soybean lecithins and compared the fractions obtained from the two sources. In a paper published in 1981, Weber (8) used column, thin-layer and gas chromatography to compare the compositions of samples of commercial corn and soybean lecithins. The sample of corn lecithin was prepared by the A. E. Staley Manufacturing Company (Decatur, IL). However, with the phenomenal growth now occurring in the demand for corn sweeteners, other products of the corn-refining industry, such as corn lecithin, may become available and competitive.

## Composition of Crude Corn Lecithin Compared to Commercial Soybean Lecithin

### Lipid Class Analysis

The physical properties of a commercial lecithin are determined by the proportions and the fatty acid compositions of the various phospholipids and other lipids that it contains (9). Therefore, a comparison of samples of commercial corn and soybean lecithins should give some insight into the properties and potential uses of corn lecithin. The samples of crude corn and soybean lecithins were provided by the A. E. Staley Manufacturing Co. (Decatur, IL). The soybean lecithin sample was a fluid type, usually ca. 65% acetone insoluble.

The lipids of the crude lecithins were separated into fractions on silicic acid columns (8). The nonpolar lipids were eluted from the column with chloroform, the glycolipids with acetone, and the phospholipids with methanol. The percentage of each lipid fraction was determined in two ways: (a) by gravimetric analysis, i.e., direct weighing of aliquots from each column fraction, and (b) by gas chromatography (GC) of the fatty acid methyl esters (FAME) with an internal standard–methyl heptadecanoate.

**TABLE 3-I**
**Silicic Acid Column Chromatography of Commercial Corn and Soybean Lecithins[a]**

| Lipids | Corn | | Soybean | |
|---|---|---|---|---|
| | Wt[b] (%) | Wt from FAME[c] (%) | Wt[b] (%) | Wt from FAME[c] (%) |
| Nonpolar lipids | 52.8 | 60.5 | 36.8 | 42.1 |
| Glycolipids | 18.4 | 12.0 | 13.2 | 8.4 |
| Phospholipids | 28.8 | 27.5 | 50.0 | 49.5 |

[a]Data from ref. 8.
[b]Gravimetric weight.
[c]Weight calculated from fatty acid methyl esters (FAME) determined by gas chromatography with an internal standard.

Both the commercial corn and soybean lecithins contained a large proportion of carrier oil, nonpolar lipids. Corn had the largest amount, 60.5% by FAME weights, compared to 42.1% for soybean (Table I). Despite the higher proportion of nonpolar lipids, corn also had a higher percentage (12.0%) of glycolipids than did soybeans (8.4%). The percentage of phospholipids

(27.5%) in corn lecithin, however, was only ca. one-half of that of soybeans (49.5%). Thus, the ratio of glycolipids to phospholipids was higher in corn than in soybean lecithin.

The largest differences between the two methods for determining the proportions of the lipids appeared to be within the glycolipid fractions (Table I). The FAME weights did not include sterylglycosides or hydroxy fatty acids from cerebrosides. Both sterylglycosides and cerebrosides gave heavy spots on thin-layer plates of the glycolipids from corn and soybean. The glycolipid fractions may also contain carbohydrates as contaminants (10,11) which would add to the gravimetric weight determinations. Alcohol-soluble proteins, amino acids and phytic acid may account for the slightly heavier gravimetric weights of the phospholipid fractions (10,11).

The nonpolar and polar lipids have been compared separately, because the proportion of carrier oil (nonpolar lipids) was not the same for the corn and soybean samples. The individual lipids were separated by thin-layer chromatography (TLC) (8). For the nonpolar lipids, the solvent system was petroleum ether (bp 60–68 C)/diethyl ether/acetic acid (80:20:1, v/v/v). Triglycerides (TG) made up over 90% of the nonpolar lipids in both samples (Table II). The percentages of free fatty acids and diglycerides were only slightly higher in corn than in soybean and were low, less than 3%, in both cases.

**TABLE 3-II**
**Lipids of the Nonpolar Lipid Fraction from Silicic Acid Column Chromatography**

| Nonpolar lipids | Corn | Soybean |
|---|---|---|
| | % by wt of fraction[a] | |
| Triglyceride | 91.0 | 92.9 |
| Free fatty acid | 2.4 | 1.1 |
| Diglyceride | 2.7 | 1.3 |

[a]Weights calculated from FAME in ref. 8.

The glycolipids were separated by TLC with chloroform/methanol/ammonia (29.3% w/v) (70:20:1) (8). The phospholipids were separated by two-dimensional TLC. The solvent systems were chloroform/methanol/ammonia (65:35:5) in the first direction and chloroform/acetone/methanol/acetic acid/water (50:20:10:10:5) in the second. The proportions of the individual phospholipids were determined by analyzing the phospholipid spots for phosphorus (8,12). When the phospholipids of corn lecithin were fractionated on a thin-layer plate, 22.3% of the total phosphorus remained at the origin. The compounds at the origin were not completely identified, but they were made up mainly of phytic acid and a small amount of phytoglycolipid (PGL) (13,14). The fact that phytate is found in crude corn lecithin is not surprising, because the lecithin was extracted from corn germ and corn germ contains 88% of the total phytate found in the corn kernel (15). From soybean lecithin, only 7% of the total phosphorus remained at the origin. This material was probably also a mixture of phytate and PGL.

PGL is a unique lipid because it has the structural features of a glycolipid and of a phospholipid. The starting material for the original isolation of PGL in 1958 (16) was called "Vodol", a commercial corn lecithin prepared by Refining Unincorporated (Chicago, IL). Structural studies also were conducted on PGL isolated from commercial lecithins prepared from soybean, flaxseed, peanut, cottonseed, sunflower and safflower (14,16,17). The basic unit of PGL was found to be an N-acyl phytosphingosine-type long-chain base attached as a phosphate ester to inositol-glucuronic acid-glucosamine (16–22). Mannose is bound to glucuronic acid and oligosaccharides with varying numbers of galactose and arabinose and fucose molecules are attached to the glucosamine. Thus, PGL includes a family of related compounds differing in the length of the carbohydrate chain. Another glycophosphoceramide that contains all the components of PGL except glucosamine also has been isolated from commercial corn, soybean, flax and safflower lecithins (14).

The phosphorus remaining at the origins after TLC of the crude corn and soybean lecithins was not included in the calculations of the proportions of the individual phospholipids. Table III shows the distributions of both the glycolipids and phospholipids of the corn and soybean lecithins. Glycolipids represented a higher proportion of the polar lipids in corn (30.3%) than in soybean lecithin (14.5%). The percentages of monogalactosyldiglycerides (MGDG) and digalactosyldiglycerides (DGDG) were only slightly higher in corn, but the sum of the percentages of sterylglycoside ester (SGE) and other glycolipids was more than twice that of soybean. Phosphatidylcholine (PC) was, by far, the most abundant phospholipid in both lecithin samples, representing 30.4% of the polar lipids in corn and 33.0% in soybean. Second was phosphatidylinositol (PI) at 16.3% in corn and 16.8% in soybean. The only other phospholipids that were found at levels above 3% for both corn and soybean were phosphatidylethanolamine (PE), N-acyl lysophosphatidylethanolamine (ALPE) and phosphatidic acid (PA). ALPE was much higher in soybean lecithin (10.4%) than in corn (3.7%). The percentage of PE nearly equalled that of PI in soybean lecithin, but in corn, the percentage of PE was only ca. one-fifth the percentage of PI. Scholfield et al. (7) also noted a much smaller amount of PE in corn than in soybean lecithin. Some degradation may have occurred during processing, because the levels of PE and PC were lower and that of PA was higher than those normally found in phospholipids isolated directly from mature corn. These differences will be discussed later in the section on polar lipids of corn germ.

**TABLE 3-III**
**Distribution of Polar Lipids in Commercial Corn and Soybean Lecithins[a]**

| | Corn | Soybean |
|---|---|---|
| | % of wt of polar lipids | |
| Sterylglycoside ester | 15.0 | 4.3 |
| Monogalactosyldiglyceride | 1.8 | 0.8 |
| Digalactosyldiglyceride | 3.7 | 3.0 |
| Other glycolipids | 9.8 | 6.4 |
| N-acyl phosphatidylethanolamine | 2.6 | 2.2 |
| N-acyl lysophosphatidylethanolamine | 3.7 | 10.4 |
| Phosphatidylethanolamine | 3.2 | 14.1 |
| Phosphatidylglycerol | 1.4 | 1.0 |
| Phosphatidylcholine | 30.4 | 33.0 |
| Phosphatidylinositol | 16.3 | 16.8 |
| Phosphatidic acid | 9.4 | 6.4 |
| Phosphatidylserine | 1.0 | 0.4 |
| Lysophosphatidylethanolamine | tr[b] | 0.2 |
| Lysophosphatidylcholine | 1.7 | 0.9 |

[a]Data calculated from FAME in ref. 8.
[b]tr = trace.

### Fatty Acid Compositions

The fatty acid compositions of the crude corn and soybean lecithins are shown in Table IV. Linoleic (18:2) was the predominant fatty acid, and the percentages were equal in the corn and soybean lecithins. Noticeable differences between the two lecithins were lower levels of stearic (18:0) and linolenic (18:3) acids and a higher level of oleic (18:1) acid in corn than in soybean.

When the lipids were separated into classes, the same trends (lower percentages of stearic and linolenic acids and a higher percentage of oleic acid in corn) were observed in the fatty acid compositions of all the classes (Table IV). However, other differences also emerged. The nonpolar lipids of corn had a higher level of linolenic acid, as is normal for corn oil compared to soybean oil. In contrast to the oils, both the glycolipids and phospholipids of corn had lower percentages of linoleic acid and were more saturated than the soybean glycolipids and phospholipids. Soybean

**TABLE 3-IV**
**Fatty Acid Compositions of Commercial Corn and Soybean Lecithins and the Lipid Fractions from Silicic Acid Column Chromatography[a]**

| | Fatty acid compositions, (mol %) | | | | |
|---|---|---|---|---|---|
| | 16:0 | 18:0 | 18:1 | 18:2 | 18:3 |
| Lecithin | | | | | |
| Corn | 17.7 | 1.8 | 25.3 | 54.2 | 1.0 |
| Soybean | 17.4 | 4.0 | 17.7 | 54.0 | 6.8 |
| Nonpolar lipid | | | | | |
| Corn | 13.1 | 1.8 | 25.8 | 58.4 | 0.9 |
| Soybean | 11.8 | 4.2 | 24.7 | 52.1 | 7.2 |
| Glycolipid | | | | | |
| Corn | 34.2 | 2.3 | 18.6 | 42.6 | 2.4 |
| Soybean | 24.2 | 4.9 | 10.8 | 48.6 | 11.5 |
| Phospholipid | | | | | |
| Corn | 22.8 | 1.5 | 26.5 | 48.5 | 0.7 |
| Soybean | 21.4 | 3.8 | 12.0 | 57.0 | 5.8 |

[a]Data from ref. 8.

glycolipids had a higher level of linolenic acid.

Each individual lipid had a characteristic fatty acid pattern (Table V), but the overall tendencies for lower stearic and linolenic percentages and higher oleic percentages in the corn lipids persisted. The characteristic fatty acid pattern of the triglycerides was indicated by a low percentage of palmitic acid and high percentage of linoleic acid. The galactosyldiglycerides were distinguished from the other lipids by higher levels of linolenic acid, particularly in soybean DGDG. The SGE had the highest percentage of palmitic acid within the corn or soybean lipids. Among the phospholipids, PC had the highest percentage of oleic acid, and PI the highest percentage of saturated palmitic acid. Within the corn phospholipids, PE had the highest level of linoleic acid. We have found the same characteristic fatty acid patterns for these individual lipids isolated from many different corn inbreds which varied in fatty acid composition from 42 to 70% in linoleic acid (23).

## Potential of Commercial Corn Lecithin

We have analyzed only one sample of crude corn lecithin and one of crude soybean lecithin. Changes in the relative proportions of the lipids could occur or could be made by changing the processing conditions. The amount of carrier oil, for example, could be adjusted easily. Elimination of some of the phytate in the corn lecithin may be desirable, because phytate binds

**TABLE 3-V**
**Fatty Acid Compositions of Lipids of Commercial Corn and Soybean Lecithins[a]**

| Lipid | Fatty acid composition, (mol %) | | | | |
|---|---|---|---|---|---|
| | 16:0 | 18:0 | 18:1 | 18:2 | 18:3 |
| Triglyceride | | | | | |
| Corn | 12.2 | 2.0 | 25.7 | 59.2 | 1.0 |
| Soybean | 11.4 | 4.3 | 25.2 | 51.9 | 7.2 |
| Free fatty acid | | | | | |
| Corn | 37.1 | 4.1 | 23.8 | 34.0 | 1.0 |
| Soybean | 31.6 | 8.5 | 16.7 | 39.0 | 4.1 |
| Diglyceride | | | | | |
| Corn | 27.0 | 2.5 | 25.7 | 44.0 | 0.8 |
| Soybean | 24.8 | 6.9 | 16.5 | 46.6 | 5.2 |
| Monogalactosyldiglyceride | | | | | |
| Corn | 20.5 | 3.3 | 23.1 | 46.6 | 6.5 |
| Soybean | 23.8 | 5.4 | 16.9 | 46.0 | 7.9 |
| Digalactosyldiglyceride | | | | | |
| Corn | 21.0 | 2.5 | 15.8 | 55.5 | 5.2 |
| Soybean | 15.7 | 4.8 | 10.2 | 47.7 | 21.6 |
| Sterylglycoside ester | | | | | |
| Corn | 54.9 | 3.1 | 15.1 | 26.1 | 0.8 |
| Soybean | 38.7 | 6.9 | 12.2 | 37.0 | 5.2 |
| Phosphatidylethanolamine | | | | | |
| Corn | 23.9 | 1.8 | 21.6 | 52.1 | 0.6 |
| Soybean | 23.2 | 2.6 | 11.6 | 57.1 | 5.5 |
| Phosphatidylcholine | | | | | |
| Corn | 21.7 | 1.7 | 31.0 | 45.0 | 0.6 |
| Soybean | 15.5 | 4.0 | 14.5 | 60.3 | 5.6 |
| Phosphatidylinositol | | | | | |
| Corn | 33.8 | 1.5 | 18.5 | 45.6 | 0.6 |
| Soybean | 35.5 | 6.9 | 7.8 | 44.3 | 5.4 |

[a]Data from ref. 8.

zinc, iron, magnesium and calcium and decreases the nutritional availability of these important minerals (15).

The physical properties, particularly the emulsifying properties of corn lecithin, may differ from those of soybean lecithin because of the higher proportion of glycolipids to phospholipids in the corn lecithin. The ratio of glycolipids to phospholipids was 0.44 for corn lecithin and 0.17 for soybean. Alcohol fractionation, which increases the ratio of PC to PE, improves the emulsifying and antispattering properties of soybean lecithin (24). Corn lecithin naturally has a higher ratio of PC to PE.

When the unsaturated fatty acids of soybean lecithin are hydrogenated, the resulting product has a lighter color, shows greater resistance to oxidative rancidity and possesses less odor and flavor than unhydrogenated lecithin. However, the hydrogenated soybean product has fewer emulsifying properties than the natural form (25). The major differences in the fatty acid compositions of the corn and soybean lecithins were a higher percentage of oleic and lower percentages of stearic and linolenic acids in corn compared to soybean. The lower level of linolenic acid in corn may be an advantage for corn lecithin over soybean lecithin, because the triunsaturated linolenic acid oxidizes readily to produce off-flavors (26,27).

Over the past 10 years, the amount of corn processed by the wet milling industry has more than doubled (28,29). Much of this remarkable growth is due to an increasing demand, particularly by the beverage industry, for high fructose corn syrups (HFCS). Although only first introduced in 1968, HFCS production at the end of 1980 was ca. 6 billion lb (30). As the corn refining industry grows, the potential for developing other by-products, such as corn lecithin, should increase.

## Polar Lipids of Corn Grain

### Lipid Class Analysis

Changes may take place in the naturally-occurring polar lipids of the corn kernel during harvesting, storage and processing of the grain to prepare commercial lecithin. Hydrolytic rancidity may result from the activity of lipases. Oxidative rancidity can be caused by lipoxygenase or autoxidation or both. Under modern agricultural practices, corn is harvested at high moisture, ca. 30%, and dried with heated air. When Gardner and Inglett (31) used temperatures ranging from 25 to 283 C to dry corn germ to 15% moisture, they observed that inactivation of the linoleic acid-hydroperoxide isomerase, lipoxygenase, peroxidase and lipase enzymes increased with increasing temperature, but, even at the highest drying temperature, slight residual activity of all the enzymes, except lipoxygenase, remained.

High temperatures during drying can damage corn kernels by causing stress cracks in the seed coats. Harvesting, cleaning and unloading operations also may damage kernels. The greatest loss encountered in processing damaged corn is in oil (32). Corn with seed coat damage is particularly susceptible to molds, and molds have very active lipases. The germ is infested selectively by molds before the endosperm is attacked. The quality of corn in storage is influenced by the mold and insect populations, temperature, handling and moisture. To be stored safely, corn should have a moisture content of not more than 13.5% (32). Above this moisture level, molds and endogenous degradative enzymes exhibit a sharp increase in activity.

The specific activity of phospholipase D in corn germ is ca. three times that of the endosperm (33). Since phospholipase D produces PA from the other phospholipids, PC, PE and PI, of the extracted oil, and PA is not precipitated out during water-degumming, the yield of lecithin is decreased (34,35). The presence of nonhydratable PA in oil extracted from field-damaged and storage-damaged soybeans has been a serious problem for soybean oil processors (36–38). However, corn was reported to have only ca. one-tenth of the phospholipase D activity of soybean (33).

Peroxidase is another degradative enzyme which has higher activity in corn germ than in the endosperm. Gardner et al. (39) found that peroxidase activity in hand-dissected germ was ca. five times that of the endosperm. Peroxidase was even more active in aleurone than in germ and the presence of aleurone tissue may account for most of the activity of the endosperm fraction.

In addition to possible enzymatic changes, commercial production conditions affect the yield and quality of lecithins (40). Differential extraction and/or separation of the polar lipids from the oil may enrich commercial lecithin in certain lipids. Earle (41) collected data from three commercial processors on the oil content of commercial corn. From 1917 to 1977, the oil content varied between 4.0 and 4.9%. No correlations were found between the year-to-year variations in oil content and factors such as temperature, rainfall or yield. The data on oil content from one company were consistently lower than those from the other two companies, because a different solvent, petroleum ether rather than hexane, was used to extract the oil.

Corn oil is extracted from germs isolated by wet or dry milling. Ca. 80% of corn oil is derived from wet milling (42). Sulfur dioxide, used to soften corn before wet milling, has been reported to inhibit lipase and phospholipase A effectively and partially inhibit lysophospholipase (43). Germ from normal corn varieties has been found to account for 10.1%–14.1% of the kernel dry weight (44,45). The lipid content, measured as FAME, represented 36%–40% of the germ (46,47), and germ contained 76%–83% of the total kernel lipids.

Examination of polar lipids extracted directly from mature corn germs will indicate the fatty acid composition and lipids of indigenous corn lecithin, and whether degradation has occurred during commercial processing. Table VI shows the percentages of nonpolar lipids, glycolipids, phospholipids and unsaponifiables found in the germ of three strains of corn. The nonpolar lipids made up the major fraction (87–92%) of the total lipids. Within the nonpolar lipid fraction, 93–96% was TG. Small quantities of diglycerides, steryl esters, FFA and monoglycerides made up the remainder of the nonpolar lipid fraction. Together, the glycolipid and phospholipid fractions accounted for 5–8% of the total lipids and the phospholipid fraction was the larger.

**TABLE 3-VI**
**Lipids of Corn Germ**

| Lipids | LG-11[a] | Waxy corn[b] | H51[c] |
|---|---|---|---|
| | | % of wt of total lipids | |
| Nonpolar lipids | 90.4 | 87.4 | 92.4 |
| Glycolipids | 1.9 | 2.6 | 2.0 |
| Phospholipids | 3.3 | 3.5 | 5.6 |
| Unsaponifiables | 4.4 | 6.5 | n.d.[d] |

[a]Seed from a triple cross hybrid forage corn from France (ref. 46).
[b]Seed from commercial waxy corn from Italy (ref. 46).
[c]Seed from inbred line grown at Urbana, IL, USA (ref. 47).
[d]n.d. = not determined.

Distribution of individual glycolipids and phospholipids in corn germ is given in Table VII. The proportions of the individual polar lipids vary among the strains. Several factors undoubtedly account for this variability. The corn strains came from different parts of the world, LG-11 from France, waxy corn from Italy and H51 from the USA. The extraction and purification procedure for the lipids of H51 (47) differed from that of the other two strains (46), although all three were extracted finally with hot, water-saturated butanol. The distribution of all the phospholipids was determined by phosphorus analyses, but Tan and Morrison (46) quantified glycolipids by GC of FAME and Weber (47) used sugar analysis. However, the data are in general agreement. The hybrid, LG-11, and the inbred, H51, probably resemble more closely present commercial hybrids than does the waxy corn. The waxy mutant gene has been shown to affect the oil content and fatty acid composition of a dent corn inbred line (48), although individual lipids were not examined.

The distribution of polar lipids in corn germ was similar in many respects to that of commercial corn lecithin. Glycolipids made up 35.8% of the weight of the polar lipids in LG-11 and 26.4% in H51 (Table VII). The percentage of glycolipids in the polar lipids of the commercial corn lecithin was 30.3% (Table III). The digalactosyl glycerides were present in larger amounts than were the monogalactosyl glycerides in both corn germ and the

**TABLE 3-VII**
**Distribution of Polar Lipids in Corn Germ**

| Lipid | LG-11[a] | Waxy corn[a] | H51[b] |
|---|---|---|---|
| | % of wt of polar lipids | | |
| Sterylglycoside ester | 3.2 | 16.8 | 0.5 |
| Monogalactosyldiglyceride | 5.7 (combined with next row) | 7.4 (combined with next row) | 2.9 |
| Monogalactosylmonoglyceride | | | ud.[d] |
| Cerebroside + sterylglycoside | n.d.[c] | n.d. | 10.7 |
| Cerebroside (OH-fatty acids) | n.d. | n.d. | 7.2 |
| Digalactosyldiglyceride | 19.3 | 6.6 | 3.9 |
| Digalactosylmonoglyceride | 7.6 | 12.4 | 0.8 (combined with next row) |
| Sulfoquinovosyldiglyceride | n.d. | n.d. | |
| Other glycolipids | — | — | 0.4 |
| N-acyl phosphatidylethanolamine | 4.9 | 4.9 | 1.0 |
| N-acyl lysophosphatidylethanolamine | 0.8 (combined with next row) | 1.2 (combined with next row) | 1.2 |
| Diphosphatidylglycerol | | | 1.4 |
| Phosphatidylethanolamine | 8.6 | 2.0 | 8.1 |
| Phosphatidylglycerol | 1.6 | 1.1 | 3.1 |
| Phosphatidylcholine | 33.1 | 24.1 | 40.2 |
| Phosphatidylinositol | 10.9 | 16.5 | 15.2 |
| Phosphatidic acid | 3.4 | 2.5 | 0.8 |
| Phosphatidylserine | — | 0.8 (combined with next row) | 0.7 |
| Lysophosphatidylethanolamine | — | | 0.2 |
| Lysophosphatidylcholine | 1.0 | 2.5 | 0.4 |
| Other phospholipids | — | 1.2 | 1.3 |

[a]See footnotes to Table VII for identification of seed. Data calculated from ref. 46.
[b]See footnote to Table VII for identification of seed. Data calculated from ref. 47.
[c]n.d. = not determined.
[d]ud. = undetected.

commercial lecithin. PC and PI were the major phospholipids in both. The most important differences were a lower percentage of PE and a higher percentage of PA in the commercial lecithin which may indicate some degradation of phospholipids during processing.

### Fatty Acid Compositions of H51

Table VIII shows the fatty acid compositions of nonpolar lipids, glycolipids and phospholipids which were isolated from the germ of inbred H51. This particular inbred has a lower level of polyunsaturated linoleic acid than most current commercial hybrids. Current hybrids produce commercial corn oil with ca. 60% linoleic acid. Thus, the nonpolar lipids from the H51 inbred had a lower level of linoleic acid than the nonpolar

TABLE 3-VIII
**Fatty Acid Compositions of Lipids from Germ of Corn Inbred H51[a]**

| Lipid | Fatty acid composition, (mol %) | | | | |
|---|---|---|---|---|---|
| | 16:0 | 18:0 | 18:1 | 18:2 | 18:3 |
| Nonpolar lipids | 16.6 | 1.3 | 31.8 | 49.5 | 0.8 |
| Glycolipids | 24.9 | 1.7 | 28.3 | 42.8 | 2.3 |
| Phospholipids | 24.9 | 1.3 | 31.1 | 42.0 | 0.7 |

[a]Data from ref. 47.

lipids of the commercial corn lecithin (Table IV).

Compared to the nonpolar lipids, glycolipids from germ had a higher percentage of linolenic acid, and both the phospholipids and glycolipids had higher levels of the saturated palmitic acid. The same trends were noted in the fatty acid compositions of the glycolipids and phospholipids of the corn lecithin. The fatty acid compositions of the individual glycolipids and phospholipids of the H51 germ were not determined.

## Influence of Genotype on Polar Lipids

Commercial seed corn producers usually do not monitor the fatty acid compositions of the hybrids that they sell. Yield potential, disease resistance and other desirable agronomic traits are the determining factors in the selection of parental lines. When superior lines are developed, the older lines are replaced. This process has been slow and gradual in the past. A commercial company sampled oil processed from midwestern corn in 1963 and in 1968 (49,50). During each sampling period, the fatty acid composition of the corn oil was relatively uniform, but in the five years between sampling periods, the average linoleic percentage increased from 58.7 to 61.9. The linoleic acid content of the private lines used as parental sources for corn hybrids during these years is unknown, but fatty acid analyses of the public lines indicated that the increase in polyunsaturation corresponded to changes in the public inbred lines used (23). Zuber (51) suggests that changes in parental sources may occur even more rapidly in the future. In his survey of US corn germplasm, he found that, due to a more intensive breeding effort, more new superior lines have been developed during the past decade than in the previous two decades.

Changes in the parental sources affect the fatty acid composition of the phospholipids as well as that of the oil (triglycerides). Reciprocal crosses were made of inbreds, C103D and B73, which differ widely in linoleic acid content (52). The lipids were isolated from the germ only, because inheritance of fatty acids was determined more easily in the diploid germ than in the triploid endosperm. Linoleic acid percentages in the $F_1$ hybrid seed of C103D and B73 were between those of the inbred parents for each class of lipid (Table IX). For this particular cross, a large maternal effect was observed. The percentage of linoleic acid tended toward the maternal parent in the reciprocal crosses. This maternal effect on linoleic acid levels was discernible in the phospholipids as well as in the TG. Genotype influenced the fatty acid compositions of the phospholipids and the TG even though each lipid class had its own characteristic fatty acid pattern. The TG had lower percentages of saturated palmitic acid and higher percentages of linoleic acid than the other lipid classes. PC had the highest level of oleic acid. PI had the highest percentage of palmitic acid. PE had the highest percentage of linoleic acid within the phospholipids. The same fatty acid patterns were observed in the lipid classes of the commercial corn lecithin.

**TABLE 3-IX**
**Fatty Acid Compositions of the Lipids of the Inbreds C103D and B73 and Their Reciprocal Crosses[a]**

| Lipid | Inbred or Cross | Fatty acid composition, (mol %) | | | | |
|---|---|---|---|---|---|---|
| | | 16:0 | 18:0 | 18:1 | 18:2 | 18:3 |
| Triglyceride | C103D | 13.3 | 2.0 | 43.4 | 40.3 | 1.1 |
| | C103D × B73 | 12.6 | 2.1 | 38.6 | 45.7 | 1.0 |
| | B73 × C103D | 12.2 | 2.2 | 33.7 | 50.8 | 1.0 |
| | B73 | 11.4 | 1.9 | 29.6 | 55.9 | 1.3 |
| Phosphatidylcholine | C103D | 20.8 | 2.3 | 49.5 | 26.6 | 0.7 |
| | C103D × B73 | 19.1 | 1.8 | 44.1 | 34.2 | 0.8 |
| | B73 × C103D | 20.1 | 1.8 | 37.6 | 39.6 | 0.9 |
| | B73 | 20.9 | 1.6 | 30.5 | 45.9 | 1.1 |
| Phosphatidylinositol | C103D | 38.9 | 4.0 | 29.8 | 26.8 | 0.5 |
| | C103D × B73 | 36.3 | 2.8 | 26.8 | 33.5 | 0.7 |
| | B73 × C103D | 37.5 | 2.5 | 20.9 | 38.4 | 0.7 |
| | B73 | 33.7 | 2.4 | 19.8 | 42.9 | 1.2 |
| Phosphatidylethanolamine | C103D | 23.0 | 3.9 | 38.4 | 34.2 | 0.5 |
| | C103D × B73 | 20.7 | 3.3 | 34.0 | 41.6 | 0.4 |
| | B73 × C103D | 20.6 | 2.2 | 26.7 | 49.8 | 0.7 |
| | B73 | 21.5 | 2.7 | 21.9 | 52.8 | 1.0 |

[a]Data from ref. 52.

A great diversity exists in fatty acid composition among corn inbreds, particularly in oleic and linoleic acid levels. Jellum (53) surveyed a large number of US and foreign lines and found a range of 14–64% for oleic acid and 19–71% for linoleic acid. These two fatty acids usually make up 80–90% of the fatty acids in corn oil and are correlated negatively with each other (53,54). Four inbreds that differ widely in fatty acid compositions are listed in Table X. The range was much larger for the TG, from 42.2% linoleic acid for H21 to 69.6% for NY16, but the effect of the genotype was also apparent in each phospholipid class. Therefore, changes in the fatty acid composition of the polar lipids of corn lecithin could occur due to the introduction of new commercial hybrids or deliberate selection of inbreds to produce a desired level of polyunsaturation.

**TABLE 3-X**
**Fatty Acid Compositions of Triglycerides and Phospholipids from Mature Kernels of Four Corn Inbreds[a]**

| | | Fatty acid composition, (mol %) | | | | |
|---|---|---|---|---|---|---|
| Lipid | Inbred | 16:0 | 18:0 | 18:1 | 18:2 | 18:3 |
| Triglyceride | H21 | 16.5 | 2.9 | 37.4 | 42.2 | 1.0 |
| | IHO | 12.9 | 2.1 | 35.4 | 48.8 | 0.8 |
| | K6 | 11.3 | 1.1 | 22.1 | 64.2 | 1.3 |
| | NY16 | 7.4 | 1.6 | 20.1 | 69.6 | 1.3 |
| Phosphatidylcholine | H21 | 20.0 | 1.5 | 42.1 | 36.0 | 0.5 |
| | IHO | 19.0 | 1.5 | 37.5 | 41.4 | 0.6 |
| | K6 | 23.9 | 1.8 | 28.3 | 45.2 | 0.8 |
| | NY16 | 18.6 | 2.2 | 25.8 | 52.3 | 1.1 |
| Phosphatidylinositol | H21 | 42.1 | 2.6 | 18.7 | 35.6 | 1.0 |
| | IHO | 39.5 | 2.2 | 20.6 | 37.4 | 0.3 |
| | K6 | 40.2 | 1.9 | 12.3 | 44.1 | 1.5 |
| | NY16 | 38.3 | 2.2 | 12.8 | 45.8 | 1.0 |
| Phosphatidylethanolamine | H21 | 20.8 | 0.8 | 23.5 | 54.2 | 0.7 |
| | IHO | 25.1 | 1.0 | 18.9 | 54.3 | 0.7 |
| | K6 | 24.0 | 1.0 | 15.3 | 59.0 | 0.7 |
| | NY16 | 20.6 | 1.9 | 18.1 | 58.3 | 1.0 |
| Phosphatidylglycerol | H21 | 36.3 | 2.0 | 19.7 | 40.3 | 1.7 |
| | IHO | 36.7 | 2.7 | 23.6 | 36.6 | 0.5 |
| | K6 | 35.5 | 2.5 | 15.6 | 42.6 | 3.8 |
| | NY16 | 34.2 | 3.6 | 16.6 | 44.3 | 1.4 |

[a]Data from ref. 23.

## Summary

The lipid and fatty acid compositions of samples of crude corn and soybean lecithins have been compared. The types of lipids were similar but the proportions varied. The ratio of glycolipids to phospholipids was 0.44 for corn lecithin and 0.17 for soybean. PC and PI were the major phospholipids in both lecithins. In soybean lecithin, the weight of PE equalled 84% of that of PI, but in corn, the weight of PE was only 20% of the PI weight. When compared with polar lipids isolated directly from corn germ, a higher level of PA and a lower level of PE in the commercial corn lecithin indicated some degradation of the phospholipids during processing.

The major differences in the fatty acid compositions of the commercial lecithins were a higher percentage of oleic acid and lower percentages of stearic and linolenic acids in corn compared to soybean. The lower level of linolenic acid should give corn lecithin greater resistance toward autoxidation and the development of off-flavors. Although corn lecithin has not been produced as a commercial product in recent years, the growth in the corn processing industry due to the increasing demand for corn sweeteners may again make corn lecithin a viable product. The effect of genotype on the fatty acid compositions of the phospholipids introduces the possibility that lecithins of selected fatty acid composition could be produced by corn breeding.

## References

1. Kraybill, H. R., P. H. Brewer and M. H. Thornton, U.S. Patent 2,353,571 (1944); Chem. Abstr. 38:P6121 (1944).
2. Scharf, A., U.S. Patent 2,678,320 (1954); Chem. Abstr. 48:10362b (1954).
3. Augustin, J., Seifensieder-Ztg. 65:633 (1938).
4. Taub, A., and M. Mattikow, U.S. Patent 2,532,206 (1950); Chem. Abstr. 45:1735f (1951).
5. Juhan, P. L., and E. W. Meyer, U.S. Patent 2,400,120 (1946); Chem. Abstr. 40:P5958p (1946).
6. Brian, R., JAOCS 53:27 (1976).
7. Scholfield, C. R., T. A. McGuire and H. J. Dutton, JAOCS 27:352 (1950).
8. Weber, E. J., JAOCS 58:898 (1981).
9. Rydhag, L., and I. Wilton, JAOCS 58:830 (1981).
10. Rouser, G., G. Kritchevsky and A. Yamamoto, in Lipid Chromatographic Analysis, edited by G. V. Marinetti, Vol. 1, Marcel Dekker, Inc., New York, NY, 1967, pp. 99–162.

11. Aneja, R., J. S. Chadha and R. W. Yoell, Fette Seifen Anstrichm. 73:643 (1971).
12. Parker, F., and N. F. Peterson, J. Lipid Res. 6:455 (1965).
13. Johnson, P., E. J. Weber and H. E. Carter, J. Lipid Res. 6:425 (1965).
14. Carter, H. E., and A. Kisic, J. Lipid Res. 10:356 (1969).
15. O'Dell, B. L., A. R. de Boland and S. R. Koirtyohann, J. Agric. Food Chem. 20:718 (1972).
16. Carter, H. E., W. D. Celmer, D. S. Galanos, R. H. Gigg, W. E. M. Lands, J. H. Law, K. L. Mueller, T. Nakayama, H. H. Tomizawa and E. Weber, JAOCS 35:335 (1958).
17. Wagner, H., W. Zofcsik and I. Heng, Z. Naturforsch. 24:922 (1969).
18. Carter, H. E., R. H. Gigg, J. H. Law, T. Nakayama and E. Weber, J. Biol. Chem. 233:1309 (1958).
19. Carter, H. E., D. S. Galanos, H. S. Hendrickson, B. Jann, T. Nakayama, Y. Nakazawa and B. Nichols, JAOCS 39:107 (1962).
20. Carter, H. E., S. Brooks, R. H. Gigg, D. R. Stroback and T. Suami, J. Biol. Chem. 239:743 (1964).
21. Carter, H. E., B. E. Betts and D. R. Stroback, Biochemistry 3:1103 (1964).
22. Carter, H. E., D. R. Stroback and J. N. Hawthorne, Biochemistry 8:383 (1969).
23. Weber, E. J., Cereal Chem. 55:572 (1978).
24. Van Nieuwenhuyzen, W., JAOCS 53:425 (1976).
25. Wolf, W. J., and D. J. Sessa, in Encyclopedia of Food Science, edited by M. S. Peterson and A. H. Johnson, Avi Publishing Co., Inc., Westport, CT, 1978, pp. 461–467.
26. Ho, C.-T., M. S. Smagula and S. S. Chang, JAOCS 55:233 (1978).
27. Smouse, T. H., JAOCS 56:747A (1979).
28. Curry, J., in Corn Annual, Corn Refiners Association, Inc., Washington, D.C., 1979, pp. 12–15.
29. Liebenow, R. C., in Corn Annual, Corn Refiners Association, Inc., Washington, D.C., 1980, p. 3.
30. Korab, H. E., in Corn Annual, Corn Refiners Association, Inc., Washington, D.C., 1981, pp. 14–15.
31. Gardner, H. W., and G. E. Inglett, J. Food Sci. 36:645 (1971).
32. Jugenheimer, R. W., Corn Improvement, Seed Production, and Uses, John Wiley & Sons, Inc., New York, NY, 1976, Ch. 14, p. 215.
33. Quarles, R. H., and R. M. C. Dawson, Biochem. J. 112:787 (1969).
34. Hvolby, A., JAOCS 48:503 (1971).
35. Kanamoto, R., Y. Wada, G. Miyajima and M. Kito, JAOCS 58:1050 (1981).

36. List, G. R., C. D. Evans, K. Warner, R. E. Beal, W. F. Kwolek, L. T. Black and K. J. Moulton, JAOCS 54:8 (1977).
37. Mounts, T. L., G. R. List and A. J. Heakin, JAOCS 56:883 (1979).
38. Nakayama, Y., K. Saio and M. Kito, Cereal Chem. 58:260 (1981).
39. Gardner, H. W., G. E. Inglett and R. A. Anderson, Cereal Chem. 46:626 (1969).
40. List, G. R., J. M. Avellaneda and T. L. Mounts, JAOCS 58:892 (1981).
41. Earle, F. R., Cereal Chem. 54:70 (1977).
42. Watson, S. A., and J. E. Freeman, Proc. Corn Sorghum Res. Conf. 30:251 (1975).
43. Bolling, H., W. Kempf, H. D. Ocker, W. Nierle and A. W. El Baya, Staerke 25:402 (1973).
44. Earle, F. R., J. J. Curtis and J. E. Hubbard, Cereal Chem. 23:504 (1946).
45. Morrison, W. R., in Advances in Cereal Science and Technology, Vol. 2, edited by Y. Pomeranz, American Association of Cereal Chemists, St. Paul, MN, 1978, pp. 221–348.
46. Tan, S. L., and W. R. Morrison, JAOCS 56:531 (1979).
47. Weber, E. J., JAOCS 56:637 (1979).
48. Flora, L. F., and R. C. Wiley, J. Am. Soc. Hort. Sci. 97:604 (1972).
49. Beadle, J. B., D. E. Just, R. E. Morgan and R. A. Reiners, JAOCS 42:90 (1965).
50. Reiners, R. A., and C. M. Gooding, in Corn: Culture, processing, products, edited by G. E. Inglett, Avi Publishing Co., Inc., Westport, CT, 1970, Ch. 13, p. 241.
51. Zuber, M. S., and L. L. Darrah, Proceedings of the Corn Sorghum Research Conference 35:234 (1980).
52. Weber, E. J., in The Resource Potential in Phytochemistry, edited by T. Swain and R. Kleiman, Plenum Press, New York, NY, 1980, Ch. 5, p. 97.
53. Jellum, M. D., J. Agric. Food Chem. 18:365 (1970).
54. Trifunovic, V., S. Ratkovic, M. Misovic, S. Kapor and J. Dumanovic, Maydica 20:175 (1975).

# Chapter Four
## Cottonseed Lecithin

**John P. Cherry**
Agricultural Research Service,
U.S. Department of Agriculture
Eastern Regional Research Center
600 E. Mermaid Ln.
Philadelphia, PA 19118

### Abstract

Industrial lecithin can be fractionated from cottonseed as pho and glycolipids after neutral lipids and protein-containing co are removed. The purity and special properties of the lecith improved by a number of solvent extraction and precipitation p Studies are determining the role of the phospholipids of lecit synthesis of triglycerides in maturing cottonseed, and the molecu the functionality of food ingredients. Lecithin, having both pola polar groups, has high surface activity and is reactive with bo protein, making it an excellent emulsifying agent in food system also slows autoxidation and enzyme hydrolysis of fats. Cottonseed low in linolenic acid and slows flavor and color deterioratior when blended with other vegetable oils. Gossypol binds to lecit oil extraction from glanded cottonseed, economically negating these phospholipids as a commercial source. New cultivars glandless or gossypol-free cottonseed, now being grown in southw western United States, have potential as commercial sources of edib

### Introduction

Phospholipids contribute to transport and metabolic reactions substances in living systems (1–5). Lecithin is the commercial name for the naturally occurring mixture of several phospholipic mainly at this time from soybeans, including lecithin (phosphati and cephalin (phosphatidylethanolamine, phosphatidylserine,

phatidylinositol) (Fig. 1; 5, 6). Diphosphatidylglycerol, lysophosphatidylcholine and a group of minor and unidentified polar lipids are present in small amounts (Fig. 2; 5, 7). Phospholipids are used as emulsifiers and antioxidants in many food systems (5, 8, 9). These applications generally take advantage of the primary property of phospholipids (polar and nonpolar groups contained within the molecular structures) to lower surface tension of liquid and dry substances into which they are incorporated, and to make homogeneous systems out of mixtures such as oil and protein, and as immiscible liquid phases.

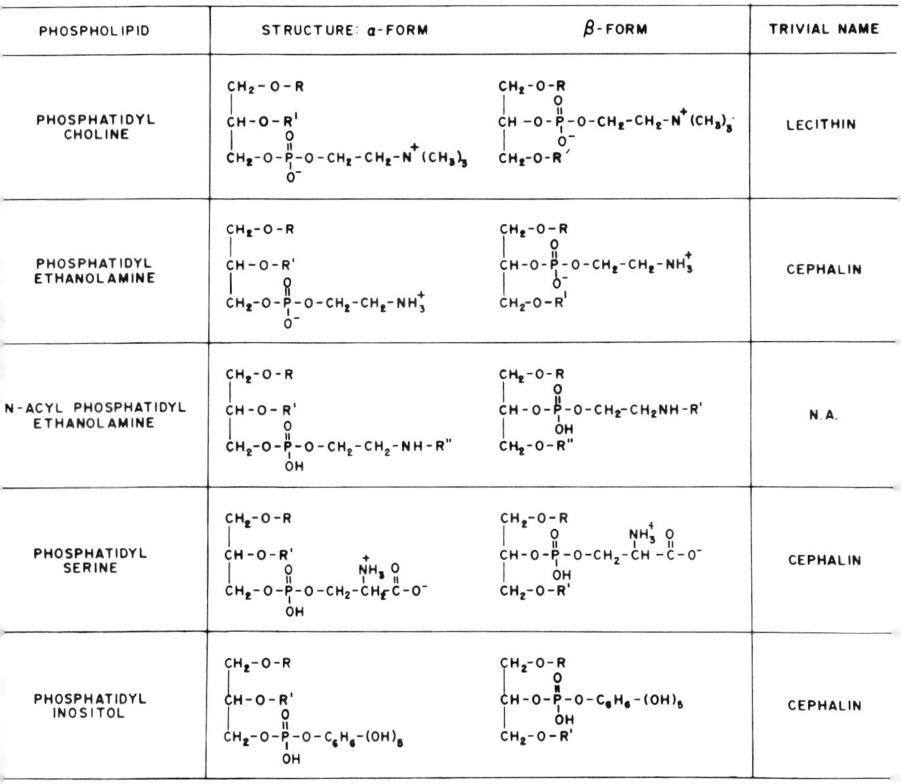


Fig. 4-1. Structures of phospholipids. R, R[1,11] = various fatty acids; N.A. = not available. (5)

The phospholipids phosphatidylcholine, phosphatidylethanolamine and phosphatidic acid were shown to contribute to antioxidant properties in

| PHOSPHOLIPID | STRUCTURE: α-FORM | TRIVIAL NAME |
|---|---|---|
| SPHINGOMYELIN | 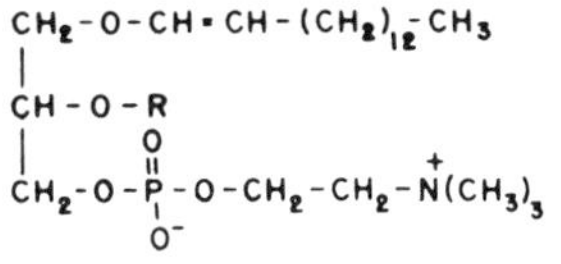 | N.A. |
| PHOSPHATIDYL GLYCEROL | 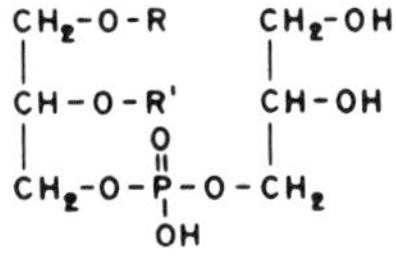 | N.A. |
| DIPHOSPHATIDYL GLYCEROL | 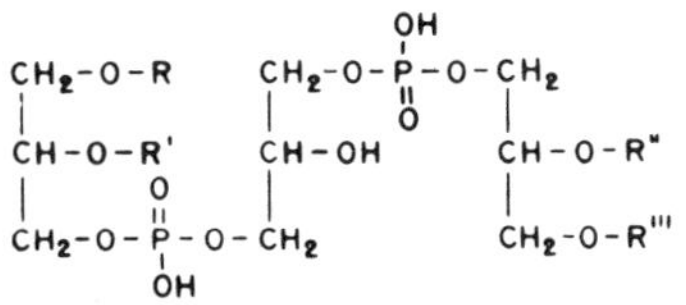 | CARDIOLIPIN |
| PHOSPHATIDIC ACID | 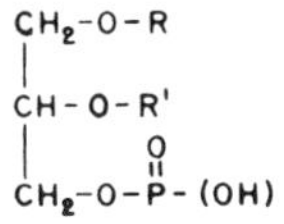 | N.A. |
| LYSOPHOSPHATIDYL CHOLINE | 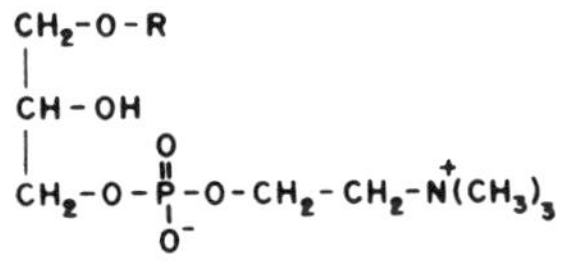 | LYSOLECITHIN |
| LYSOPHOSPHATIDYL ETHANOLAMINE | 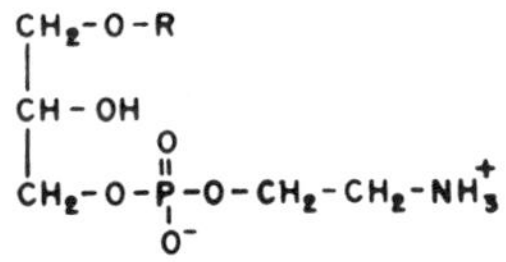 | LYSOCEPHALIN |
| LYSOPHOSPHATIDYL SERINE | 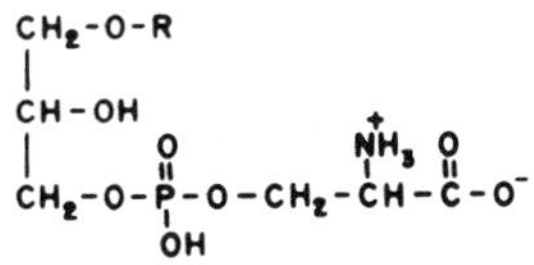 | N.A. |
| LYSOPHOSPHATIDYL INOSITOL | 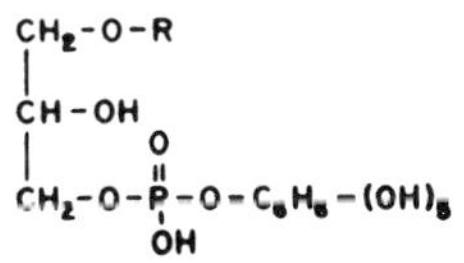 | N.A. |

Fig. 4-2. Structures of minor phospholipids. R, $R^{1,11,111}$ = various fatty acids; N.A. = not available. (5)

stored foods (10–12). Phosphatidylethanolamine gives the best protection against oxidation of unsaturated fatty acids. Addition of phosphatidylethanolamine and phosphatidylcholine reduced lipolysis, probably by interacting with the lipase system. The fatty acid portion of the phospholipid molecules played no role in protecting against food oxidation, although a lack of linolenic acid in cottonseed lecithin contributed to storage stability.

## Phospholipid Extraction and Composition

During the 1930s and 1940s, cottonseed phospholipids were, to a small extent, commercially available (13). Most of the phospholipids were in the nonoil materials (1.0–2.0%) separated from hydraulically pressed oil by alkali refining or water washing. The phospholipids could also be extracted from the nonoil material with either ether, petroleum naphtha or benzene. They were shown to be relatively insoluble in acetone, and once purified could be further separated into alcohol-soluble lecithins and alcohol-insoluble cephalins.

From a commercial cottonseed product, Olcott (14) obtained a bright yellow-colored powdery solid of purified phospholipid (phosphorus, 2.9%; nitrogen, 1.2%) and noted that it was relatively storage-stable. The fatty acid composition of these phospholipids from cottonseed was palmitic acid, 17.3%; stearic acid, 7.3%; arachidic acid, 2.8%; palmitoleic acid, 1.5%; oleic acid, 20.3%; linoleic acid, 44.4%; and unsaturated $C_{20-22}$ fatty acids, 6.4%. The absence of the triunsaturated linolenic acid contributed to the storage stability of cottonseed phospholipids.

Since highly pure lecithin with equivalent fatty acids as crude preparations would yield the considerably different analysis of about 4.0% phosphorus, 1.8% nitrogen, 70% fatty acid and no ash, Woolley (15) suggested that the phospholipid fraction of cottonseed oil did not consist predominantly of lecithins and cephalins, but contained others like "lipositol" found in soybean oil. In lipositol (phosphatidylinositol; Figs. 1, 2), the polyhydric alcohol with which the fatty acids are esterified, was suggested to consist of the cyclic alcohol inositol. Also esterified with the alcohol, or otherwise combined, are phosphoric acid, ethanolamine tartrate and galactose.

Lishkevich (16) found that the phospholipid content of various oilseeds was highest in cottonseed, followed by soybean, sunflower, flax, castor bean and peanut seed. He separated cottonseed phospholipids into three fractions including (a) acetone soluble, 16.5%, of which 46.2% was lecithin, and 53.8% was cephalin; (b) acetone insoluble, 76.5%, of which 53.2–59.4% was lecithin, and 40.6–46.8% was cephalin; and (c) benzene soluble, 7.0%, which was almost entirely lecithin. In other studies, associated impurities were removed from the phosphorus-containing residues by washing them with acidified water (<pH 3.0) or with 10% sodium chloride solution (17–21). The acetone insoluble portion was shown to contain bound gossypol (20).

The method of cottonseed crushing and extraction affects the phospholipid content in oil (22–25). Hydraulically pressed oils were richer in acetone insoluble materials, including phospholipids, than those extracted with either screw prepress-solvent or solvent methods (Table I). The greatest percentage of phospholipid recovered was from the solvent extracted oil (Table II). Total phospholipid content was highest in the screw prepress-solvent extracted oil. Phospholipids (% of total) extracted varied little among extraction methods and included phosphatidylinositol, lysolecithin, phosphatidylserine and phosphatidylethanolamine. The infra-red (IR) spectrum of lecithin revealed a band at 5.8 $\mu$ corresponding to the carbonyl ester of glycerophosphatide, and 10.3 $\mu$, a property of phospholipids with choline (Fig. 3; 25). Bands at 6.8, 8.1, 8.6, and 9.1 represent $CH_{1-3}$, P=O, C–O–C, and P–O– groups, respectively. Fatty acid composition of phospholipids from cottonseed oil varied according to the extraction method (Table III). Neutral lipids from cottonseed contained more polyunsaturated fatty acids than the phospholipids; the reverse was true for the saturated fatty acids. Phospholipids extracted by hexane contained the highest percentage of linoleic acid. The highest percentage of oleic acid was present in phospholipids of chloroform-extracted oil. Ether-extracted phospholipids were highest in unsaturated fatty acids.

For comparison, Alter and Gutfinger (26) presented percentages of fatty acids of prepress-solvent and solvent extracted cottonseed neutral and polar lipid fractions (Table IV). The content of saturated fatty acids was significantly higher in the phospholipid fractions than in the corresponding neutral lipid portions. The increase in saturated fatty acids was associated with a corresponding reduction in the dienoic fatty acid, linoleic acid. Overall, neutral lipids from cottonseed contained more polyunsaturated fatty acids than their phospholipids.

Gas liquid chromatographic (GLC) analysis of fatty acid methyl esters derived from total lecithin of cottonseed showed approximately 80% unsaturated fatty acids (27). The fatty acids in the $\beta$-position of cottonseed lecithin that are liberated by phospholipase-$A_2$ hydrolysis consist mainly of oleic (29.5%) and linoleic (61.9%) acids. Cottonseed lecithin also contains 8.6% palmitic acid in the $\beta$ position. The $\alpha$ position consists mainly of palmitic, oleic and linoleic acids, and a small amount of stearic acid. The unsaturated palmitic and stearic acids make up 60% of the $\alpha$-position.

The observations showing that percentage of individual phospholipids in

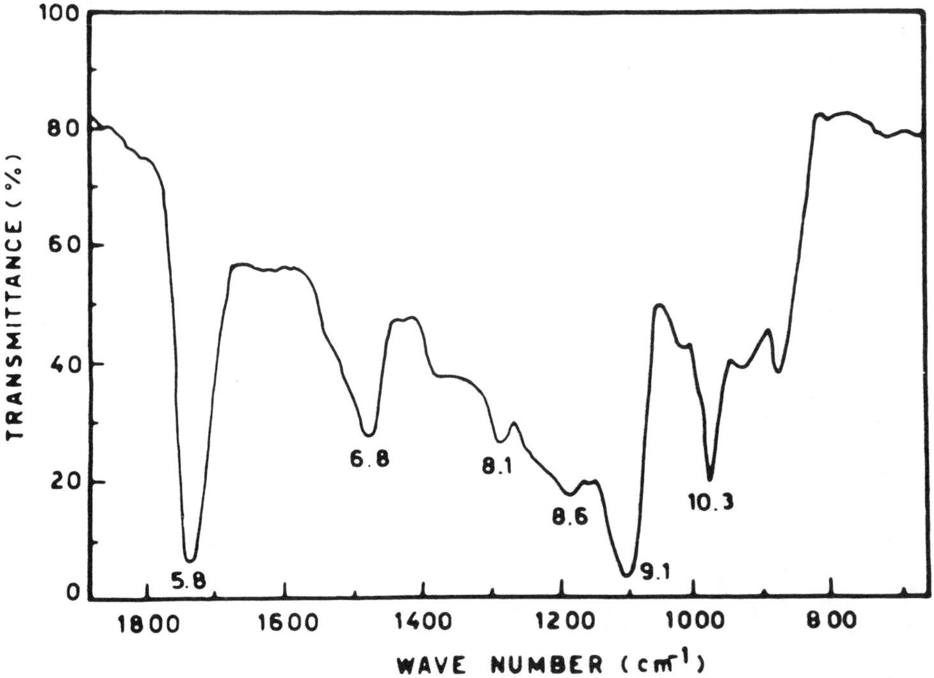


Fig. 4-3. Infra-red spectrum of cottonseed lecithin. (25)

oil varies with the processing technique were emphasized (27, 28). Such differences are less prevalent when the individual phospholipid composition is calculated as percentage of total phospholipids. For example, cooking crushed cottonseed enhances the extraction of phospholipids with oil. In hydraulic and screw press methods, the greater amount of phospholipids remains in the cake than oil. Therefore, solvent extracted oil contains higher phospholipid content than oil from the press procedures. The average amounts of individual components in the phospholipids are 26.3% phosphatidylcholine, 20.2% phosphatidylserine, 17.7% phosphatidylethanolamine, 9.3% phosphatidylinositol, 5.7% lysolecithin and 12.7% unknown substances.

Vijayalakshmi et al. (29) showed that 92% of the cottonseed phospholipids could be eluted from a silicic acid column and contained 33% phosphatidylcholine, 20% phosphatidylethanolamine and 37% phosphatidylinositol. Other phospholipids (8%) consisted of phytoglycolipid, phytosphingosine and phosphatidylserine (Figs. 1, 2). This study also showed that cottonseed phospholipids are rich in simple inositol phospholipids (Fig. 2). The sum of

**TABLE 4-I**
**Composition (%)[a] of Cottonseed Soapstock Prepared by Different Oil Extraction Methods (23)**

| | Extraction method | | |
|---|---|---|---|
| Analysis | Hydraulic press | Screw prepress-solvent | Solvent |
| Moisture, volatiles | 46.03 | 43.47 | 44.11 |
| Free fatty acids | 0.17 | 0.27 | 0.20 |
| Total fatty acids | 27.99 | 28.88 | 27.68 |
| Neutral oil | 10.45 | 13.67 | 10.99 |
| Ash | 3.05 | 3.15 | 3.05 |
| Successive extractions: | | | |
| Hexane soluble | 0.54 | 0.64 | 1.47 |
| Ether soluble | 2.68 | 3.37 | 3.14 |
| Chloroform soluble | 8.07 | 8.19 | 8.25 |
| Successive extractions: | | | |
| Acetone insoluble from hexane soluble | 8.73 | 6.49 | 4.60 |
| Acetone insoluble from ether soluble | 15.97 | 8.78 | 7.42 |
| Acetone insoluble from chloroform soluble | 89.99 | 51.21 | 47.20 |

[a]Averages of two lots.

phosphatidylethanolamine and phosphatidylinositol, about 59%, was in fair agreement with values of alcohol-extracted cephalin, at 71% (Fig. 1; 30). The phospholipids contained gossypol, probably bound to phosphatidylethanolamine and phosphatidylinositol (Fig. 4). This interaction was disrupted by conditions of silicic acid chromatography (29), or mildly acidic pH (31).

In summary, the saturated:unsaturated fatty acid ratio of cottonseed phospholipids is approximately 1:2 (Table V; 5, 22). Palmitic acid constitutes 90% of the total saturated fatty acids, while linoleic acid is approximately 80% of the total unsaturated fatty acids. Isolated cottonseed crude phospholipids contain approximately 9.13% total gossypol, some of which is in the free form. Most of the gossypol is associated with phosphatidylethanolamine and phosphatidylserine. Yields of phosphatidylcholine, phosphatidylethanolamine and phosphatidylserine are 34.85%, 20.05% and 16.2%, respectively, of the acetone-insoluble residue. Total saturated fatty acid composition is higher in phosphatidylethanolamine than in phosphatidylcholine or phosphatidylserine. Phosphatidylethanolamine has the lowest amount of unsaturated fatty acids of the three phospholipids. Palmitic acid

TABLE 4-II
**Phosphulipid Composition[a] (%) of Cottonseed Soapstock Prepared by Different Oil Extraction Methods (23)**

| Phospholipid | Hydraulic press: In soapstock | Hydraulic press: Of total phospholipid | Screw prepress-solvent[b]: In soapstock | Screw prepress-solvent[b]: Of total phospholipid | Solvent[b]: In soapstock | Solvent[b]: Of total phospholipid |
|---|---|---|---|---|---|---|
| Phosphatidylinositol | 0.03 | 5.91 | 0.05 | 5.72 | 0.35 | 5.96 |
| Lysolecithin | 0.05 | 9.23 | 0.07 | 9.39 | 0.06 | 9.25 |
| Phosphatidylserine | 0.17 | 29.92 | 0.27 | 28.50 | 0.21 | 28.63 |
| Phosphatidylethanolamine | 0.15 | 34.17 | 0.22 | 35.24 | 0.17 | 35.13 |
| Unknown | 0.06 | 12.61 | 0.10 | 12.99 | 0.08 | 13.22 |
| % Recovery | — | 91.83 | — | 91.83 | — | 92.18 |
| Total phospholipid | 0.51 | — | 0.78 | — | 0.59 | — |

[a]Averages of two lots.
[b]Hexane.

TABLE 4-III
**Fatty Acid Composition (%) of Cottonseed Soapstock and Phospholipids (23)**

| Fatty acid | Hexane soluble soapstock | Phospholipid of hexane soluble fraction | Ether soluble soapstock | Phospholipid of ether soluble fraction | Chloroform soluble soapstock | Phospholipid of chloroform soluble fraction |
|---|---|---|---|---|---|---|
| $C_{14:0}$ | 0.36 | 0.24 | 0.26 | 0.21 | 0.31 | 0.54 |
| $C_{16:0}$ | 20.68 | 22.26 | 20.16 | 34.26 | 24.96 | 10.87 |
| $C_{18:1}$ | 14.26 | 16.57 | 10.47 | 8.57 | 9.34 | 35.33 |
| $C_{18:2}$ | 64.70 | 60.93 | 69.11 | 56.96 | 65.39 | 53.26 |

is the major saturated fatty acid, and oleic and linoleic acids make up most of the unsaturated fatty acids.

## Varietal Phospholipids

Shustanova et al. (32) presented yields of free and bound lipids of cottonseed from two *Gossypium hirsutum* varieties, S-4727 and S-6029 (Table VI). The lipids were extracted by successively treating full fat meals with a number of

**TABLE 4-IV**
**Fatty Acids (%) in Phospholipids and Neutral Lipids of Cottonseed Oil (26)**

| | Extraction technique | | | |
|---|---|---|---|---|
| | Prepress-solvent[b] | | Solvent[b] | |
| Fatty acid[a] | Phospholipids | Neutral lipids | Phospholipids | Neutral lipids |
| $C_{14:0}$ | 1.4 | 1.2 | 1.4 | 1.3 |
| $C_{16:0}$ | 31.5 | 25.6 | 30.0 | 25.3 |
| $C_{18:0}$ | 4.0 | 2.4 | 3.8 | 2.3 |
| $C_{18:1}$ | 20.8 | 18.8 | 21.0 | 18.8 |
| $C_{18:2}$ | 42.3 | 52.0 | 43.8 | 52.3 |
| Saturated fatty acids | 36.9 | 29.2 | 35.2 | 28.9 |
| Polyunsaturated fatty acids | 42.3 | 52.0 | 43.8 | 52.3 |

[a]$C_{16:1}$, trace; $C_{18:3}$, none detected.
[b]Hexane.

solvents and fractionating the components by silica gel two-dimensional chromatography. The main components of the phospholipid fraction were identified as phosphatidylcholine (52.0%), phosphatidylinositol (21.8%), and phosphatidylethanolamine (11.4%). Minor components were polyglycerophosphatides (two components, 2.8% and 5.8%), lysophosphatidylcholine (1.7%) and unidentified polar phospholipids (4.5%).

On the whole, the total phospholipid composition was similar for the two cotton varieties (Table VII). Variety S-4727 contained more unsaturated fatty acids than variety S-6029. This was particularly noticeable in the total phospholipid fraction. Thus, the potential exists for breeding for select fatty acid composition in cottonseed phospholipids. The fatty acid compositions of the total phospholipid fraction of both varieties were more saturated than the other lipid extracts; this was due mostly to the high palmitic acid composition. Of the unsaturated fatty acids, linoleic and oleic acids predominated. Overall, the free lipids extracted by acetone contained more oleic acid and less linoleic acid than the petroleum ether extract.

## Minor Phospholipids

Minor phospholipids, cerebroside and phytoglycolipid, are isolated by extensive sequential fractionation with solvent (acetic acid, benzene insoluble), silicic acid column chromatography and preparative thin layer chromatography in separate analyses, (93:7 and 30:70 chloroform:methanol separate cerebroside and phytoglycolipid, respectively) (33). The sugar moieties of

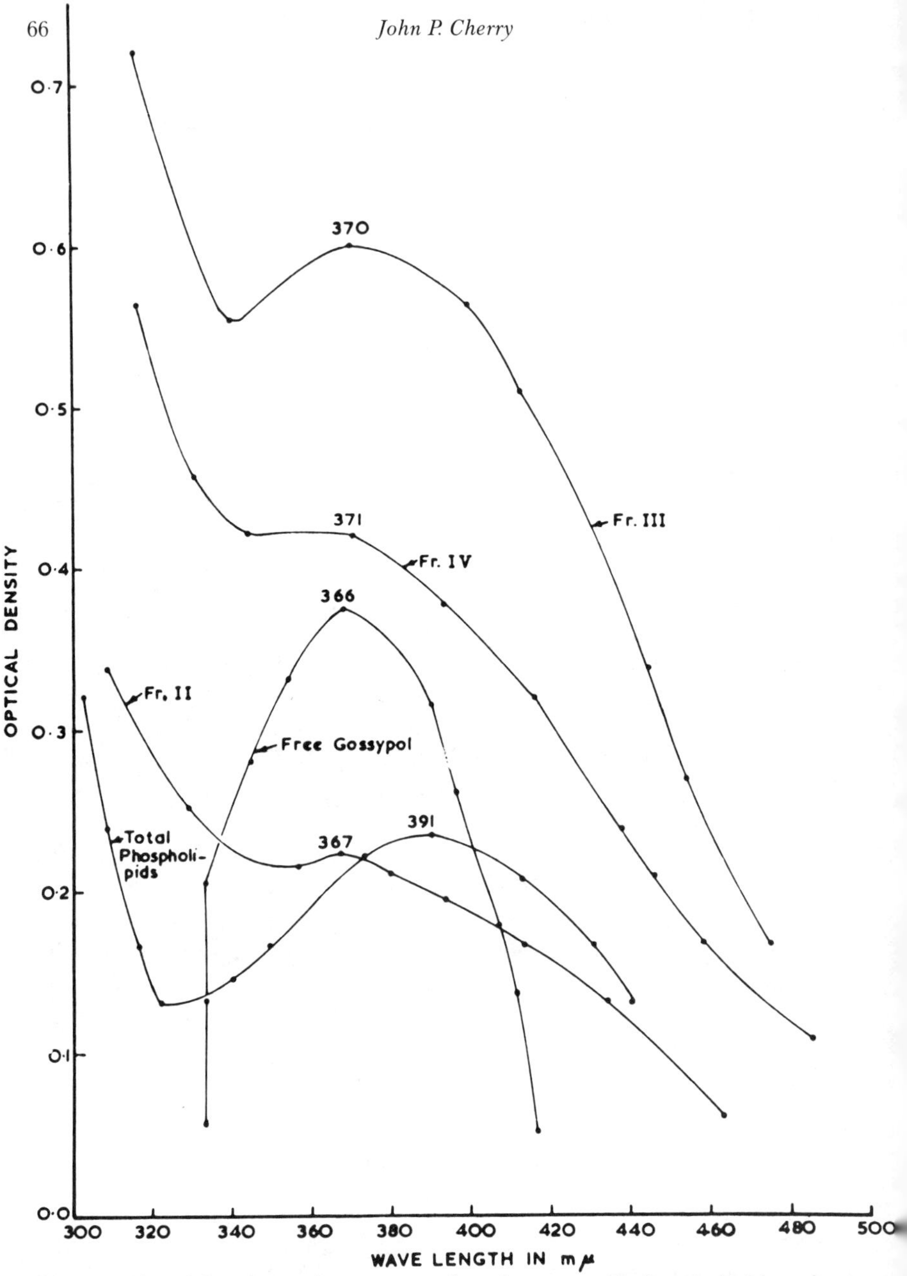


Fig. 4-4. Ultraviolet absorption spectra of total cottonseed phospholipids, column chromatographic fractions and free gossypol; given is the wavelength of λ maximum. Fr. (Fraction) II = glycerides + gossypol; Fr. III = phosphatidylethanolamine + gossypol; Fr. IV = phosphatidylinositol + gossypol. (29)

**TABLE 4-V**
**Composition of the Acetone-Insoluble Fraction of Cottonseed Oil** (5, 22)

| Component | Crude phospholipids[a] | Phosphati-dylcholine[b] | Phosphatidyle-thanolamine[b] | phosphati-dylserine[b] |
|---|---|---|---|---|
| Extracted | 2.24 | 34.85 | 20.05 | 16.20 |
| Nitrogen | 1.19 | 1.58 | 1.34 | 1.19 |
| Phosphorus | 2.49 | 3.37 | 2.67 | 2.21 |
| Total gossypol | 9.13 | 2.34 | 22.43 | 19.90 |
| Free gossypol | 0.02 | 2.24 | 0.05 | 0.01 |
| Fatty acid | | | | |
| $C_{8:0}$ | — | — | 0.4 | 0.3 |
| $C_{12:0}$ | — | — | 0.4 | 0.4 |
| $C_{14:0}$ | 0.4 | 0.3 | 0.4 | 0.6 |
| $C_{16:0}$ | 32.9 | 31.1 | 33.7 | 33.3 |
| $C_{16:1}$ | 0.5 | 0.3 | 0.3 | 0.6 |
| $C_{18:0}$ | 2.7 | 2.8 | 2.2 | 0.3 |
| $C_{18:1}$ | 13.6 | 11.0 | 11.5 | 14.4 |
| $C_{18:2}$ | 50.0 | 54.0 | 49.0 | 50.4 |
| Unknown | — | — | 2.3 | — |
| Saturated fatty acids | 36.0 | 34.2 | 37.1 | 34.9 |
| Unsaturated fatty acids | 64.1 | 65.3 | 60.8 | 65.4 |

[a]Percentage of crude oil.
[b]Percentage of crude phosopholipids.

cerebroside are glucosamine, glucose and rhamnose. Those of phytoglycolipid are glucuronic acid, galactose, glucose and rhamnose. The IR spectra of these components confirm their structures. Specifically, in addition to the bands noted for lecithin (Fig. 3), cerebroside has bands at 6.04 and 6.4 $\mu$ corresponding to amides I and II, which are characteristic of sphingolipids, while phytoglycolipid has a band at 5.7 $\mu$, typical of glycerophosphatides. These phospholipids contain substantial amounts of hydroxylated fatty acids. Hydroxy:nonhydroxy fatty acid ratios in cerebroside and phytoglycolipid are 1:9 and 1:16, respectively. Oleic acid ($C_{18:1}$) makes up most of the nonhydroxy acids (45.9% and 60.4% for cerebroside and phytoglycolipid, respectively). Nonhydroxy linoleic acid ($C_{18:2}$) is the main polyunsaturated fatty acid, making up only 0.6% of the fatty acids in cerebroside and 10.0% in phytoglycolipid. The total saturated nonhydroxy acids in the minor phospholipids are 53.2% and 19.5%, respectively. The major saturated nonhydroxy acid in cerebroside is $C_{24:0}$ (25.6%); others include stearic, palmitic and arachidic acids. No hydroxy $C_{24}$ is present in either of the minor phospholipids. Hydroxy $C_{18:1}$ is a major fatty acid in both minor phospholipids

TABLE 4-VI

**Successive Extraction (%) with Various Solvents of Free and Bound Lipids from Seeds of Two Cotton Varieties (32)**

| | | | Yield | | | Distribution of phosphorus | |
|---|---|---|---|---|---|---|---|
| Solvent | Extracted compounds | Variety | Kernel basis | Fraction basis | Phosphorus content | Kernel basis | Total phosphorus |
| | | Free lipids | | | | | |
| Petroleum ether | Glycerides, fatty acids, | S-4727 | 31.3 | 72.4 | 0.0045 | 0.0014 | 2.5 |
| | gossypol, etc. | S-6029 | 26.3 | 72.4 | 0.0038 | 0.0010 | 1.7 |
| Acetone | Gossypol, glycerides, fatty | S-4727 | 6.1 | 14.2 | 0.050 | 0.003 | 5.2 |
| | acids, phospholipids, etc. | S-6029 | 5.0 | 13.5 | 0.046 | 0.002 | 3.4 |
| | | Bound lipids | | | | | |
| Chloroform-methanol (2:1)[a] | Phospholipids, carbo- | S-4727 | 6.1 | 13.4 | 1.20 | 0.0525 | 92.3 |
| | hydrates, glycerides, etc. | S-6029 | 5.3 | 14.4 | 1.12 | 0.0550 | 94.9 |

[a]Best extraction technique.

**TABLE 4-VII**
**Fatty Acid Composition (%) of Free and Bound Lipids from Seeds of Two Cotton Varieties (32)**

| | Extract | | | | | |
|---|---|---|---|---|---|---|
| | Petroleum ether | | Acetone | | Total phospholipid | |
| Fatty acid | S-4727 | S-6029 | S-4727 | S-6029 | S-4727 | S-6029 |
| $C_{10:0}$ | 0.5 | 2.2 | 0.4 | 1.2 | 0.6 | 3.5 |
| $C_{12:0}$ | 0.4 | 0.9 | 0.3 | 0.6 | 0.5 | 0.7 |
| $C_{14:0}$ | 0.5 | 0.7 | 0.5 | 0.5 | 0.5 | 0.5 |
| $C_{16:0}$ | 24.0 | 23.4 | 22.8 | 23.5 | 25.4 | 27.9 |
| $C_{16:1}$ | trace | 1.3 | 0.5 | 1.2 | 0.7 | 0.9 |
| $C_{18:0}$ | 0.7 | 0.8 | 1.4 | 1.1 | 2.3 | 0.9 |
| $C_{18:1}$ | 18.1 | 13.6 | 25.4 | 21.6 | 17.3 | 13.2 |
| $C_{18:2}$ | 54.3 | 57.1 | 48.1 | 50.3 | 52.0 | 52.4 |
| $C_{18:3}$ | 1.5 | 0 | 0.7 | 0 | 0.7 | 0 |
| Total saturated | 26.1 | 28.0 | 25.3 | 26.9 | 29.3 | 33.5 |
| Total unsaturated | 73.9 | 72.0 | 74.7 | 73.1 | 70.7 | 66.5 |

(40.9% and 50.7%, respectively). Hydroxy $C_{18:2}$ is present in amounts of 9.0% and 10.0%, respectively. Total saturated hydroxy acid composition, mainly stearic acid, is 44.3% and 30.5%, respectively. Cerebroside contains 1.43% N and no P; phytoglycolipid has 1.37% N and 2.68% P. The presence or absence of phosphorus is the major difference of the two minor lipids. Gossypol is not present in either fraction; no available amine groups are present for reaction with gossypol. Chromatographs of cerebroside and phytoglycolipid acid hydrolysates show the presence of a long chain component that is thought to be a phytosphingosine.

## Phospholipids in Maturing Cottonseed

Changes in phospholipid composition during maturation of cottonseed are presented in Table VIII (34). The total phospholipid composition of five- and ten-day old cottonseed was very high (30.5% and 23.7% of oil, respectively). Between 15 and 40 days, the composition dropped to between 5.71% and 4.06%, and at 50 and 60 days, 3.20% and 2.38%. During the growing interval, lecithin increased from 25.8% at five days to 38.2% of total phospholipids at 60 days. Combined cephalin (phosphatidylinositol, phosphatidylserine and phosphatidylethanolamine) percentages were higher than those of lecithin up to 50 days, then lessened as the seed matured to 60 days. Overall, the individual phospholipids in the cephalin fraction vary during the maturing interval; only phosphatidylethanolamine decreases to a lower amount at 60 days compared to 5 days. Variations also were noted for phosphatidylcholine.

Except for the higher total saturated fatty acid content, most fatty acid amounts of the phospholipid fraction in the maturing cottonseed are similar to those in the oil from which they were separated (Table IX). The percentage of total saturated fatty acids, especially palmitic acid which makes up most of these fatty acids, decreases from 68.4% to 24.7% during the 60-day growing period; palmitic acid decreases from 59.0% to 23.1%. The unsaturated fatty acids, oleic and linoleic acids, increase significantly, 15.1% to 28.2% and 14.7% to 45.8%, respectively.

## Gossypol-Phospholipid Reactions

The heat and moisture of the old hydraulic press method of extracting oil from gossypol-containing glanded cottonseed caused most of the pigment to bind to constituents in the meal rather than those in the oil. Changes in oil extraction processes produced oils that contained considerable amounts of free gossypol pigments that bound to crude phospholipids and caused color and toxicity problems (5, 35). Cottonseed products, or blends containing gossypol, intended for human use in the United States must contain no more than 0.045% free gossypol (Food and Drug Administration). The Protein Advisory Group of the United Nations has set limits of 0.6% free gossypol and 1.2% total gossypol for human consumption in its programs (36). The dark-brown color caused by gossypol in cottonseed meal, oil and phospholipid extracts also limits their use in foods.

In general, cottonseed processing involves the following steps: (a) eliminate leaves, twigs, pieces of bolls and sand; (b) remove linters; (c) dehull; (d) screw press, prepress-solvent or solvent extract; and (e) desolventize, toast, and grind or pelletize products (Fig. 5; 37–39). The crude oil is purified by heating and treating with sodium hydroxide to remove soapstock, or foots, which contains most of the phospholipids. It is this refining process that removes the darker coloring materials such as gossypol by binding them to phospholipids to leave a clear yellow-colored oil. Bleaching clay is used to remove any remaining colored substances in the oil.

The ultraviolet spectrum (391 $\mu$) of the total phospholipid extract and fractionation studies show that gossypol in cottonseed oil is present in a bound form with phosphatidylethanolamine and phosphatidylserine (Table X, 40). These authors also presented the IR spectra of the two phospholipids and showed that both of them had a band at 6.2 $\mu$ which is typical of a C=N

**TABLE 4-VIII**

**Phospholipid Composition (%) During Cottonseed Development and Maturity (34)**

| Day: | 5 | | 10 | | 15 | | 20 | | 30 | | 40 | | 50 | | 60 | |
|---|---|---|---|---|---|---|---|---|---|---|---|---|---|---|---|---|
| Phospholipid | In O[a] | In P[b] | In O | In P | In O | In P | In O | In P | In O | In P | In O | In P | In O | In P | In O | In P |
| PI[c] | 4.19 | 13.7 | 2.90 | 12.3 | 0.64 | 11.2 | 0.56 | 11.5 | 0.57 | 13.3 | 0.56 | 13.8 | 0.44 | 13.8 | 0.33 | 13.9 |
| LS | 4.96 | 16.3 | 3.62 | 15.3 | 0.75 | 13.1 | 0.61 | 12.5 | 0.56 | 13.0 | 0.52 | 12.8 | 0.41 | 12.8 | 0.30 | 12.6 |
| PS | 4.47 | 14.7 | 3.04 | 12.9 | 0.76 | 13.3 | 0.60 | 12.3 | 0.54 | 12.6 | 0.52 | 12.8 | 0.40 | 12.5 | 0.28 | 11.8 |
| LE | 7.87 | 25.8 | 7.19 | 30.4 | 1.73 | 30.3 | 1.58 | 32.4 | 1.40 | 32.6 | 1.31 | 32.3 | 1.09 | 34.1 | 0.91 | 38.2 |
| PE | 5.25 | 17.2 | 3.57 | 15.1 | 1.02 | 17.9 | 0.78 | 16.0 | 0.68 | 15.8 | 0.61 | 15.0 | 0.40 | 12.5 | 0.26 | 10.9 |
| U | 3.76 | 12.3 | 3.33 | 14.1 | 0.81 | 14.2 | 0.74 | 15.2 | 0.55 | 12.8 | 0.54 | 13.3 | 0.46 | 14.4 | 0.30 | 12.6 |
| Total % | 30.50 | | 23.65 | | 5.71 | | 4.87 | | 4.30 | | 4.06 | | 3.20 | | 2.38 | |

[a]O = oil.

[b]P = total phospholipids.

[c]PI: phosphatidylinositol; LS: lysolecithin; PS: phosphatidylserine; LE: lecithin; PE: phosphatidylethanolamine; U: unknown phospholipid.

**TABLE 4-IX**
**Fatty Acid Composition (%) of Mixed Phospholipid during Development and Maturity of Cottonseed (34)**

| | Days after flowering | | | |
|---|---|---|---|---|
| Fatty acid[a] | 5 | 15 | 30 | 60 |
| $C_{12:0}$ | 2.4 | 1.2 | 0.5 | – |
| $C_{14:0}$ | 5.9 | 1.8 | 0.9 | 0.6 |
| $C_{16:0}$ | 59.0 | 37.2 | 36.4 | 23.1 |
| $C_{16:1}$ | 1.6 | 0.6 | 1.5 | 1.3 |
| $C_{18:0}$ | 1.1 | 2.2 | 1.0 | 1.0 |
| $C_{18:1}$ | 15.1 | 22.2 | 23.6 | 28.2 |
| $C_{18:2}$ | 14.7 | 34.6 | 36.0 | 45.8 |

[a]Relative areas under the curves for the methyl esters of the component fatty acids as estimated from GLC tracings.

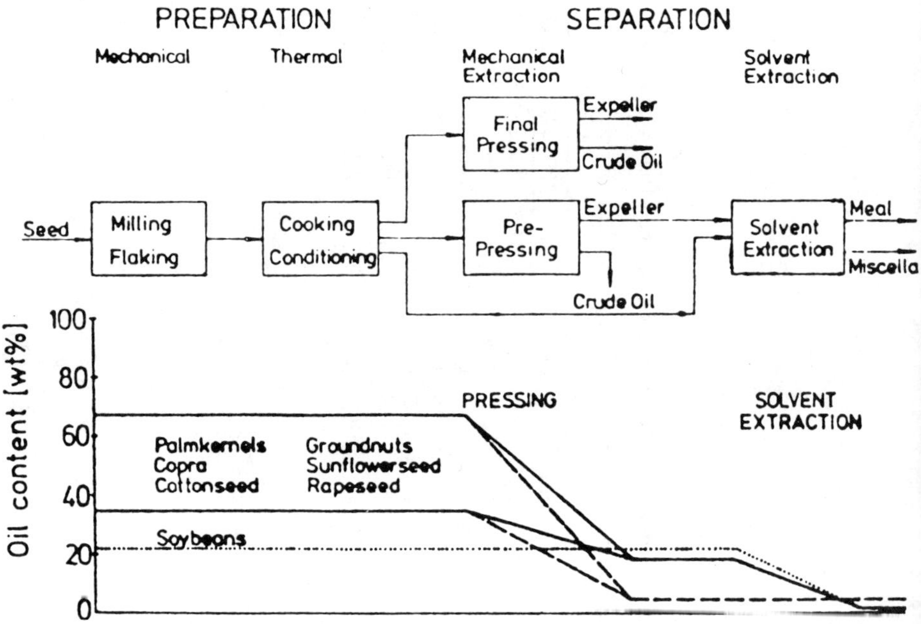


Fig. 4-5. Oil recovery processes. (38)

bond reactive with gossypol, i.e., interaction of the free amino group of the phospholipids with the aldehydes of gossypol. The components formed include mono- and diphosphatidylethanolamine-monogossypol, mono- and diphosphatidylserine-monogossypol, and monophosphatidylethanolamine monophosphatidylserine-monogossypol (Fig. 6). Phosphorus, nitrogen, and gossypol IR spectra analyses revealed that N:P:gossypol ratios are 2.14:1.94:1 to 2.32:2.11:1 for phosphatidylethanolamine and 2.19:1.89:1 to 2.28:1.88:1 for phosphatidylserine (Table X; 22).

**TABLE 4-X**
**Gossypol Composition (%) of Cottonseed Phospholipids from Two *Gossypium* Species (40)**

| | Nitrogen (N) | Phosphorus (P) | Gossypol | | Ratio of N:P: Gossypol |
|---|---|---|---|---|---|
| | | | Total | Free | |
| Phosphatidylethanolamine | | | | | |
| *G. barbadense* | 1.47 | 2.93 | 25.64 | 0.06 | 2.14:1.94:1 |
| *G. hirsutum* | 1.21 | 2.41 | 19.21 | 0.04 | 2.32:2.11:1 |
| Phosphatidylserine | | | | | |
| *G. barbadense* | 1.27 | 2.31 | 20.95 | 0.04 | 2.28:1.88:1 |
| *G. hirsutum* | 1.10 | 2.11 | 18.85 | 0.03 | 2.19:1.89:1 |

## Glandless Cottonseed Phospholipids

Glandless or gossypol-free cottonseed now provides potential sources of commercial food-grade oil, lecithin or phospholipids, and meals or flours that are free of gossypol pigments, color and toxicity problems (5, 37–39).

Glandless cottonseed can be processed much more efficiently than those containing gland-filled gossypol to high quality edible kernel, oil and vegetable protein products (Fig. 7; 37–39). Kernels and meat fines can be flaked and extracted with hexane to produce high quality oil and defatted meal. With glandless cottonseed, there is no need for expensive heat treatments during processing to bind gossypol in the phospholipid fraction and meal. A refined oil from glandless cottonseed requires less purification processing, and as a result, less loss of neutral oil. The advent of glandless, or gossypol-free, cottonseed provides an opportunity to produce a food-grade phospholipid fraction as a by-product of edible oil products. Glandless cottonseed oil and lecithin products thus become more economically attractive by increasing revenues, decreasing waste-disposal costs and reducing emulsion problems that occur during processing.

The phospholipid and fatty acid compositions of the lecithin fraction

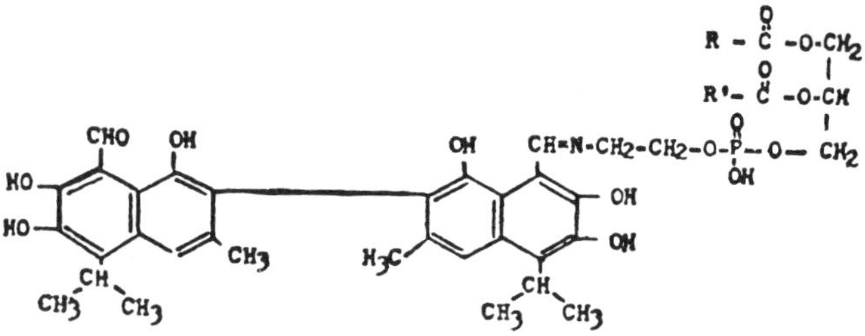

I-A : Monophosphatidyl ethanolamine-monogossypol

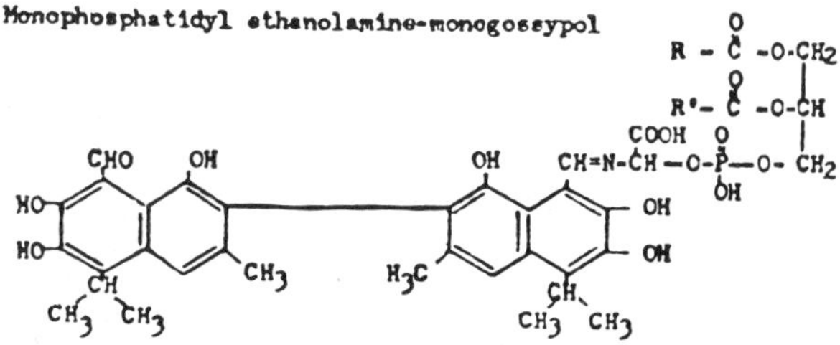

I-B : Monophosphatidyl serine-monogossypol

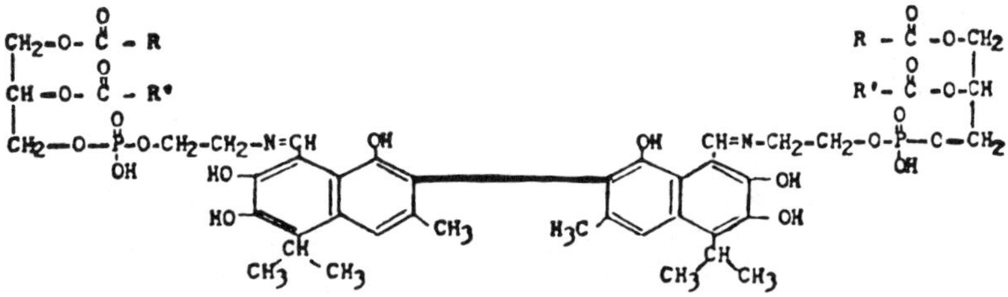

II-A : Diphosphatidyl ethanolamine-monogossypol

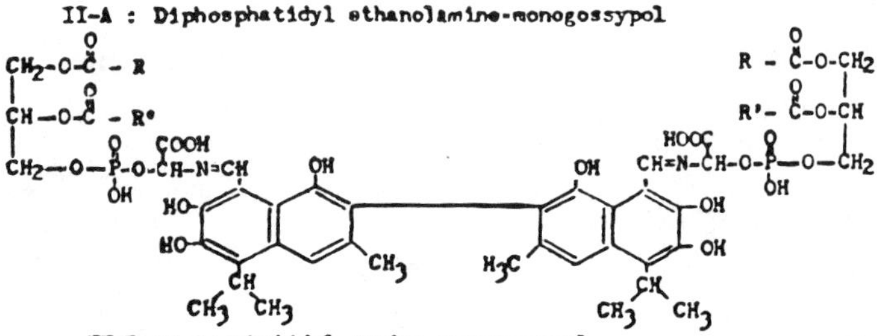

II-B: Diphosphatidyl serine-monogossypol

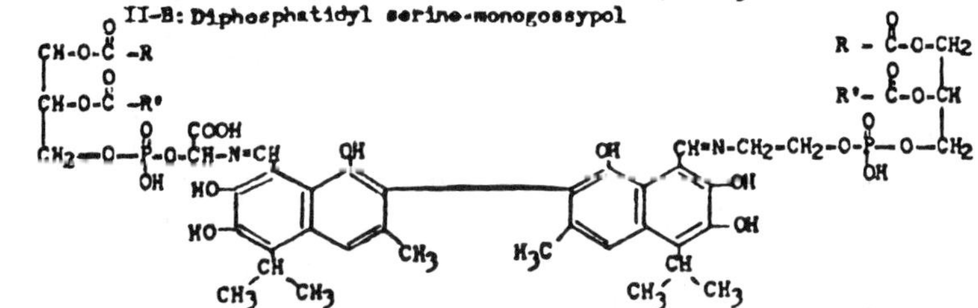

III: Monophosphatidylethanolamine monophosphatidylserine -monogossypol

Fig. 4-6. Possible structures of cephalin-gossypol interactions. (40)

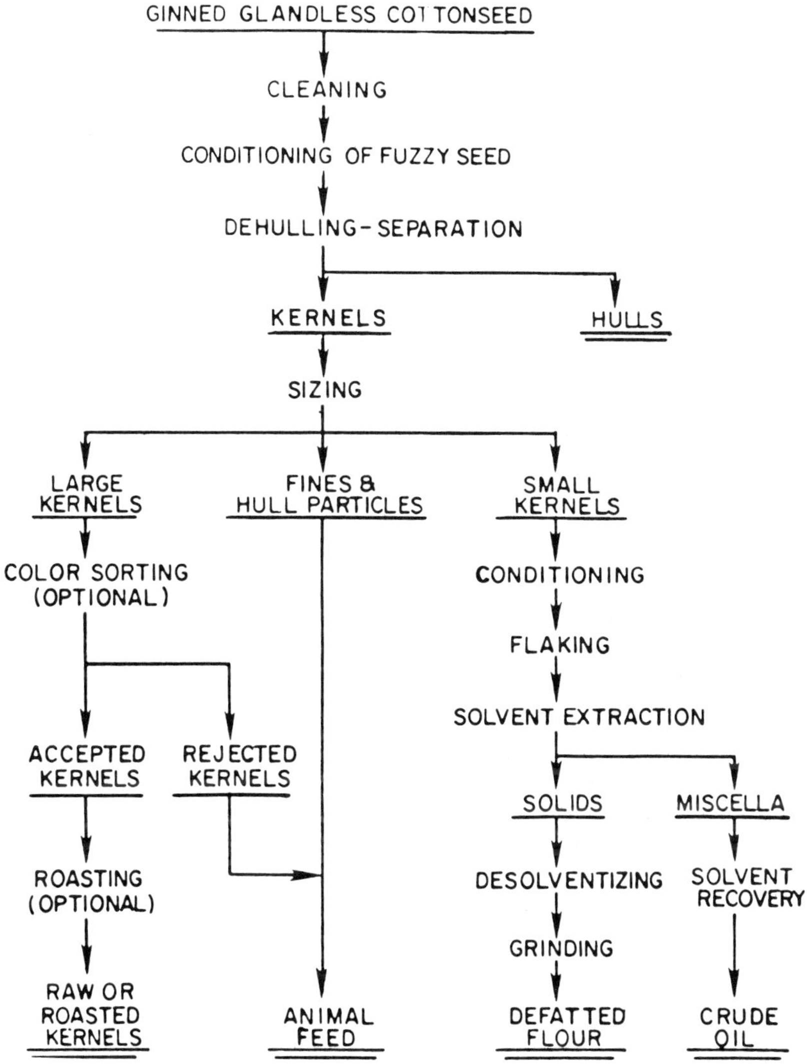


Fig. 4-7. Flow chart for production of glandless cottonseed kernels, flour, and oil. Solvent = hexane. (38)

from hexane-defatted glandless cottonseed oil are summarized in Tables XI and XII. Cottonseed phospholipids are superior to those of other oilseeds, especially soybeans, which presently are the main source of commercial lecithin (5). Since cottonseed oil contains only trace amounts of fatty acids with more than two double bonds (linolenic acid), it is more stable to oxidation and rancidity processes, first indicated in earlier studies by Olcott (14). Other sources of phospholipids, e.g., soybeans, contain linolenic acid in amounts that can cause flavor, color and odor problems. Blending phospholipids from glandless cottonseed with those from other sources improves their flavor, color and odor properties during storage and use in food processing. Cottonseed phospholipids also contain high percentages of lecithin components that improve their functionality as food emulsifiers.

**TABLE 4-XI**
**Composition of Phospholipid Fraction from Glandless Cottonseed Oil** (37, 38)

| Phospholipid[a] | Composition (% of total phosphorus) |
|---|---|
| Origin | 4.12 |
| Lysophosphatidylcholine | 2.56 |
| Phosphatidylinositol | 13.41 |
| Phosphatidylserine | 2.38 |
| Phosphatidic acid | 8.76 |
| Phosphatidylcholine | 23.16 |
| Phosphatidylethanolamine | 13.46 |
| Phosphatidylglycerol | 7.62 |
| Lysophosphatidylserine | ND[b] |
| Lysophosphatidylethanolamine | ND |
| Unknown (sum: 6 TLC spots) | 25.30 |

[a]Water (2–4%) was added to hexane-extracted glandless cottonseed oil, the resulting mixture stirred 30 min at 70°C, and centrifuged to separate the oil and phospholipid-containing fraction (41). The phospholipids were separated by 2-dimensional thin layer chromatography (TLC) on Silica gel-60 plates. Dimension I = chloroform: methanol:7N $NH_4OH$ (65:30:1); Dimension II = chloroform:methanol:acetic acid:water (170:25:25:4). Quantitation of the phospholipids was according to El-Sebajy et al. (42).
[b]ND = not detected.

## References

1. Gurr, M. I., and A. J. James, Lipid Biochemistry: An Introduction, Cornell University Press, Ithaca, NY, 1972, p. 125.
2. Galliard, T., in Form and Function of Phospholipids, edited by G. B. Ansell, J. N. Hawthorne and R. M. C. Dawson, Elsevier Scientific Publishing Co., New York, 1973, p. 253.
3. Finean, J. B., Ibid., p. 171.
4. Zahler, P., and E. R. Weibel, Biochim. Biophys. Acta 219:320 (1970).

**TABLE 4-XII**
**Fatty Acid Composition of Phospholipid Fraction from Glandless Cottonseed Oil (37, 38)**

| Fatty acid | Composition (% of total fatty acids) |
|---|---|
| Myristic, $C_{14:0}$ | 0.72 |
| Palmitic, $C_{16:0}$ | 24.19 |
| Palmitoleic, $C_{16:1}$ | 0.72 |
| Stearic, $C_{18:0}$ | 2.96 |
| Oleic, $C_{18:1}$ | 16.94 |
| Linoleic, $C_{18:2}$ | 53.58 |
| Linolenic, $C_{18:3}$ | 0.23 |
| Arachidic, $C_{20:0}$ | 0.41 |
| Gadoleic, $C_{20:1}$ | 0.08 |
| Lignoceric, $C_{24:0}$ | 0.18 |
| Percentage of fatty acids recovered | 49.90 |


5. Cherry, J. P., M. S. Gray and L. A. Jones, JAOCS 58:903 (1981).
6. Brian, R., JAOCS 53:27 (1976).
7. Shustanova, L. A., A. Sh. Isamukhamedov and S. T. Akramov, Khimiya Prirodnykh Soedinenii 4:449 (1975).
8. Kamat, V., G. Graham and M. Davis, Cereal Chem. 55:295 (1978).
9. Kinsella, J. E., JAOCS 56:242 (1979).
10. El-Tarras, M., Fayez, M., M. Ezzat Moetz Abdel, S. K. Abdel Ahmad and M. M. Amer, Oleagineux 34:139 (1979).
11. Bhatia, I. S., N. Kaur and P. S. Sukhija, J. Sci. Food Agri. 29:747 (1978).
12. El-Tarras, M. Fayez M., M. Ezzat Moetz Abdel, S. K. Abdel Ahmad and M. M. Amer, Riv. Ital. Sostanze Grasse 53:15 (1976).
13. Bailey, A. E., Cottonseed and Cottonseed Products, Wiley Interscience Publishers, Inc., New York, NY, 1948, p. 377.
14. Olcott, H. S., Science 100:226 (1944).
15. Woolley, D. W., J. Biol. Chem. 147:581 (1943).
16. Lishkevich, M. I., Maslobovino Zhirovoe Delo. 15:6 (1939).
17. Thurman, B. H., U.S. Patent 2,150,732 (1939).
18. Thurman, B. H., U.S. Patent 2,182,767 (1940).
19. Thurman, B. H., U.S. Patent 2,201,061 (1940).
20. Thurman, B. H., U.S. Patent 2,201,063 (1940).
21. Thurman, B. H., U.S. Patent 2,206,210 (1940).
22. El-Shattory, Y. A., Studies on Cottonseed Phospholipids, Ph.D. Dissertation, National Research Center, Cairo, U.A.R., 1971, 169 pp.
23. El-Shattory, Y. A., Rev. Franc. Corps Gras 26:187 (1979).

24. El-Nockrashy, A. S., and Y. El-Shattory, Rev. Franc. Corps Gras 19:13 (1972).
25. El-Nockrashy, A. S., and Y. El-Shattory, Rev. Franc. Corps Gras 20:217 (1973).
26. Alter, M., and T. Gutfinger, La Rivista Italiana Delle Sostanze Grasse 59:14 (1982).
27. Vijayalakshmi, B., and S. Venkob Rao, Fette, Seifen, Anstrichm. 74:404 (1972).
28. Armour's Neo-Fat: New Fatty Acids and Oils for Industrial Uses, Armour and Company (renamed Akzo Chemie America), Bulletin #1303-R, unpaged, Neofat Division, Chicago, IL, June 1940.
29. Vijayalakshmi, B., S. Venkob Rao and K. T. Achaya, Fette, Seifen, Anstrichm. 71:757 (1969).
30. Rewald, B., Biochem. J. 36:822 (1942).
31. Pons, W. A., J. Pominski, W. H. King, J. A. Harris and T. H. Hopper, JAOCS 36:328 (1959).
32. Shustanova, L. A., A. Sh. Isamukhamedov, and S. T. Akramov, Khimiya Prirodnykh Soedinenii 2:449 (1975).
33. El-Nockrashy, A. S., and Y. Shattory, Rev. Franc. Corps. Gras 19:233 (1972).
34. El-Nockrashy, A. S., H. M. Mostafa, M. M. El-Fouly, and Y. El-Shattory, Die Nahrung 20:125 (1976).
35. Berardi, L. C., and L. A. Goldblatt, in Toxic Constituents of Plant Foodstuffs, edited by I. E. Liener, Academic Press, New York, NY, 1980, p. 183.
36. Milner, M., Food and Nutr. Bull. 2:51 (1980).
37. Cherry, J. P., JAOCS 60:360 (1983).
38. Cherry, J. P., and H. R. Leffler, in Cotton Monograph, edited by R. J. Kohel and C. F. Lewis, Amer. Soc. Agron., Crop Sci. Soc. Amer., and Soil Sci. Soc. Amer., Madison, WI, 1984, p. 511.
39. Cherry, J. P., and L. C. Berardi, in CRC Handbook of Processing and Utilization in Agriculture, Vol. II:Part 2, Plant Products, edited by I. A. Wolff, CRC Press, Inc., Boca Raton, FL, 1982, p. 187.
40. El-Nockrashy, A. S., and Y. El-Shattory, Rev. Franc. Corps Gras 19:89 (1972).
41. Carr, R. A., JAOCS 53:347 (1976).
42. El-Sebajy, L. A., A. R. El-Mahdy, E. K. Moustafa and M. S. Mohamed, Food Chem. 5:217 (1980).

# Chapter Five
## Rapeseed Lecithin

**F. W. Hougen[1], V. J. Thompson[1] and J. K. Daun[2]**
[1]Department of Plant Science
University of Manitoba
Winnipeg, Manitoba
Canada R3T 2N2

[2]Canadian Grain Commission
Grain Research Laboratory
1404–303 Main Street
Winnipeg, Manitoba
Canada R3C 3G8

### Introduction

The term rapeseed, used in the most general sense, refers to seeds from several different plant species (Table I). These seeds have all been grown and harvested for their oil content in different parts of the world for thousands of years. On a worldwide production basis, rapeseed ranks fifth among vegetable oils (after soybeans, sunflowerseed, peanuts and cottonseed). The major producing countries, Canada, India, China and France, accounted for 71% of the world's production of rapeseed in 1981.

Both winter and summer types are cultivated, the winter types being preferred where possible because of substantially higher yields. There is a considerable variation in composition between and within species of rapeseed. Major variations in composition within species became commercially important about 1970 when the western world began converting to low erucic acid varieties of *Brassica napus* and *Brassica campestris.* The terms "Canbra" and "LEAR" have been used to refer to varieties of rapeseed with low levels of erucic acid in the oil. More recently, varieties of rapeseed with low levels of glucosinolates have been introduced. In Canada the name "canola", a trademark of the Canola Council of Canada, refers to seed and seed products from *Brassica napus* and *Brassica campestris* with low levels of erucic acid and glucosinolates.

Rapeseed lecithin is the "gums" or the lipid fraction of crude rapeseed oil

**TABLE 5-I**
**Definition of Rapeseed**[a]

| Botanical name | | Synonyms |
|---|---|---|
| *Brassica napus* ssp. *oleifera* *forma biennis* *forma annua* | | Summer rape, winter rape, oil rape, Argentine rape, swede rape, oilseed rape, Colza (Fr.), Raps (Ger.) |
| *Brassica campestris* ssp. *oleifera* *forma biennis* *forma annua* | | Winter turnip rape, summer turnip rape, rapeseed, oil turnip, Polish rape, Navette (Fr.), Rübsen (Ger.) |
| *Forma annua* | var. *chinensis* | Summer turnip rape, chinese mustard, Pak-choi (Fr.) |
| | var. *pekinensis* | Summer turnip rape, celery cabbage, chinese kale, Pet-sai (Fr.) |
| | var. *dichotoma* | Summer turnip rape, Toria |
| | var. *trilocularis* | Summer turnip rape, Sarson |
| *Brassica juncea* | | Brown mustard, leaf mustard, Indian mustard, oriental mustard |

[a]Ref. 1.

which is removed by degumming (or desliming) techniques. This chapter will discuss the composition of rapeseed lecithin, processes related to its manufacture, and its utilization.

Rapeseed lecithin has been reviewed (2, 3) and an extensive bibliography on rapeseed and other oilseed crops has been compiled and computer indexed (4). A selection of about 3000 titles from this bibliography has been published (5).

## Chemical Composition of Rapeseed Lecithin

Rapeseed lecithin, as obtained by the traditional degumming process, is an aqueous emulsion of phospholipids which also contains triglycerides and small amounts of other lipid and nonlipid materials. The emulsion usually contains 25–35% water, with the remainder being mainly phospholipids and triglycerides in a 2:1 ratio (3,6,7), as well as small amounts of free fatty acids, mono- and diglycerides, tocopherols, sterols, pigments, terpene hydrocarbons, waxes, glycolipids, aliphatic aldehydes and ketones, sugars, other carbohydrates, carbohydrate derivatives, proteins, protein fragments, amino acids, iron compounds, sulfur compounds, calcium and magnesium salts and various resins and mucilaginous materials. The complexity of the composition of rapeseed lecithin is well demonstrated by the study of Zajac and Niewiadomski (8) who showed, by a combination of column and thin layer chromatography, the presence of 29 classes of compounds in rapeseed gum. Some of these were not identified.

**TABLE 5-II**
**Composition of Rapeseed and Soybean Gums (%)**

| | Rapeseed | | | | | Soybean |
|---|---|---|---|---|---|---|
| Components | (9)[a] | (9)[b] | (10)[b] | (11)[b] | (8)[c] | (8)[c] |
| Water | 24 | | | | | |
| Nonlipid | 9 | | | | | |
| Triglycerides or neutral lipids | 16 | | 38.1 | 5.6 | 29 | 35 |
| Phospholipids (total) | 51 | | | | | |
| Phosphatidylcholine | | 22 | 16.2 | 24.6 | 20 | 21 |
| Phosphatidylethanolamine | | 15 | 17.5 | 22.1 | 16 | 8 |
| Phosphatidylinositol | | 18 | 7.6 | 14.7[d] | 8 | 20 |
| Lysophosphatidylcholine<br>Lysophosphatidylethanolamine | | | <br>2.0 | 19.4 | 1 | 0 |
| Sterol glycosides or glycolipids | | 9[e] | 7.9 | 13.6 | 11 | 0 |
| Unidentified | | 36[f] | 10.7 | | 15 | 11 |
| Unaccounted for | | | | | | 5 |

[a]Complete gum sample, including water.
[b]Acetone precipitate from gum sample.
[c]Dried gum sample.
[d]Tentatively identified.
[e]Tentatively: phytoglycolipids.
[f]Tentatively: 16% acidic phospholipids (plus 20% unidentified).

Canadian, Swedish and Polish workers have reported the composition of rapeseed lecithin (Table II). Direct comparison of the percentage figures in these studies may be somewhat misleading since some of the analyzed gums had been subjected to various treatments to remove water and triglycerides, as indicated in the footnotes. The table, nevertheless, gives an indication of the major substances which have been found in rapeseed lecithin. The composition of a sample of soybean lecithin included in the table indicates the overall similarity between rapeseed and soybean lecithin.

### Phospholipids

The major phospholipids present in rapeseed lecithin are phosphatidyl choline, phosphatidyl ethanolamine and phosphatidyl inositol. The relative proportions of these three components (Table III) do not differ greatly among the rapeseed lecithin samples investigated, nor between rapeseed and soybean lecithin.

**TABLE 5-III**
**Major Phospholipids in Rapeseed and Soybean Gum (%)[a]**

| | Rapeseed | | | | |
|---|---|---|---|---|---|
| | Reference | | | | Soybean |
| Components | (9) | (10) | (11) | (8) | (8) |
| Phosphatidylcholine | 40 | 38 | 40 | 46 | 43 |
| Phosphatidylethanolamine | 27 | 43 | 36 | 36 | 16 |
| Phosphatidylinositol | 33 | 19 | 24 | 18 | 41 |

[a]Calculated from the percentage figures in Table 2; each phospholipid is expressed as a percentage of the total amount of the 3 phospholipids.

Although phosphatidyl serine was not reported in rapeseed lecithin by the authors referred to in Table II, its presence in rapeseed gum has been inferred by Sulser (12) who identified small amounts of serine on paper chromatograms of the hydrolysis products of rapeseed phospholipids. Larsson et al (13) also detected small amounts of phosphatidylserine in laboratory extracted lecithin from two out of three rapeseed varieties studied. Phosphatidylserine is reported to be widely distributed (14), but in small amounts, in higher plants, animals and microorganisms.

The phosphatidic acid content of lecithin has been reported to vary greatly depending on processing conditions (13). About 5% of phosphatidic acid was found in lecithin prepared in the laboratory from two varieties of rapeseed. Lecithin prepared from the same seed in an industrial process contained 20% phosphatidic acid. The increase in phosphatidic acid appeared to mainly be derived from hydrolysis of phosphatidylcholine during processing. This production of phosphatidic acid during processing may be an advantage when considering lecithin utilization since the emulsifying properties of lecithin have been shown to depend on the amounts of negatively charged phospholipids present, namely, phosphatidic acid, phosphatidylinositol and possibly phosphatidylserine.

The large variation in the reported amounts of lysophospholipid (Table II) may be a reflection of the reported tendency of phospholipids to hydrolyze to lyso compounds during processing, with the degree of hydrolysis depending on the processing conditions and the treatment of the oils (15).

**Fatty Acids**

Most reports on rapeseed gums deal with older seed types with high erucic acid contents (20–40%) in the oil. For high erucic acid oils, the fatty acid composition of the lecithin differs considerably from that of the triglyceride portion. The main difference is that the long chain fatty acids (C20 and C22) are present only in small amounts in the lecithin (Table IV). For low erucic acid oils (including canola oils), the fatty acid composition of the lecithin does not differ substantially from that of the parent oil except for a somewhat higher content of C16:0 fatty acid in the gums and somewhat higher C18:1 and C18:3 fatty acids in the oils.

The individual phospholipids in the gums differ somewhat in fatty acid composition particularly in the amount of palmitic acid (C16:0) present (Table IV). It should be noted, however, that the data presented for many of the samples in Table 4 are the results of the work of different authors who used rapeseed oils of different fatty acid composition as starting material for their studies. Thus the data are not directly comparable. Studies on phospholipids isolated from seed (Table V) show similar differences in fatty acid composition similar to those noted in Table 4 for gum phospholipids.

**Other Components**

Since rapeseed gum contains ca. 30% entrained oil, minor components normally associated with the oil, such as sterols, tocopherols and pigments, would also be expected to be present in the gum although there are no reports in the literature detailing their presence. Small amounts of sulfur compounds are present in crude rapeseed oil (ca. 25 ppm in conventional crude oils and 12 ppm in crude canola oils) and studies have shown that ca. 50% of these are removed during degumming (22,23). This would result in relatively high levels of sulfur compounds in rapeseed lecithin. One analysis of a gum fraction of a conventional rapeseed oil showed 385 ppm sulfur (24).

Carbohydrates in both free and combined form have been associated with phospholipids in vegetable oils (6,7). A number of sugars and related compounds have been found in rapeseed lecithin, including glucose (1.3% of the lecithin) (25), glycolipids (monogalactosyldiglyceride, 0.16%, and digalactosyldiglyceride, 0.65%) (26), and inositol (3). One study (27) has reported the presence of phytin (the mixed calcium and magnesium salt of phytic acid) in rapeseed lecithin. It is conceivable that rapeseed phytin may be extracted to a limited extent with the crude oil and subsequently separated out with the lecithin fraction.

**TABLE 5-IV**
**Fatty Acid Composition of Rapeseed Oil and Its Fractions**

| Sample | Fatty acids (%) | | | | | | | | | | | Reference |
|---|---|---|---|---|---|---|---|---|---|---|---|---|
| | 16:0 | 16:1 | 18:0 | 18:1 | 18:2 | 18:3 | 20:0 | 20:1 | 22:0 | 22:1 | Others | |
| High-euric rapeseed | | | | | | | | | | | | |
| Oil | 3.3 | 0.2 | 1.4 | 30.5 | 21.5 | 6.6 | 0.4 | 10.0 | 0.1 | 25.5 | 0.5 | (16) |
| Gum | 17.0 | 1.5 | 2.0 | 27.6 | 31.3 | 7.9 | 0.5 | 3.3 | trace | 7.0 | 1.9 | (17) |
| Gum phospholipids | 16.2 | 1.0 | 0.9 | 24.8 | 45.2 | 9.3 | | 1.9 | | | 0.7 | ( 9) |
| Phosphatidylcholine | 8.8 | 1.3 | 0.8 | 34.5 | 44.9 | 6.2 | | 2.4 | | | 1.1 | ( 9) |
| Phosphatidylethanolamine | 12.9 | 1.2 | 1.0 | 23.1 | 51.2 | 8.2 | | 1.7 | | | 0.7 | ( 9) |
| Phosphatidylinositol | 34.4 | 2.8 | 1.5 | 15.0 | 36.4 | 8.5 | | | | | 1.4 | ( 9) |
| Gum glycolipids | 17.4 | 1.6 | 3.1 | 17.4 | 31.3 | 13.5 | | 4.0 | | 6.4 | | (18) |
| Low-erucic rapeseed | | | | | | | | | | | | |
| Oil | 4.2 | 0.2 | 2.0 | 60.5 | 19.6 | 9.0 | 0.6 | 2.2 | 0.3 | 1.4 | | (16) |
| Gum | | | | | | | | | | | | |
| Tower | 4.3 | 0.3 | 1.6 | 57.5 | 22.1 | 11.0 | | 1.8 | 0.5 | 0.7 | 0.2 | (19) |
| Midas | 9.3 | 0.5 | 1.2 | 47.5 | 31.7 | 8.3 | | 0.9 | 0.3 | 0.3 | | (19) |
| Canola | | | | | | | | | | | | |
| Pressed oil | 4.9 | 0.3 | 1.7 | 57.4 | 20.9 | 11.4 | 0.7 | 1.8 | 0.4 | 1.8 | 0.1 | (20) |
| Gum[b] | 13.4 | 1.0 | 3.3 | 46.3 | 24.5 | 5.0 | 0.7 | 1.8 | 0.6 | 2.0 | 1.4 | (20) |
| Solvent-extracted oil | 3.9 | 0.3 | 1.7 | 56.4 | 21.6 | 11.3 | 0.7 | 1.8 | 0.4 | 1.8 | 0.1 | (20) |
| Gum[b] | 13.2 | 0.8 | 2.2 | 46.0 | 29.3 | 5.5 | 0.5 | 1.0 | 0.5 | 0.8 | 0.2 | (20) |
| Crude oil | 4.2 | | 1.9 | 61.1 | 21.1 | 8.9 | 0.7 | 1.0 | 0.3 | 0.6 | 0.2 | (20) |
| Gum[b] | 10.4 | 0.8 | 1.8 | 52.6 | 25.0 | 5.8 | 0.6 | 1.6 | 0.4 | 0.8 | 0.2 | (20) |

[a]Other fatty acids, and material unaccounted for.
[b]Prepared by acetone precipitation from the oils.

**TABLE 5-V**
**Fatty Acid Composition of Phospholipids and Neutral Lipids from Rapeseed Oil[a]**

| | Major fatty acids (%) | | | | | | |
|---|---|---|---|---|---|---|---|
| Sample | 16:0 | 18:0 | 18:1 | 18:2 | 18:3 | 20:1 | 22:1 |
| High-erucic seed | | | | | | | |
| Triglycerides[b] | 4.3 | 0.9 | 18.1 | 16.6 | 9.8 | 11.9 | 36.9 |
| Phospholipids[c] | 21.7 | 1.1 | 23.1 | 38.0 | 9.4 | 1.1 | 2.6 |
| Low-erucic seed | | | | | | | |
| Triglycerides[b] | 6.3 | 0.9 | 40.2 | 37.4 | 12.1 | 1.6 | |
| Phospholipids[c] | 18.3 | 0.6 | 22.3 | 47.9 | 7.4 | | |

[a]Ref. 21.
[b]i.e., Neutral oil, mainly triglycerides.
[c]Including sulfoquinovosyl diglyceride.

Glycine has been reported in rapeseed lecithin (25) but the high concentration reported (25%) is obviously in error.

Substantial amounts of metal ions are removed from vegetable oils in the refining process, mostly in the degumming and caustic refining (28). Lecithin might thus be expected to be relatively rich in metal ions, particularly iron and copper. The amounts of metal ions present in the gums depend on the particular degumming procedure used—acidulation treatment being particularly effective in removing metal ions from the oil. The copper content of rapeseed gum has been reported as 1.2 ppm (25).

## Lecithin Content in Crude Rapeseed Oils

Whereas Kaufmann (29) reports only 0.1% lecithin in rapeseed oil (40 ppm phosphorus), other workers (Table VI) have reported considerably higher values ranging from 0.8–3.5% (320–2400 ppm phosphorus). A recent study of the phosphorus content of crude oils produced at Canadian canola crushing plants gives a range of 0.8–3.3% lecithin (Table VII) which is similar to the range 1.8–3.5% reported by Teasdale (36).

Solvent-extracted oils have been found to contain the highest levels of phosphorus. For this reason it is a common practice to degum solvent extracted oils but not pressed oils.

### Effect of Processing and Other Conditions

The contents of phospholipids or lecithin in crude oils is dependent on many factors, such as the species of plant, the maturity of the seed, the

**TABLE 5-VI**
**Phospholipid Content of Crude Rapeseed Oils**

| Crude oil | | Press oil | | Solvent-extracted oil | | |
|---|---|---|---|---|---|---|
| Lecithin (%) | Phosphorus (ppm) | Lecithin (%) | Phosphorus (ppm) | Lecithin (%) | Phosphorus (ppm) | Reference |
| 0.1 | — | — | — | — | — | (29) |
| 0.8 | — | 0.2–0.5 | — | 1.2 | — | ( 3) |
| — | — | — | — | 2 | — | (30) |
| — | 70–963 | — | — | — | — | (31) |
| — | — | — | 125, 128 | — | 338, 343 | (32) |
| — | — | — | 77–113 | — | 132–386 | (33) |
| — | 133–844 | — | — | — | — | (34) |
| — | — | — | 88–605 | — | 79–167 | (35) |
| 1.8–3.5 | — | — | — | — | — | (36) |

**TABLE 5-VII**
**Lecithin Content of Crude Oils from Western Canadian Canola Crushing Plants (1981-1982)[a]**

| Crushing plant | % Lecithin in oil[b] | | | | | |
|---|---|---|---|---|---|---|
| | Solvent-extracted | | Expelled | | Crude | |
| | 1 | 2 | 1 | 2 | 1 | 2 |
| A | 1.7 | 2.0 | 0.6 | 0.8 | (1.0) | (1.2)[c] |
| B | — | — | — | — | 0.9 | 1.1 |
| C | 5.3 | 4.6 | 2.4 | 1.4 | (3.3) | (2.4) |
| D | — | — | — | — | 1.8 | 2.0 |
| E | — | — | — | — | 0.8 | 0.9 |
| F | — | — | — | — | 0.9 | 1.1 |

[a] J. K. Daun and B. A. Misener, unpublished data.
[b] Based on two samples spaced at least one month apart. Acetone insolubles accord to ref. 49.
[c] Bracketed crude oil results calculated (0.7 × expelled + 0.3 × solvent extraction).

method of extraction of the oil, the climate and the nature of the soil (29,37,38). The influence of some of these factors is discussed in the following.

*Moisture content of the seed.* Nielson (10) has shown that an increase in the moisture content of soybeans results in an increased phosphorus content in the extracted oil. Solomon (2) has similarly reported that the phospholipid content in rapeseed is reduced with a decrease in moisture content of the seed to be extracted. Thus, with seeds of 5% and 9–11% moisture content, the extracted oils contained more than two and three times as much phospholipids, respectively, as with seeds of 2.5% moisture content.

*Heating of the seed.* Sawicki et al. (15) have studied the effect of heating of rapeseed on the phospholipid content in the extracted oil. Generally, increased temperature produced increased amounts of phospholipids in the oil. In industrial experiments, they found 150 ppm phosphorus in the oil from nonheated seed and 330 ppm for seed heated at 90 C for one hour.

Solomon (2) similarly reported that the pressing of seeds at reduced temperature reduced the quantity of aphospholipids obtained in the oil.

Boekenoogen (38) states that while a cold-pressed oil is free from foots (i.e., lecithin), a wet heating to 45 of the meal decomposes the insoluble protein-phospholipid complex, thus releasing the free phospholipid into the oil. Palaveeva and Gurdev (39) found as much as a 60% (relative) increase in phospholipids in the extracted oil from seeds treated with 10–12% water before extraction. The nonhydratable portion of the phospholipids increased from an original 16% to 32% after the water treatment. Heating the seed also increased the amount of phospholipids in the oil.

*Expelled versus extracted oil.* The phosphorus content is generally lower in pressed oils than in extracted oils. Franzke et al. (40), in Germany, found that pressed rapeseed oil contained 77–113 ppm phosphorus and extracted oil 132–386 ppm phosphorus. Persmark (3), in Sweden, found somewhat higher values, namely, ca. 200 ppm phosphorus in pressed oils and ca. 500 ppm in extracted oils values which compare well to results for canola oils (Table VII).

*Solvent used in extraction.* The type of solvent used in extraction has a marked influence on the composition of the extracted oil. A relatively nonpolar solvent like hexane will extract preferentially the nonpolar lipids, i.e., triglycerides, whereas a more polar solvent like alcohol will extract more of the more polar lipids such as phospholipids. This is a well known principle that applies to the extraction of lipids from any organic tissue.

*Completeness of extraction.* The amount of phospholipids passing into the oil during extraction is known to increase during the last phases of the extraction. This increase in extraction of phospholipids is reported to begin after 40–70% of the total oil has been extracted (2). Niewiadomski (41) showed that the residual lipids in the meal after industrial extraction of rapeseed "contained five times more" unsaponifiables (Presumably meaning a concentration five times higher than that in the extracted oil), three times more sterols, and 20 times more phosphorus than in the extracted oil. Persmark (3) similarly reports that the less available fatty material from the crushed seed contains considerably higher concentrations of phosphorus than the readily extractable oil. Defromont et al. (42) support these findings.

*Delayed extraction.* Experiments with soybeans have shown that if the oil extraction is not carried out immediately following the crushing, but 24 hours later, the quality of the oil is much reduced. There was a threefold increase in the phosphorus content of the oil, from 93 ppm with immediate extraction, to 299 ppm with delayed extraction (43).

*Damaged seeds.* Poor weather and early frost may cause severe damage to oil-bearing seed crops, resulting in poor quality oils. Damage to soybeans is known to result in increased amounts of free fatty acids, iron, and non-hydratable phospholipids in the crude oil, and to result in a lower flavor quality of the refined oil. The increased iron content presumably arises from the free fatty acids reacting with the steel equipment used in the processing (44,45).

Ong (43) has compared the quality of oil from mechanically damaged soybeans and sound soybeans. The phosphorus content in the oil was considerably higher for the damaged seed (539 ppm) than for the sound seed (352 ppm).

## The Degumming Process

Although lecithin is the major component separated from the oil through the degumming process, the procedures used in this process affect the composition and quantity of the lecithin removed from the oil.

Degumming is an important refining step because the gum, if left in the oil, can cause increased refining losses, difficulties in the splitting of soapstock, deposits in storage tanks, oxidation and polymerization of the oil, development of undesirable color, odors and flavors in the oil, and excessive foaming and smoking of the heated oil (6,40, 46,47). Residual phospholipids in refined oil can deactivate bleaching clay leading to greater clay usage and higher triglyceride losses. Phospholipids can also act as catalyst poisons in hydrogenation reactions.

### Water Degumming

In its simplest form, degumming involves treating the crude oil with water or steam in order to hydrate the phospholipids and thus make them insoluble in the oil. Hydration occurs at the ionic part of the molecule. The crude oil is usually treated with 2–4% water at moderate temperature and the precipitated gums are removed from the oil by centrifugation.

Optimization of the oil/water ratio is important for efficient degumming. Too little water will cause insufficient hydration and incomplete degumming; excess water may result in the formation of emulsions with accompanying refining losses (37). In laboratory degumming experiments on rapeseed oil (48) the optimum water content was found to be 4–5%, and even at this optimum only 30% of the phosphorus was removed. Other reports claim that 70–85% of the phosphorus can be removed by water degumming (3,40). In order to meet the CGSB standard of a maximum of 0.6% acetone insolubles (49), the crude oils sampled from Western Canadian crushing plants (Table 7) would have to be degummed with an efficiency of at least 40%.

*Hydratable and nonhydratable phospholipids.* Water degumming may leave as much as 200 ppm residual phosphorus in rapeseed oil (3,30). This is due to the presence of nonhydratable and poorly hydratable phospholipids, mainly the calcium and magnesium salts of phosphatidic acids, and possibly phosphatidylserine and phosphatidylinositol. The hydratable phospholipids are believed to be mainly phosphatidylcholine and phosphatidylethanolamine. Some nonhydratable phospholipids occur in intact oilseeds, but an additional amount of phosphatidic acid is formed from the hydratable phospholipids during processing (50,51).

Carr (47) reported that certain phospholipids were hydratable or nonhydratable according to whether the phosphoric acid molecule were attached at the outer position ($\alpha$) or middle position ($\beta$) of the glycerol molecule. The natural occurrence of $\beta$-phospholipids, however, has not been confirmed in recent literature; Morrison (52) states that the phosphoric acid residue is always attached at the $\alpha$-carbon of the glycerol molecule.

**Degumming Agents**

Degumming agents have recently been used in western Canada for the production of "semi-refined canola oils" (53). These oils contain as little as 20 ppm phosphorus and, after bleaching with 1–2% of acid activated earth, are suitable for "physical" or "steam refining". The low level of phosphorus in these oils is due to the use of degumming agents, either in conjunction with normal water degumming or in a second process. The use of degumming agents has been reviewed (6,37,46).

To remove nonhydratable phospholipids, water degummed rapeseed oils are often further degummed with phosphoric acid, usually at a refinery. This treatment has resulted in lower alkali refining losses, easier hydrogenation and filtration, better color of the oil, and better oxidative and flavor stability (54). Phosphoric acid also combines with copper, iron and other metals which may act as prooxidative catalysts, thus facilitating their removal from the oils (46).

Ohlson and Svensson (55) studied phosphoric, acetic, citric, nitric, oxalic and boric acids as degumming agents for rapeseed oil. Phosphoric and oxalic acids were found to be the most efficient degumming agents. Franzke et al. (33) compared sulfuric and phosphoric acids in degumming rapeseed oil. They found sulfuric acid to be superior to phosphoric acid. The sulfuric acid caused extensive degradation of the phospholipids into lyso compounds and polar hydrolysis compounds; these were only partially removed with the gum, but almost completely by the subsequent alkali refining and bleaching.

Guillaumin et al. have reported an extensive laboratory study on the degumming of rapeseed oil (48). Agents tested included salts, salt mixtures, alkali, acids, acids plus salts, acids plus active clays and others. The best results were obtained with dilute nitric acid, used alone or in conjunction with activated clay. Dilute nitric acid removed 80% of the original 142 ppm phosphorus in the oil. Optimal concentrations of the nitric acid were found to be 5–7% for use with extracted oils and 2–3% for pressed oils; the acid solutions were used in an amount of 5% of the oil. A temperature of 80 C with 30 min contact time and medium agitation was found to be the most efficient. A wash with saturated NaCl was recommended, especially if steam refining were to follow, in order to remove traces of acid and prevent ester formation. Defromont et al. (42) confirmed the effectiveness of nitric acid as a degumming agent for rapeseed oil on a semiindustrial scale.

Although the above studies on the use of nitric acid as a degumming agent gave high quality oils with practically no color, Wettström (56) found that deodorization imparted a dark color to nitric acid degummed oil.

Diosday et al. (66) recently evaluated 54 reagents for chemical degumming of canola oil. Best results were obtained with citric acid and maleic anhydride, which gave residual phosphorus levels of less than 30 ppm in the oil.

## Utilization of Rapeseed Lecithin

Lecithin is used in a number of industrial processes (57) and as a food additive. Although soybean is the main source of industrial lecithin, rapeseed lecithin has found some use as an animal feed additive. Further uses of rapeseed lecithin may be anticipated provided the quality of the product is improved.

### Additive in Animal Feed

At most crushing plants in western Canada, the gums removed from the rapeseed oil during degumming are added back to the meal, usually during the desolventization process. This provides an outlet for the lecithin in locations where it is not economically feasible to refine and market a finished product. The addition of lecithin (and acidulated soapstock) reduces the dustiness of the meal and increases its metabolizable energy (58).

Rapeseed meals with added lecithin have been extensively tested in diets for livestock and poultry. Although gums from high glucosinolate rapeseed significantly increased the metabolizable energy of the feed, results from studies with turkey chicks and poults (60) suggest that these gums may contain some factor which reduces feed consumption and growth. Lecithin from low glucosinolate types of rapeseed (canola), however, was found to give results similar to soybean lecithin in feed trials (61). Broilers and layers have not been adversely affected by addition of up to 6% canola lecithin to the canola meal used in their rations (58). Usually canola meal produced in western Canada contains ca. 1.5% added lecithin. Addition of canola gums to canola meal at levels greater than this norm produced no adverse effects in nutritional studies involving beef cattle, dairy cattle and swine (59).

## Quality of Rapeseed Lecithin

If rapeseed lecithin were to compete in the general market with soybean lecithin, only rapeseed lecithin of the highest quality could be considered (41).

The gross composition of rapeseed lecithin does not differ substantially from that of soybean lecithin (3). The main reason why rapeseed lecithin has not achieved the same widespread use in the food industry as soybean lecithin is that rapeseed lecithin has been found inferior to soybean lecithin in color, flavor, taste and general appearance (3,62).

Rapeseed lecithin of comparable quality to soybean lecithin may be produced by the use of suitable refining treatments (3); rapeseed lecithin has thus been successfully utilized by margarine and chocolate manufacturers in England and Germany. Polish researchers described a degumming process which produces a high grade lecithin from rapeseed, with a moisture content of less than 2%, an acid number of 4–8 a and good flavor. This lecithin required no further refining (3).

Various factors in the oil extraction and processing also influence the quality of the lecithin. These factors include proper attention to the seed cooking conditions, the distillation temperature of the miscella, the water content of the sludge, the temperature and pressure used in drying the lecithin, and the use of sound seed (3,63,64).

When properly prepared, rapeseed lecithin has a light color, an appropriate consistency and a pleasant taste and smell (65).

## References


1. Bengtsson, I. A. von Hofsten and B. Lööf in Rapeseed; Cultivation, Composition, Processing and Utilization, edited by L.A. Appelqvist and R. Ohlson, Elsevier Publishing Co., Amsterdam, 1972, p. 37.
2. Solomon, B., Rev. Fr. Corps Gras. 20:567 (1973).
3. Persmark, U. in Rapeseed; Cultivation, Composition, Processing and Utilization, edited by L. A. Appelqvist and R. Ohlson, Elsevier Publishing Co., Amsterdam, 1972, p. 288.
4. For computerized bibliography of rapeseed and associated oilseed crops (*Brassica* species), contact the Librarian, Agriculture Canada Research Station, Saskatoon, Saskatchewan, Canada S7N OX2.
5. McPherson, A., Rapeseed and Mustard: A Selected Bibliography, Publication No. 52, Agriculture Canada Research Station, Saskatoon, Saskatchewan, Canada, 1978. For copies, contact the Canola Council of Canada, 301–433 Main Street, Winnipeg, Manitoba, Canada R3B 1B3.
6. Norris, F. A., in Bailey's Industrial Oil and Fat Products, 3rd edn., edited by D. Swern, Interscience Publishers Inc., New York, 1964, pp. 28–34 and p. 732.
7. Eckey, E. W., Vegetable Fats and Oils, Reinhold Publishing Corp., New York, 1954, pp. 54–57.
8. Zajac, M., and H. Niewiadomski, Acta Aliment. Pol. 1:63 (1975).
9. Weenik, R. O., and A. P. Tulloch, JAOCS 43:327 (1965).
10. Persmark, U., JAOCS 45:742 (1968).
11. Bratkowska, I., and H. Niewiadomski, Acta Aliment. Pol. 1:339 (1975).
12. Sulser, H., Mitt. Lebensm. Hyg. 45:251 (1954).
13. Larsson, I., L. Rydhag and I. Wilton in Proceedings of the 5th International Rapeseed Conference, Malmö, Sweden, 1978, Vol. 2, p. 187.
14. Gurr, M. I., and A. T. James, Lipid Biochemistry: an Introduction, 2nd edn., Chapman and Hall, London, 1975, p. 127.
15. Sawicki, J., I. Bratkowska, H. Niewiadomski, H. Ziminska and A. Gorska, Acta Aliment. Pol. 1:323 (1975).
16. Cheadle, A. E., in Proceedings of the Annual Meeting of the Rapeseed Association, Rapeseed Association of Canada, Publication No. 38, 1975, p. 81.
17. Zajac, M., H. Niewiadomski and B. Sobolewska, Tluszcze Jadalne 18:107 (1974).
18. Ackman, R. G., in Proceedings of the Symposium on Rapeseed Oil, Meal and By-product Utilization, edited by J. M. Bell, Rapeseed Association of Canada, Vancouver, B.C., Canada, Publication No. 45, 1977, p. 12.

19. Hougen, F. W., Unpublished results, Dept. of Plant Science, University of Manitoba, Winnipeg, Canada R3T 2N2.
20. Daun, J. K. and B. A. S. Misener, Unpublished results, Canadian Grain Commission, Grain Research Laboratory, 1404–303 Main Street, Winnipeg, Manitoba, Canada R3C 3G8.
21. Appelqvist, L. A., in Rapeseed; Cultivation, Composition, Processing and Utilization, edited by L. A. Appelqvist and R. Ohlson, Elsevier Publishing Co., Amsterdam, 1972, p. 132.
22. Franzke, C., R. Göbel and E. Hollstein, Nahrung 16:873 (1972).
23. Daun, J. K., and F. W. Hougen, JAOCS 53:169 (1976).
24. Daun, J. K., Sulphur in Rapeseed Oil, M.Sc. Thesis for the Department of Plant Science, University of Manitoba, Winnipeg, Canada, 1973, p. 42.
25. Linow, F., and G. Mieth, Nahrung 11:663 (1967).
26. Appelqvist, L. A., JAOCS 49:151 (1972).
27. Ackman, R. G., and A. D. Woyewoda, J. Chromatogr. Sci. 17:514 (1979).
28. Franzke, C., Nahrung 16:859 (1972).
29. Kaufmann, H. P., Fette, Seifen, Anstrichm. 48:53 (1941).
30. Anjou, K., in Rapeseed; Cultivation, Composition, Processing and Utilization, edited by L. A. Appelqvist and R. Ohlson, Elsevier Publishing Co., Amsterdam, 1972, p. 214.
31. Devinat, G., S. Biasini and M. Naudet, Rev. Fr. Corps Gras 27:333 (1980).
32. Babuchowski, K., and R. Zadernowski, Tluszcze Jadalne 16:238 (1972).
33. Franzke, C., P. Brandt and E. Hollstein, Nahrung 17:653 (1973).
34. Daun, J. K., L. D. Davidson, J. A. Blake and W. Yuen, JAOCS 58:914 (1981).
35. Denecke, P., Öl ung Margarine 25:450 (1978).
36. Teasdale, B. F., in Grains and Oilseeds, Production, Processing and Utilization, Canadian International Grains Institute, 100–303 Main Street, Winnipeg, Manitoba, Canada (in press).
37. Andersen, A. J. C., in Refining of Oils and Fats for Edible Purposes 2nd edn., edited by P. N. Williams, Pergamon Press, New York, 1962, pp. 28–42.
38. Delvaux, E. L., and J. E. Bertrand in Analysis and Characteristics of Oils, Fats and Fat Products, Vol. 1, edited by H. A. Boekenoogen, Interscience Publishers, London, 1964, pp. 281–307.
39. Palaveeva, T., and M. Gardev, Maslo-Sapunena Promst. 17:15 (1981); Chem. Abstr. 96:12111C (1982).
40. Franzke, C., E. Hollstein and P. Brandt, Lebensm. Ind. 19:347 (1979).
41. Niewiadomski, H., in Proceedings of the International Conference

of Science and Technology, Marketing Rapeseed and Rapeseed Products, Ste. Adèle, Quebec, 1970, Rapeseed Association of Canada, p. 223.
42. Defromont, C., R. Guillaumin, D. Delahaye and N. Drouhin, Rev. Fr. Corps Gras 9:486 (1962).
43. Ong, I. T. L., The necessity for introducing new quality requirements for crude soybean oil, presented at the Deutsche Gessellschaft für Fettforschung, Aachen, West Germany, October, 1978.
44. Evans, C. D., G. R. List, R. E. Beal and L. T. Black, JAOCS 51:444 (1974).
45. List, G. R., C. D. Evans, L. T. Black and T. L. Mounts, JAOCS 55:275 (1978).
46. Brae, B., JAOCS 53:353 (1976).
47. Carr, R. A., JAOCS 53:347 (1976).
48. Guillaumin, R., and M. Boulot, Rev. Fr. Corps Gras 7:506 (1960).
49. Canadian Government Specifications Board, National Standard of Canada, Rapeseed Oil Low Erucic Acid Crude and Crude, Degummed, CAN 2–32.300M-76, Canadian Government Specifications Board, Ottawa, (1976).
50. Bratkowska, I., and H. Niewiadomski, Acta Aliment. Pol. 3:39 (1977).
51. Hvolby, A., JAOCS 48:503 (1971).
52. Morrison, W. R., in Advances in Cereal Science and Technology, Vol. 2, edited by Y. Pomeranz, American Assoc. of Cereal Chemists Inc., 1978, p. 222.
53. Smythe, J., United Oilseeds Products Ltd. Lloydminster, Alberta, Canada, Press Release October 9, 1981.
54. Teasdale, B. F., Zesz. Probl. Postepow Nauk Roln. 91:243 (1970).
55. Ohlson, R., and C. Svensson, JAOCS 53:8 (1976).
56. Wettström, R., in Rapeseed; Cultivation, Composition, Processing and Utilization, edited by L. A. Appelqvist and R. Ohlson, Elsevier Publishing Co., Amsterdam, 1972, p. 218.
57. Van Nieuwenhuyzen, W., JAOCS 58:886 (1981).
58. Salmon, R. E., Can. J. Anim. Sci. 50:211 (1970).
59. Clandinin, D. R. (ed.), Canola Meal for Livestock and Poultry, Publication No. 59, Canola Council of Canada, Winnipeg (1981).
60. Lall, S. P., and S. J. Slinger, J. Sci. Food Agric. 25:423 (1974).
61. March, B. F., in 5th Progress Report, Publication No. 50, Rapeseed Association of Canada, 1977 p. 95.
62. Ohlson, R., in Proceedings of the 5th International Rapeseed Conference, Malmö, Sweden, 1978.
63. Niewiadomski, H., in Proceedings of the 4th International Rapeseed

Conference, Giessen, Germany, 1974, p. 307.
64. Bratkowska, I., Acta Aliment. Pol. 4:255 (1978).
65. Subbiah, M. T. R., Am. J. Clin. Nutr. 26:219 (1973).
66. Diosady, L. L., P. Sleggs and T. Kaji, JAOCS 59:313 (1982).

# Chapter Six

## Sunflower Lecithin

**W. Herbert Morrison, III**
USDA, ARS, Richard B. Russell Agricultural Research Center
P.O. Box 5677
Athens, GA 30613

### Introduction

The increase in production of sunflowers in the USA to slightly over five million acres in 1979 has increased the domestic availability of the oil and byproducts of sunflower seed. Although the acreage of oil-type sunflower seed in 1981 was ca. 3.4 million acres, the increase in domestic use has increased from 10 to 40% of the crop. For the 1980–81 crop, 125,000 metric tons of oil was crushed for domestic consumption (1).

Although sunflower lecithin currently is not used to any great extent in the USA, nevertheless, as sunflower oil production is increased, the availability and, consequently, the use of lecithin should increase. Table I shows the content and composition of the phospholipids in sunflower oil. The percentage of phospholipids ranges from .02 to 1.5% with the average at ca. 0.75%. The composition of the phospholipids varies greatly with phosphatidylcholine, ranging from 12.7 to 64.2%. At present, there have been no studies which investigate the relationship between phospholipid composition and hybrid differences, growing conditions and location. However, from the data in Table I, several points are suggestive. The data of Popov (8) and Boroduliua (9) indicate that seed of different varieties grown at the same location will have similar phospholipid content and composition. The three entries for Peredovik indicate that the same variety grown at different locations will have different phospholipid composition.

### Lipid and Fatty Acid Composition

The composition of the fatty acids associated with the phospholipids is shown in Table II. Their composition parallels that of the oil itself except for there being considerably more palmitic acid associated with the phos-

**TABLE 6-I.**
**Phospholipid Content (%) of Sunflower Seed Oils**

| Source | Shcherbabov (2) | Chapman (3) | Litvinova (4) | El-Sabaiy (5) | Galliard (6) | Grewal Grewal (7) | Popov Popov (8) | Popov Popov (8) | Borodulina Borodulina (9) | | | | |
|---|---|---|---|---|---|---|---|---|---|---|---|---|---|
| Seed type | Peredovik | Master farmer 700 | Unknown | Meak | Unknown | Sunrise | Peredovik | A-41 | VNIIMK 8931 | Peredovik | Amer virsky 3497 | Saliut | Enisey |
| Phosphatidylcholine | 12.7 | 52 | 14.4 | 56.2 | 51 | 26.8 | 42.3 | 44.7 | 55.4 | 64.2 | 58.8 | 56.1 | 61.0 |
| Phosphatidylethanol | 46.6 | 19.7 | 23.6 | 11.7 | 23 | 9.9 | 15.7 | 29.4 | 18.2 | 19.5 | 19.6 | 17.0 | 13.1 |
| Phosphatidylinositolamine | 3.7 | 26 | 13.4 | 7.4 | 22 | 18.5 | 36.6 | 19.8 | 24.0 | 15.2 | 20.6 | 22.4 | 21.4 |
| Phosphatidic acid | 32.6 | 2.2 | 7.3 | 4.3 | 4 | — | 5.2 | 6.0 | 2.2 | 1.2 | 1.1 | 4.5 | 4.4 |
| Total phospholipids | — | — | — | .02 | — | 1.5 | 0.80 | 0.77 | 0.82 | 0.75 | 0.86 | 0.72 | 0.74 |

pholipids than in the seed oil. A further breakdown of fatty acids and phospholipids at various stages of seed maturity was conducted by Grewal (7) and the results are shown in Table III. Phosphatidylcholine increased in concentration with seed development as did phosphatidylinositol. Phosphatidyl ethanol and phosphatidic acid remained essentially unchanged throughout seed development. The fatty acids followed the same pattern as the oil fatty acid except for palmitic which has a content four times greater in the phosphatides than in seed oil.

**TABLE 6-II.**
**Fatty Acid Composition (%) of Neutral (Oil) and Polar (Phospho) Lipids**

| Fatty acids | Grewal (7) | | Chapman (3) | | Vicentini (20) | | | |
|---|---|---|---|---|---|---|---|---|
| | Neutral | Polar | Neutral | Polar | Neutral | Polar | Neutral | Polar |
| 16:0 | 4.6 | 14.8 | 5.6 | 31.9 | 6.1 | 11.1 | 6.6 | 13.0 |
| 16:1 | — | — | — | — | — | — | 0.1 | — |
| 18:0 | 3.4 | 3.0 | 3.2 | 7.9 | 4.8 | 4.1 | 5.4 | 4.6 |
| 18:1 | 40.3 | 13.3 | 44.7 | 17.3 | 30.9 | 14.9 | 30.3 | 16.0 |
| 18:2 | 51.7 | 68.7 | 45.3 | 42.8 | 57.4 | 68.0 | 57.3 | 67.3 |

**TABLE 6-III.**
**Phospholipid and Fatty Acid Composition (%) of Sunflower Seed Oils at Various Stages of Maturity (6)**

| | Phospholipids[a] | | | | Fatty acids of the polar lipids | | | |
|---|---|---|---|---|---|---|---|---|
| DAF[b] | PC | PE | PA | PI | 16:0 | 18:0 | 18:1 | 18:2 |
| 10 | 8.7 | 11.0 | 13.6 | 6.9 | 53.3 | 7.1 | 14.3 | 15.9 |
| 20 | 18.3 | 15.5 | 10.1 | 11.2 | 28.0 | 2.0 | 42.0 | 26.4 |
| 30 | 28.0 | 11.6 | 9.0 | 14.1 | 10.6 | 1.7 | 37.4 | 50.3 |
| Mature | 26.8 | 9.9 | 9.9 | 18.5 | 14.8 | 3.0 | 13.3 | 68.7 |

[a]PC, phosphatidylcholine; PE, phosphatidylethanol; PA, phosphatidic acid; PI, phosphatidylinositol.
[b]Day after flowering.

## Effects of Processing on Quality

Removal of the phospholipids or degumming is dependent on the amount of hydratable phospholipids in the oil. This is governed by a number of different factors and can change at various stages of processing. Arutyunyan

(11) has studied the effects of seed defects on phospholipid content and ease of removal from the oil. He found that as the degree of defects increased, the amount of hydratable phospholipids decreased, and as a result, fewer phospholipids could be removed during degumming.

## Drying Seeds

Drying is the first important step in processing. Seed dried to a moisture below 9.5% can be stored for extended periods without resulting in oil deterioration (13). Generally, seed heated to 70–75 C during drying to 7.0% moisture results in maximum conversion of phospholipids into the oil and a minimum of nonhydratable phospholipids (12). However, the stage of seed maturity or ripening rate also plays an important role in considering the temperature which should be used during drying to insure maximum phospholipid production and removal from the oil. Arutyunyan (12) determined the optimum heating needed for drying unripe (seed reaching physiological maturity 26 days after the end of seed flowering, DAF), harvest ripe (harvested around 35 DAF), and ripe (harvested about 56 DAF) seed. Unripe seed could only be heated to 70–75C, harvest ripe to 75–80C, and mature seed to 85–90C.

Shcherbakov (2) studied the effects of drying at 60, 80, and 120 C on seed harvested at three different maturity levels and stored for eleven weeks. Samples were evaluated weekly after drying at each temperature. Cephlin, lecithin and phosphatidic acid levels were measured It was found that the greatest change in the levels of the phosphatides during drying occurred in the least ripe seed. Regardless of stage of maturity or drying temperature, phosphatidic acid and lecithin exhibited little change. Cephlin generally decreased with the length of storage, drying temperature and seed maturity. Phosphatidic acid and lecithin remained fairly constant in the mature seed showing little change with increased drying temperature. Interestingly, after one and two weeks storage, there was a sharp inflection either positive or negative in a plot of phospholipid concentration after drying versus weeks of storage of the mature seed. Phosphatidic acid reached a maximum value at one week storage, then dropped the next week only to reach close to its initial maximum value after 11 weeks storage. With lecithin, an initially high concentration in the ripe seed was followed by a drop which after two weeks slowly increased over the remaining storage period. The opposite effect was seen in the concentration of cephlin. No explanation was offered for this occurrence.

## Drying Crude Oil

The type of heat used in the drying operation has been shown to have an important effect on the phospholipid content of the oil. Arutyunyan (12) conducted studies in which the amount of hydratable phospholipids in the oil was affected by the use of either steam or dry heat. This study showed that, as the heat was increased from 40 to 100 C, the amount of phospholipids remaining in the oil nearly tripled. It was suggested that humid heat increased bonding strength for the more polar phospholipids and decreased bonding strength for the less polar, such as lecithin. This would promote increased transfer of the less polar phospholipids into the oil. These results are consistent with the findings of Shcherbakov (2). In addition to decreased nonhydratable phospholipids, seed dried with steam heat retained its properties for 5–7 months.

Other workers (14) ascribed increased amounts of nonhydratable phospholipid to enzymic fermentation during slow heating of the ground meal from 20 to 70 C during worm gear transport of the press to the vat. This could be reduced by preheating the meal to 80–85 C at 8.5–9% moisture. It was suggested that phospholipases and lipase were inactivated, thus causing the production of less nonhydratable phospholipids. Arutyunyan (11) later demonstrated that with sunflower oil, this is not the case. Samples of sunflower meal were treated with a solution containing 1–2% phospholipase D and heated. The results showed that there was no change in the amount of nonhydratable phospholipids in the oil spiked with phospholipase D as compared to the control.

## Degumming

Once sunflower oil has been prepressed, it is important to degum the oil as soon as possible in order to achieve maximum removal of phospholipids. Arutyunyan (11) showed that an oil of 0.16% moisture which was stored under laboratory conditions for 24 hr developed a sediment and the moisture of the oil decreased. The phospholipids in the oil decreased from 0.56 to 0.30% over the 24-hr period. Longer holding periods in storage tanks reduced oil phospholipid content from 0.55 to 0.20% as a result of absorption of moisture from the atmosphere.

The effect of oil moisture was also demonstrated in an experiment in which an oil, dried to 0.02% moisture, was stored under laboratory conditions for 24 hr. During this period, oil phospholipids decreased from 0.56 to 0.52% as compared to the drop from 0.56 to 0.30% for the "wet" oil (9). This is important because as the amount of phospholipids in the oil decreases, so does the ease and efficiency of removal.

The use of either mineral acids such as phosphoric acid or alkali refining will also reduce the amount of phospholipids remaining in the oil. However, even at high concentrations of base, there remains a small percentage of material in the oil (9).

Phosphoric acid and some of its derivatives can reduce the amount of nonhydratable phospholipids to a negligible amount. This technique has been incorporated into the dewaxing process by a major vegetable oil manufacturer (15). Cold alkali refining, followed by the addition of phosphoric acid, has been used by another major producer to refine and dewax in one step (16). Unfortunately, phospholipids removed using mineral acid cannot be used in food products.

A Russian investigator (17) has achieved removal of phospholipids by mixing 3% ground sunflower hulls with hydrated oil (0.5% water) followed by filtration using a disc filter. The filtered oil contained 0.09–0.13% phospholipids and 0.04–0.07% moisture. The sediment contained 6–16% phospholipids and 59–68% oil.

## Isolation

Once the sludge (or foots) has been isolated, oil and moisture are removed in order to produce a powdered product. Use of the powdered lecithin is hampered by its tendency to undergo oxidative deterioration. The effectiveness of the stabilization of lecithin is determined by the method of dehydration and fat removal of the sludge. Popov (18) found that removal of fat by multiple washings with acetone (5:1 to 1:1 acetone/product) produced an oil-free, dehydrated lecithin product with color and stability superior to those of powdered lecithin obtained from crude lecithins using commercial conditions. Another approach was employed by German investigators (19). Their approach was to use methanol, ethanol, and isopropanol to extract pure lecithin from sludge.

## References

1. Hayenga, R., A Partial Interview with Ralph Hayenga, The Sunflower, Vol. 7:9, Dec. 1981, Fargo, ND, pp. 22–26.
2. Shcherbakov, V. G., and R. B. Malkhasyan, Maslo-Zhir Prom. 35(12):2 (1969).

3. Chapman, G. W., Jr., JAOCS 57:299 (1980).
4. Litvinova, E. D., E. A. Arisheva and N. S. Arutyunyan, Maslo-Zhir. Prom. 37(9):13 (1971).
5. El-Sabaiy, L. A., A. R. El-Mahdy, E. K. Mustofa and M. S. Mohamed, Food Chem. 5:217 (1980).
6. Galliard, T., Form and Function of Phospholipids, Vol. 3, edited by G. B. Ansell, J. N. Hawthorne and R. M. C. Dawson, Elsevier Scientific Publishing Co., Amsterdam, London, New York, 1973, pp. 253–288.
7. Grewal, S. S., P. S. Sukhija and L. S. Bhatia, Biochem. Physiol. Pllanzen. 173:11 (1978).
8. Popov, P., Proceedings of the 7th International Sunflower Conference, Krasnodar, USSR, 1976, p. 420.
9. Borodulina, A. A., P. S. Popov and I. N. Harchenko, Proceedings of the 6th International Sunflower Conference, Bucharest, Romania, 1974, p. 239.
10. Biacs, P., A. Erdélyi, E. Kuracz-Lusztig and J. Holló, Acta Alimentaria 7(3):181 (1978).
11. Arutyunyan, N. S., Maslo Zhir. Prom. 40;11 (1974).
12. Arutyunyan, N. S., V. M. Kopeikovsky and L. A. Mkhitaryants, Proceedings of the 7th International Sunflower Conference, Krasnodar, USSR, 1976, pp. 478–483.
13. Schuler, R. T., and D. C. Zimmer, Trans. ASAE 16:520 (1973).
14. Matsuk, Y. P., V. V. Beloborodov, F. A. Semenov and G. T. Semyakina, Trudy Vsesoyuznogo Nauchno-Isstedovatel'skoga Inst. Zhir. 28:41 (1971).
15. Gribble, W. P., and J. S. Rhee, U.S. Patent 3,994,943 (1976).
16. Young, H. T., U.S. Patent 3,943,155 (1976).
17. Ikhno, P. N., G. E. Mel'nik, G. S. Garmush, G. N. Lozheshnik and T. T. Gun'ko, Maslo-Zhir. Prom. 43:423 (1977).
18. Popov, A., M. Gardev, N. Yanishleiva and L. Hristova, JAOCS 48:365 (1971).
19. Liebing, H., and J. Lau, Fette Seifen Anstrichm. 78(3):123. (1976).
20. Vicentini, G. E., C. Galoppini, G. Angelli and R. Fiorentini, Proceedings of the 7th International Sunflower Conference, Krasnodar, USSR, 1976, p. 438.

# Chapter Seven
## Animal Lecithins

**A. Kuksis**
Banting and Best Department of Medical Research University of Toronto, Toronto, Canada M5G 1L6

## Introduction

Commercial animal lecithin is a complex mixture of several phospholipids, principally phosphatidylcholine, phosphatidylethanolamine and sphingomyelin, but other minor phospholipids along with free fatty acids and neutral lipids including free cholesterol may also be present in small amounts. The exact composition depends on the source of the lecithin as well as on the method of extraction and purification.

The present review describes the published industrial methods of lecithin preparation along with those employed in biological research of tissue lipids. Detailed lipid class and molecular species composition is provided for the main actual or potential animal sources of industrial lecithin as well as for the major food sources. Essentially all of the analyses reported have been performed on material prepared by biological research methodology and therefore may not represent the composition of any of the phospholipid classes actually present in a given commercial preparation of lecithin. Nevertheless, the provided data should serve as an informative guide to the potential complexity of any industrial preparation of animal lecithins.

## Extraction and Isolation

The emphasis is placed upon those extraction and isolation methods which yield preparative amounts of phospholipids, although very few references to truly industrial scale preparations were found in the published literature. Precipitation and solvent partition techniques appear to have been preferred for this purpose, but certain chromatographic routines also have been used as a purification step. Essentially all analytical isolations and quantitations have been performed by thin-layer chromatography. In view of the large differences in the lipid content and composition of different animal tissues, each source of phospholipid requires a special modification of the overall extraction and isolation technique. In most instances mixtures of chloroform-methanol have been used as extraction solvents.

**Egg Phospholipids**

The classical method for isolation of egg phosphatides is the cadmium chloride procedure (1). In this technique the choline-containing phospholipids are precipitated as the cadmium chloride complexes, which are sparingly soluble in acetone. Upon treatment with acetone the egg yolks are dehydrated at the same time as the neutral lipids are dissolved and the phospholipids precipitated. This process requires repeated treatment with acetone, which is tedious and time consuming, and yields a crude product (2–7). Lyophilization before extraction with acetone has also been used (8). Ethanol (3, 4, 9–11) and chloroform-methanol mixtures (8, 12, 13) have also been employed to extract crude egg yolk phospholipids.

Recently Ramesh et al (14) have described a selective extraction of phospholipids from egg yolk, which is applicable to large scale preparation. In this method egg yolk is suspended in methanol and filtered by decantation through a cotton-plugged funnel. The methanol extracts are concentrated under reduced pressure, when two layers are formed. The top layer is methanol diluted with water from egg yolk, while the bottom layer is phospholipid in methanol. The concentrate is treated with dilute salt solution and chloroform, and the solution thoroughly mixed. This yields a lower layer containing phospholipids and an upper layer of aqueous methanol containing non-lipids extracted from egg yolk. The non-phosphorus lipids (19% of total methanol-solube lipid) is made up of triacylglycerols and free cholesterol, which can be removed by low temperature crystallization. The method gives egg yolk phospholipids almost free of triacylglycerols in one step and does not require the pre-extraction of total lipids as in conventional methods (6, 15, 16).

Further purification of egg yolk phosphatidylcholine is achieved by adsorption chromatography but it requires caution. The initial efforts of Hanahan et al. (17) and Rhodes and Lea (8) separated the choline and non-choline fractions by elution with 95% ethanol or chloroform-methanol 1:1 on an alumina column, but were unable to obtain phosphatidylcholine free from lysophosphatidylcholine, which required a second fractionation of the choline fraction on a column of silicic acid. Renkonen (18) found that synthetic phosphatidylcholine and lysophosphatidylcholine could be successfully separated on alumina column with a solvent system consisting of chloroform-methanol 9:1. This ratio of solvents is much less polar than those referred to above. With the more polar solvents, about 1% of the phosphatidylcholine was degraded to lysophosphatidylcholine in 1 hour at 22 C. The production of lysophosphatidylcholine was markedly reduced if the phosphatidylcholine was eluted at 4 C. Likewise, Singleton et al (9) did not detect lysophosphatidylcholine formation on alumina column eluted with chloroform-methanol 9:1, which gave pure phosphatidylcholine. The lysophosphatidylcholine, sphingomyelin and phosphatidylethanolamine were retained on the column from which they could be eluted with chloroform-methanol-water 2:5:2. Marsh and Holzbach (19) have described improved yields of phosphatidylcholine by starting from commercial crude egg yolk lecithin.

Recently Nielsen (11) has purified egg phosphatidylcholine from $CdCl_2$-phosphatidylcholine complex by column chromatography on alumina. In this routine yolks from fresh hen eggs are stirred for a few seconds in a Waring blender. The yolks are extracted using 96% ethanol and the mixture is filtered. The clear yellow filtrate is set aside for 1 hour at 0 C in order to sediment impurities and then refiltered. To this primary extract is slowly added 50% $CdCl_2$ until no further precipitate occurs. The mixture is centrifuged after 1 hour at room temperature. The precipitate is washed with 96% ethanol, dissolved in water-saturated diethyl ether and reprecipitated with 99.9% ethanol and centrifuged. The purified $CdCl_2$ complex is dissolved in chloroform and applied to the alumina column. The lecithin is recovered by washing the column with dichloromethane-methanol 25% aqueous ammonia 925:70:5. This method for purification of egg yolk lecithin gives chemically and chromatographically pure phosphatidylcholine. The yield is comparable to that obtained by other methods (2.8 g from 5 eggs).

Rouser et al (20) have reported that the silicic acid-silicate column is the only one that separates phosphatidylcholine and sphingomyelin quantitatively, and Lundberg (21) has incorporated this technique in a large scale preparation of pure yolk phosphatidylcholine. Instead of the customary ethanol or chloroform-methanol, diethyl ether is used for extraction of the crude phosphatides in this method. The chromatography is carried out on a silicic acid-silicate column with 6:1 and 3:1 mixture of chloroform and methanol, pure methanol and pure chloroform as eluting solvents. To avoid degradation and oxidation of the product, the solvents are deoxygenated and BHT is added as antioxidant. The chromatography is carried out at 4 C under nitrogen. Lyophilized egg yolk or dried egg yolk powder can be used with diethyl ether and the phosphatidylcholines precipitated with cold acetone. The yolk must be completely dry, however. Much time is saved by evaporating the solvent by using diethyl ether instead of ethanol or a chloroform-methanol mixture for the extraction. At the same time most of the sphingomyelin, which is insoluble in diethyl ether is separated. Phosphatidylethanolamine can be separated by extraction with ethanol, but this is of lesser importance because it can be easily removed by chromatography (21). A similar type of fractionation has been obtained on the silicic acid column by either gradient or stepwise elution with chloroform and methanol (15) or with petroleum ether and diethyl ether (16). In the former instance, the non-polar lipids (triacylglycerols) are eluted ahead of the polar lipids (phospholipids), while in the latter the polar lipids are eluted ahead of the non-polar lipids because of the formation of high molecular weight micelles of phospholipids in hydrocarbon solvents.

Ramesh et al (14) have described a simultaneous extraction and preparative fractionation of egg yolk lipids. In this method, egg yolk is adsorbed on activated thin-layer-grade silicic acid and subsequently extracted with different solvents to get a crude fractionation of its lipids into non-polar and polar components. The crude concentrates help achieve larger yields of high purity lipids per unit column load during subsequent chromatographic subfractionation.

Radin (10) has suggested the replacement of the toxic chloroform and methanol in the isolation procedure by the relatively non-toxic ethanol, isopropanol, hexane and water. In this procedure commercial crude egg phosphatidylcholine is partially purified by a single step passage through an alumina column in ethanol. It is then purified by a two-step passage through a pre-packed commercial silica gel column (Lobar size B, EM Laboratories, Elmsford, NY), which is operated with a chromatographic pump. The Lobar and guard columns are equilibrated with same hexane-isopropanol-water 60:80:12 and 2.5 g of the phosphatidylcholine dissolved in the same solvent is slowly injected into the column system. After pumping in an additional 100 ml of the same solvent, the column was eluted with 1400 ml of hexane-isopropanol-water 60:80:15. The phosphatidylcholine was eluted between approximately 300 and 1300 ml and was collected in a single flask. The method does not allow removal of sphingomyelin, but both sphingomyelin and lysolecithin can be removed readily by a lighter loading of the column. The overall yield of phosphatidylcholine from a commercial sample was 40–43%. With the above procedure, one can prepare about 5.4 g of high quality phosphatidylcholine in two days.

Collins (22) and Davenport (23) have suggested that considerable chemical changes occur in phosphatidylcholines from natural sources during silicic acid chromatography. In contrast polystyrene chromatography represents an extremely mild method for separating neutral lipid from phospholipid and affords 92% recovery of the latter in a native form (23). In this method polar and non-polar lipids are resolved by exclusion chromatography no columns of divinyl cross-linked polystyrene beads by elution with benzene (24). Davenport recommends that 1% methanol is included as an antioxidant in the eluting solution. Berry and Kaye (25), however, have experienced difficulties with the use of polystyrene columns, which suggests that there may be some variation in the polystyrene samples.

Several workers have recognized the great potential of HPLC for the fast separation of phospholipids in large quantities (26–28). Recently, Guerts van Kessel et al (29) have developed a truly preparative HPLC method for the purification of a number of phospholipids, including egg phosphatidylcholine. The lipid separation is performed on a Poligosyl column and the individual lipid classes are monitored directly by refractive index detection. Chloroform-methanol mixtures are used as eluant systems. For the purification of the egg yolk phosphatidylcholine, a Polygosil 60-1525 column 50 cm in length and 5.0 cm in diameter was used along with 4.5 to 5 g of crude egg phosphatidylcholine in 7 ml of solvent. The solvent system was chloroform-methanol 60:40 to 70:30. Hanson et al (30) also have described an HPLC method for the rapid separation of egg yolk phospholipids. Direct UV detection and column packing material resistant to water (Ultrasil-$NH_2$) was used thereby providing the advantage of non-destructive elution and convenient use of water in the mobile phase. The column was eluted with hexane-isopropanol 5.5:8.0 and hexane-isopropanol-methanol-water 5.5:8.0; 1.0:1.5 gradient. Complete separations were obtained for cholesterol, ceramide, phosphatidylcholine, sphingomyelin, lysophosphatidylcholine, phosphatidylethanolamine and lysophosphatidylethanolamine. It should be noted that the amine packed column and silicic acid column differ in their selectivity in separating phospholipids.

Complete separation of phosphatidylethanolamine, lysophosphatidylethanolamine, phosphatidylcholine, sphingomyelin and lysophosphatidylcholine has been achieved by Gross and Sobel (31) by HPLC on silicic acid using acetonitrile-methanol-water 400:100:34 as the eluant. This system, however, readily inactivates the silica gel by the water component of the mobile phase, and must be regenerated, which takes time.

Table I gives the phospholipid class composition of various total lipid extracts of egg yolk, including some large scale preparations. It is seen that phosphatidylcholine makes up about 70%, phosphatidylethanolamine 24%, sphingomyelin 4% and phosphatidylserine and phosphatidylinositol about 1% each, with the lysophosphatidylcholine and lysophosphatidylethanolamine contributing the remaining 2% of the total phospholipid (13).

**TABLE 7-I.**
**Distribution of phospholipid classes in egg yolk**

| Phospholipid Classes | Analyses (1) | (2) | (3) | (4) | (5)[a] | (5)[b] |
|---|---|---|---|---|---|---|
| | | | Weight % | | | |
| % Total Lipid | | | | 23 | | |
| PC | 66–76 | 69.1 | 77.0 | 69 | 82.6 | 87.1 |
| PE | 15–24 | 23.9 | 18.0 | 24 | 9.1 | 7.8 |
| PS | 1 | 2.7 | | trace | | |
| CL | 1 | 3.2 | | trace | | |
| SPH | 3–6 | 1.0 | 2.3 | 3 | 1.8 | 2.5 |
| LPC+LPE | 3–6 | | 2.5 | 3 | 6.5 | 2.5 |
| UNIDENTIFIED | | | | | | |

[1]Privett et al (111).
[2]Noble and Moore (103).
[3]Cook and Martin (112).
[4]Gornall and Kuksis (113).
[5]Connelly and Kuksis (114).
[a]Commercial sample.
[b]Intralipid.

## Milk Phospholipids

The common methods for the extraction of milk fat recover both neutral lipids and phospholipids. The early methods based on solvent partition with predominantly ethyl or petroleum ether in the upper phase and basic (ammonia) aqueous ethanol mixtures in the lower phase retain the acidic phospholipids as ammonium salts and underestimate the phospholipid content (32, 33). Walstra and De Graaf (34) reported that by adding 1.5% NaCl to a sample of milk, cream or buttermilk, before the extraction, they could recover an additional 10–15% of lipid phosphorus. Various workers have attempted to obtain a selective extraction of the milk phospholipids by combining solvent extraction with adsorption chromatography. Duthie and Patton (35) described a routine for the extraction of milk phospholipids by combining whole milk with silicic acid, placing the mixture in a glass column and eluting the phospholipid fraction with 20% formic acid in ethyl ether. The neutral lipids were recovered with chloroform during the initial wash and adsorption of the milk-silicic acid sample to the column. Thin-layer chromatographic examination of the eluates showed that the new method gave comparable concentration of sphingomyelin, phosphatidylcholine and phosphatidylethanolamine, but less lysophosphatidylcholine than the other methods.

Cerbulis (36) has shown that complete recovery of milk phospholipids requires extraction with chloroform-methanol 2:1 by continuous shaking for 3 to 4 hours using 20 ml of solvent per gram of dry milk solids. According to Cerbulis (36) cream-bound lipids contained 90% phospholipids, skim-milk 44%, casein 21.9%, whey 25.5% and separation slime 72.7%. Other estimates of the phospholipid composition of various milk fractions have been reported by Morrison et al (37), Privett et al (38) and Patton and Keenan (39).

Milk phospholipids (phosphatidylcholines and phosphatidylethanolamines along with sphingomyelins) can be conveniently prepared from spray-dried buttermilk powder (37, 40). In this procedure total lipids are extracted with a monophasic chloroform-methanol-water system of Bligh and Dyer (41) and are purified by partitioning between chloroform and methanol-water. The neutral and polar lipids are separated by chromatography on a column of silicic acid. The glycerolipids and phospholipids are eluted with chloroform-methanol 1:1 and methanol. The polar lipid classes were resolved by thin-layer chromatography. The glycolipids, ceramide mono and dihexosides were isolated as described by Morrison and Smith (42, 43). The ethanolamine, choline and serine phosphatides and sphingomyelin were precipitated by acetone.

Santha and Narayanan (44) determined the phospholipid composition of the butteroil residue (ghee). The total lipids were extracted by the method of Folch et al (50). Butter serum was obtained by melting butter at 50–60 C. Individual phospholipids were isolated by thin-layer chromatography. The ghee residue contained the same phospholipids as the butter serum, but there were small differences. Since the ghee is prepared by heating butter to 120 C, it is possible that some partition or precipitation complexes in the residue may be demonstrated and removal of moisture may help to dissolve the phospholipids in the oil phase.

Nutter and Privett (45) prepared milk phosphatidylcholine by chloroform-methanol extraction of dried milk serum (the residual milk serum obtained in the course of the preparation of anhydrous milk fat by a creamery packaging process). The phosphatidylcholines were prepared by a combination of DEAE cellulose column and silicic acid-silicate aqueous ammonia column chromatography as described by Rouser et al (46). The phosphatidylcholine was separated from neutral lipids and concentrated in a fraction eluted from a DEAE cellulose with 9:1 chloroform-methanol. Phosphatidylcholine was isolated from this fraction by chromatography on silicic acid-silicate aqueous ammonia column. The phosphatidylcholine was eluted with chloroform-methanol 4:1 containing 1.5% water.

Cho et al (47) used butter serum from fresh, unsalted butter as a convenient source of milk phospholipids. Butter serum was frozen as a thin-layer in polyethylene bags and freeze dried. The lyophilized material was then extracted with chloroform-methanol 2:1. The phospholipids were fractionated by DEAE cellulose and two-dimensional TLC a described by Rouser et al (48).

Morrison et al (37) and Privett et al (38) have reported the results of detailed TLC analysis of the polar lipids of various milk fractions. Skim milk and milk serum have the highest proportion of polar lipids as per cent of total lipid, while whole milk and cream have the least. Of the polar lipids, phosphatidylethanolamine constitutes the major component, with phosphatidylcholine and sphingomyelin being present in about equal proportions in most milk fractions, but at significantly lower levels than phosphatidylethanolamine. It should be noted, however, that the relative proportions of the different phospholipid classes vary considerably among the different milk fractions. According to Morrison et al (37), milk contained also significant amounts of phosphatidylserine, but Privett et al (38) noted significant amounts of phosphatidylserine only in nonfat-skim milk and to a lesser extent in milk serum. In addition, to the common phospholipids, the various milk fractions also contained significant amounts of ceramide mono and dihexoside, as well as ganglioside. Privett et al (38) pointed out, however, that several of the phospholipid classes were incompletely resolved from certain unknowns and require further quantitative assessment. Table II gives the quantitative composition of various bovine milk fractions. Other tabulations have been provided by Morrison et al (37) and Santha and Narayanan (44).

### Brain Phospholipids

Another common source of animal lecithins or phospholipids for various applications is the brain tissue (1). The original methods of phospholipid preparation from brain involved freeing the brain tissue from adhering foreign tissue and mincing in a chopping machine. The tissue was dried by extraction several times with cold acetone. The dry residue was then extracted with hot alcohol. The alcohol extracts were concentrated, refrigerated overnight at 0 C and filtered. The filtrate was finally treated with a cold saturated solution of cadmium chloride sufficient to completely precipitate the crude phosphatidylcholine. The principal impurity in the cadmium chloride precipitate is phosphatidylethanolamine, which could be removed by subsequent extraction with diethyl ether and chloroform. Alternatively, crude phosphatidylcholine from brain may be obtained by covering the finely divided brain with diethyl ether and letting it stand in the cold for 48 to 72

**TABLE 7-II.**
**Distribution of phospholipid classes in various milk fractions[a]**

| Lipid Classes | Whole Milk | Plastic Skim | Nonfat Skim | Serum | Cream | Milk[b] | Serum[c] |
|---|---|---|---|---|---|---|---|
| | | | | Weight % | | | |
| % Total Lipid | 2 | 19 | 32 | 22 | 1–2 | | |
| Ceramide monohexoside | trace | 19.5 | 14.1 | 7.8 | 17.9 | 3 | |
| Ceramide dihexoside | trace | 10.1 | | | | 3 | |
| Phosphatidylethanolamine | 36.4 | 27.5 | 45.1 | 32.4 | 25.3 | 30 | 26.9 |
| Plasmalogen | | | | | | 1 | |
| Phosphatidylcholine | 27.0 | 7.3 | 16.4 | 23.2 | 19.9 | 28 | 29.3 |
| Plasmalogen | | | | | | 3 | |
| Sphingomyelin | 29.0 | 18.2 | 14.5 | 26.9 | 21.6 | 19 | 31.3 |
| Phosphatidylserine | trace | 1.9 | 7.6 | 3.5 | trace | 8 | 5.0 |
| Phosphatidylinositol | trace | trace | trace | 1.1 | trace | 5 | 5.9 |
| Ganglioside + (unknown) Lysophospholipids | 7.6 | 15.5[4] | 2.3[3] | 5.1[4] | 15.3[4] | 0 | 1.0 |

[a]Privett et al (38).
[b]Morrison et al (37).
[c]Santha and Narayanan (44).

hours. The cold diethyl ether extracts crude phosphatidylcholine and cholesterol. The phosphatidylcholine is purified by precipitation with acetone. Both of the above methods can be readily applied to working up several kg of brain tissue and have been employed for both industrial and semiindustrial preparations (49).

More recently the brain lipids have been extracted by means of chloroform-methanol using some modification of the procedure of Folch et al (12, 50). The major drawback of these methods is that they also extract substantial quantities of non-lipid substances (51, 52), which must be removed by lengthy solvent extraction or gel chromatography. Phillips and Privett (53) have described a method for the quantitative extraction of lipids from brain tissue by means of chloroform-methanol that does not require subsequent purification by gel chromatography or aqueous washing of the lipid extracts. In this procedure, the non-lipid substances are removed in a pre-extraction of the tissue with dilute (0.25%) acetic acid. The residual tissue is extracted twice with 40 volumes of chloroform-methanol 1:1. Approximately 97% of the lipid present is removed in these extractions. There was no evidence that treatment of the tissue with dilute acetic acid would hydrolyze ether lipids. However, the vinyl ethers (plasmalogens) would be split if the full routine involving final extraction with dilute HCl were applied. Presumably this procedure can be applied to brain tissues other than that of the pig for a rapid and quantitative recovery of the polar lipids.

The lipid class composition of brain phospholipids has been extensively investigated by adsorption column and thin-layer chromatography, and, more recently, high pressure liquid chromatography. Usually aliquots of the total brain lipid extract have been chromatographed on DEAE cellulose as described by Rouser et al. (48) to segregate the choline-containing phospholipids from the ethanolamine-containing phospholipids. A chloroform-methanol 12:1 fraction, which contains phosphatidylcholine and sphingomyelin, is combined an applied to silicic acid column and the column eluted in succession with chloroform-methanol 6:1, 3:1 and 1:1. Phosphatidylcholine is eluted in the 3:1 fraction and chromatographs as a single band on TLC. The chloroform-methanol 4:3 fraction from the DEAE cellulose contains the ethanolamine phospholipids, and must be treated separately to resolve the diacyl, alkylacyl and alkenylacyl species. The thin-layer chromatography of the phospholipid classes is usually performed in the two-dimensional form as described by Rouser et al. (51) and Skipski (54). The phosphatidylcholine is recovered from the silica gel by repeated extraction with chloroform-methanol-acetic acid-water 50:39:1:10 as described by Arvidson (55). Recently, several workers (56–58) have described HPLC methods for the resolution and semi-preparative isolation of phosphatidylcholine from brain lipids. Table III lists the major classes of brain phospholipids from different animal species. Several workers have given separate estimates for the phospholipids of the myelin, endothelial cells and brain cell nuclei, which have been reproduced here. Others have distinguished between the normal glycerophospholipids and plasmalogens.

### Phospholipids of Liver, Kidney and Muscle

Other major tissue sources of dietary phospholipids and phosphatidylcholine are the organ meats, liver, and kidney. These sources are also of importance for the industrial production of lecithin concentrates. The phospholipids from these tissues are usually extracted by the chloroform-methanol mixtures proposed by Folch et al (50) or by Bligh and Dyer (41). According to Rouser et al. (48) the tissue is first ground to a uniform paste by passage through a household-type meat grinder or a Potter-Elvehjem homogenizer, depending on the size of the sample. The extraction is accomplished by homogenizing in a Waring blender for 5 min with 20 volumes of chloroform-methanol 2:1 containing 1 mg of BHT per liter of solvent. The suspension is filtered through a medium porosity sintered glass filter and the residue is re-extracted three times (3 min each) with 10 volumes each of chloroform-methanol 2:1, chloroform-methanol 1:2, and finally chloroform-methanol

**TABLE 7-III.**

**Distribution of phospholipid classes in brain of different animal species**

| Lipid Classes | Human[a,e] | Bovine[a,e] | Bovine[b,e] | Human[b,f] | Bovine[b,g] | Human[b,g] | Sheep[c] | Rat[d] |
|---|---|---|---|---|---|---|---|---|
| | | | | Weight % | | | | |
| % Total Lipid | | | | | | | | |
| PC | 21.8 | 18.4 | 32.4 | 33.2 | 48.2 | 47.6 | 37.3 | 36.8 |
| Plasmalogen | | | | | | | 0.9 | |
| PE | 35.4 | 36.1 | 23.5 | 25.2 | 24.2 | 17.8 | 7.7 | 36.4 |
| Plasmalogen | | | | | | | 16.5 | |
| PS | 18.8 | 18.0 | 11.0 | 10.7 | 6.7 | 9.3 | 9.2 | 11.8 |
| PI | 1.8 | 1.8 | 4.3 | 4.8 | 7.1 | 5.0 | 2.1 | 3.1 |
| PA | 1.1 | 1.7 | 0.9 | 0.3 | 1.3 | 1.2 | 2.6 | 1.2 |
| CL | | | 1.2 | 1.0 | ND | 0.3 | 2.0 | 2.2 |
| PG | | | 0.9 | ND | 0.4 | 0.6 | | |
| LPC (LPE) | 2.0 | 0.2 | 0.9 | 1.0 | 1.0 | 2.5 | | |
| LbisPC | 1.0 | 0.4 | 2.1 | 0.2 | 0.7 | 0.2 | | |
| SPH | 16.3 | 15.0 | 20.4 | 17.0 | 4.9 | 10.7 | 12.9 | 5.7 |
| UNIDENTIFIED | | | | | | | | |

[a]Siakotos et al (116).
[b]Siakotos and Fleischer (115).
[c]Scott et al (117).
[d]Wuthier (1966) (118).
[e]myelin.
[f]endothelial cells.
[g]nuclei.

7:1 saturated with ammonium hydroxide. The extracts are pooled and the solvent removed under vacuum, methanol and water being removed by repeated additions of chloroform and evaporation to small volume, keeping the temperature below 15 C and avoiding complete dryness. Following solvent evaporation, the lipids are transferred to a Sephadex column in chloroform-methanol 19:1, saturated with water, and the lipids separated from water-soluble non-lipid contaminants as described by Siakotos and Rouser (59). The lipids in fraction 1 from the Sephadex column were concentrated and diluted to a known volume of chloroform-methanol 9:1 and aliquots taken for two-dimensional TLC as described by Rouser et al (60).

Recently HPLC techniques have been introduced for the resolution of the phospholipid classes of liver lipids (61, 62). According to Patton et al (62) the total lipid extract is separated into phospholipid classes by chromatography on a 250 × 4.6 mm Hibar II column packed with 10 $\mu$ LiChrospher Si-100 (EM Laboratories Inc., Elmsford, NY). The eluting solvent was prepared by the addition of 490 ml of HPLC-grade 2-propanol to 62 ml of 25 mM potassium phosphate (pH 7.0). After mixing, 367 ml of HPLC-grade hexane and 100 ml of absolute ethanol were added. After filtration, 0.6 ml of glacial acetic acid was added. The final solvent composition was hexane-2-propanol-25 mM phosphate buffer-ethanol-acetic acid 367:490:62:100:0.6. Complete separations were obtained for neutral lipid, phosphatidylethanolamine, phosphatidylserine, phosphatidylinositol, cardiolipid, phosphatidylcholine, sphingomyelin and lysophosphatidylcholine. Conventional HPLC, however, is capable of handling only small amounts of material, but scaled up systems have been proposed (29). Plasmalogens have not as yet been resolved from the normal glycerophospholipids by HPLC systems. The phospholipids of heart, kidney and skeletal muscle are rich in the alkenylacyl and alkylacyl species of phosphatidylcholine and especially of phosphatidylethanolamine (63–65). The principal phospholipid classes of the liver, kidney, heart and skeletal muscle are shown in Tables IV, V, VI, and VII, respectively.

## Phospholipids of Spleen and Erythrocytes

The phospholipid composition of the spleen and the red blood cells is of interest because of its complexity. Furthermore, the plasma membrane of the red blood cell, which contains most of the phospholipid, is frequently considered as a model of the composition of other cell membranes and their physico-chemical behaviour. Total lipid and phospholipid extracts from spleen can be readily prepared by the method outlined for the kidney and liver tissue by Rouser et al (48). The preparation of total phospholipid

**TABLE 7-IV.**
**Distribution of phospholipid classes in liver of different animal species**

| Phospholipid Classes | Rat[a] | Man[a] | Ox[a] | Mouse[a] | Frog[a] | Rat[b] | Sheep[b] |
|---|---|---|---|---|---|---|---|
| | | | | % total lipid P | | | |
| PC | 48.2 | 43.6 | 52.9 | 45.6 | 46.5 | 54.6 | 39.7 |
| Plasmalogen | | | | | | | 0.8 |
| PE | 23.6 | 27.9 | 24.4 | 25.1 | 25.6 | 25.8 | 27.9 |
| Plasmalogen | | | | | | | 3.6 |
| PS | 3.4 | 3.1 | 4.1 | 3.9 | 3.9 | 2.8 | 2.3 |
| PI | 8.3 | 8.6 | 8.3 | 7.7 | 7.3 | 6.9 | 8.2 |
| PA | 0.1 | 0.7 | 0.4 | 0.1 | 0.2 | 0.4 | 0.8 |
| CL | 4.5 | 3.7 | 2.7 | 4.7 | 2.0 | 4.6 | 1.1 |
| PG | 0.2 | ND | 0.2 | 0.1 | 0.2 | | 5.3 |
| LbisPA | 0.1 | 1.0 | 0.4 | 0.1 | 0.4 | | |
| LPC | 0.7 | 1.3 | 0.9 | 0.7 | 0.8 | 0.9 | |
| LPE | ND | ND | ND | ND | ND | | |
| SPH | 4.2 | 4.6 | 4.3 | 4.7 | 5.1 | 3.2 | 5.0 |
| UNIDENTIFIED | | | | | | 0.9 | |

[a]Rouser et al (48).
[b]Patton et al (62).

extracts from the erythrocytes, however, requires a modified approach, although any of the common chloroform-methanol ratios may be utilized. It is essential that an effective lysis of the cells is achieved early in the extraction process. According to Dodge and Phillips (66) 15 ml of methanol containing 5 mg of BHT per 100 ml is added to 10 ml of packed cells and the suspension allowed to stand for 30 min. with occasional stirring. This is followed by the addition of 15 ml chloroform and the mixture again is allowed to stand for 30 min. The suspension is then centrifuged at 1500 xg for 5 min and the supernatant decanted into a separatory funnel. A second and third extraction is performed starting with 15 ml of methanol added to the residue, standing for 10 min. and followed by 15 ml chloroform and again standing for 10 minutes. After centrifugation, the supernatants are combined in the separatory funnel and 48 ml of chloroform and 28 ml of 0.05 M KCl added and mixed. After phase separation (overnight standing in the cold) and warming to 40 C, the lower phase is collected, evaporated to small volume in a rotary vacuum evaporator, taken up in chloroform, and diluted to volume for further analyses.

Rose and Oaklander (67) have recommended a chloroform-isopropanol method of extraction of red blood cell lipids. In this routine 1 ml of packed cells is treated with 1 ml of distilled water, vortexed and left standing for 15

**TABLE 7-V.**
**Distribution of phospholipid classes in kidney of different animal species**

| Phospholipid Class | Animal Species | | | | | |
|---|---|---|---|---|---|---|
| | Rat[a] | Man[a] | Ox[a] | Mouse[a] | Frog[a] | Sheep[b] |
| | | | | % total lipid P | | |
| PC | 34.3 | 33.1 | 32.4 | 35.6 | 34.3 | 35.8 |
| Plasmalogen | | | | | | 3.2 |
| PE | 27.1 | 27.4 | 28.6 | 26.4 | 30.0 | 10.6 |
| Plasmalogen | | | | | | 7.4 |
| PS | 7.3 | 6.3 | 7.5 | 7.0 | 6.0 | 3.2 |
| PI | 5.9 | 5.5 | 7.2 | 5.7 | 5.6 | 4.6 |
| PA | 0.3 | 0.2 | 0.2 | 0.2 | 0.3 | 2.3 |
| CL | 6.5 | 4.2 | 6.5 | 7.3 | 4.8 | 5.9 |
| PG | 0.3 | 0.5 | 0.4 | 0.5 | 0.3 | |
| LbisPA | 0.1 | 0.1 | 0.2 | 0.2 | 0.4 | |
| LPC | 1.0 | 2.8 | 0.7 | 0.9 | 0.2 | |
| LPE | 0.1 | 2.9 | 0.3 | ND | ND | |
| SPH | 12.1 | 12.0 | 13.7 | 10.4 | 12.2 | 12.1 |
| UNIDENTIFIED | | | | | | |

[a]Rouser et al (48).
[b]Scott et al (117).

min. Eleven ml of isopropanol is then added slowly with mixing and the mixture allowed to stand for one hour. Then 7 ml of chloroform is added and mixed and the mixture again allowed to stand for one hour. Finally the suspension is centrifuged at 500 xg for 30 min. The supernatant can be sampled directly for further analysis. The latter procedure is simple and reproducible, affords almost quantitative extraction of cholesterol and phospholipid, uses a single extraction tube, and yields an extract only slightly contaminated by heme. Tables VIII and IX give the distribution of the phospholipid classes in the spleen and the red blood cells, respectively, of different animal species.

### Phospholipids of Other Tissues

The lung and bile are two sources of phospholipids of relatively simple chemical composition. The lung and bile secrete phosphatidylcholines as natural surfactants. Both lung and bile phospholipids can be isolated readily by any one of the conventional methods of phospholipid extraction. Since these phospholipids contain very little neutral lipid, there is no need for further purification prior to lipid class analysis. Table X gives the distribution of the phospholipid classes in the total lung extracts from several animal species. The phospholipid composition of the bile

**TABLE 7-VI.**
**Distribution of phospholipid classes in heart muscle of different animal species**

| Phospholipid Classes | Rat[a] | Man[a] | Ox[a] | Mouse[a] | Frog[a] | Rabbit[b,d] | Rabbit[b,e] | Sheep[c] |
|---|---|---|---|---|---|---|---|---|
| | % total lipid P | | | | | | | |
| PC | 36.0 | 40.0 | 38.1 | 40.6 | 39.6 | 38.4 | 42.2 | 25.5 |
| Plasmalogen | | | | | | | | 20.4 |
| PE | 29.7 | 26.3 | 29.7 | 28.9 | 31.5 | 34.0 | 32.2 | 12.4 |
| Plasmalogen | | | | | | | | 11.0 |
| PS | 3.2 | 2.7 | 3.5 | 2.8 | 5.1 | 3.3 | 3.4 | 2.2 |
| PI | 3.7 | 6.1 | 5.3 | 4.3 | 4.2 | 3.1 | 3.2 | 4.7 |
| PA | 0.2 | 0.2 | trace | 0.1 | 0.1 | 0.9 | 0.4 | 0.4 |
| CL | 11.2 | 9.0 | 12.6 | 11.2 | 9.3 | 14.5 | 12.7 | 5.8 |
| PG | 1.0 | 0.6 | 0.2 | 0.6 | 0.2 | 1.2 | 1.1 | |
| LbisPA | ND | ND | ND | ND | ND | — | — | |
| LPC | 0.8 | 3.5 | 0.2 | 1.6 | 0.3 | 0.8 | 0.5 | |
| LPE | ND | 1.5 | ND | ND | ND | — | — | |
| SPH | 3.1 | 4.9 | 4.7 | 3.5 | 6.5 | 3.8 | 4.3 | |
| UNIDENTIFIED | | | | | | | | |

[a]Simon and Rouser (119).
[b]Osanai and Sakagani (120).
[c]Scott et al (117).
[d]male.
[e]female.

of different animal species is given in Table XI.

The adipose tissue contains a high proportion of neutral lipid, that must be removed during the isolation of the phospholipids. Christie and Moore (68) have extracted the dissected adipose tissue with chloroform-methanol 2:1 by the usual procedure. The separation of the phospholipids from the neutral lipids (largely triacylglycerols) was then accomplished by solvent partition between hexane and 87% ethanol using the procedure of Galanos and Kapoulas (69). All the complex lipids were retained in the ethanol layer together with some free fatty acids and small amounts of partial acylglycerols. Pure phospholipids were obtained by chromatography on a column of silicic acid, chloroform being used to elute neutral lipids and methanol to recover the phospholipids. Table XII gives the phospholipid class distribution in the adipose tissue of different animal species.

Virtually complete extraction of oil and phospholipids may be obtained with isopropanol during the production of fish protein concentrates from herring (70). Weihrauch and Son (71) have published an extensive compilation of data on the phospholipid class composition of various edible fish and shellfish products.

**TABLE 7-VII.**
**Distribution of phospholipid classes in skeletal muscle of different animal species**

| Phospholipid Classes | Man[a] | Ox[a] | Rat[a] | Mouse[a] | Frog[a] | Lobster[a] | Rabbit[b] | Chicken[b] |
|---|---|---|---|---|---|---|---|---|
| | | | | % total lipid | | | | |
| PC | 48.0 | 46.5 | 51.1 | 52.4 | 55.2 | 54.6 | 65.9 | 58.4 |
| PE | 26.4 | 26.6 | 22.2 | 25.8 | 29.0 | 23.6 | 17.4 | 19.5 |
| PS | 3.3 | 4.1 | 3.7 | 4.0 | 3.4 | 6.2 | 0.8 | 1.0 |
| PI | 8.8 | 5.6 | 8.9 | 6.7 | 5.5 | 4.9 | 10.7 | 3.8 |
| PA | 1.2 | 0.3 | 0.8 | 0.1 | 0.1 | 0.2 | | |
| CL | 6.6 | 8.9 | 1.4 | 5.8 | 2.3 | 1.3 | 0.4 | 2.3 |
| PG | 1.0 | 0.3 | 0.9 | 0.7 | 0.2 | trace | | |
| LbisPA | trace | 0.1 | 0.3 | trace | 0.1 | trace | | |
| LPC | trace | 0.7 | 2.7 | 0.6 | 0.4 | 1.1 | | |
| LPE | ND | ND | 1.3 | trace | trace | ND | | |
| SPH | 4.0 | 4.5 | 2.7 | 3.6 | 3.1 | 8.4 | 2.8 | 8.8 |
| UNIDENTIFIED | | | | | | | | |

[a]Simon and Rouser (119).
[b]Marai and Kuksis (64, 65).
[c]Sarcotubular membranes of skeletal muscle.

**TABLE 7-VIII.**
**Distribution of phospholipid classes in spleen of different animal species**

| Phospholipid Classes | Rat[a] | Man[a] | Ox[a] | Mouse[a] | Sheep[b] |
|---|---|---|---|---|---|
| | | | % total lipid P | | |
| PC | 41.9 | 41.2 | 36.0 | 41.5 | 40.5 |
| Plasmalogen | | | | | 3.0 |
| PE | 24.0 | 24.5 | 24.5 | 24.9 | 25.0 |
| Plasmalogen | | | | | 9.8 |
| PS | 8.0 | 8.3 | 12.0 | 7.9 | 2.1 |
| PI | 5.5 | 4.4 | 4.3 | 5.8 | 1.6 |
| PA | 0.6 | 0.2 | 0.6 | 0.2 | 1.5 |
| CL | 2.0 | 1.0 | 0.3 | 2.0 | 1.5 |
| PG | 0.8 | 0.3 | ND | 0.3 | |
| LbisPA | 0.5 | 0.3 | ND | 0.7 | |
| LPC | 1.3 | 1.5 | 0.8 | 1.5 | |
| LPE | ND | ND | 0.3 | ND | |
| SPH | 6.6 | 12.8 | 14.5 | 6.6 | 12.5 |
| UNIDENTIFIED | | | | | 9.1 |

[a]Rouser et al (48).
[b]Getz et al (121).

**TABLE 7-IX.**
**Distribution of phospholipid classes in red blood cells of different animal species**

| Phospholipid Classes | Man[a] | Rat[b] | Pig[b] | Dog[b] | Horse[b] | Sheep[b] | Cow[b] | Hen[c] |
|---|---|---|---|---|---|---|---|---|
| | Animal Species | | | | | | | |
| | Mole % | | | | | | | |
| PC | 29.2 | 47.5 | 23.3 | 46.9 | 42.2 | — | | 42.3 |
| PE | 27.5 | 21.5 | 29.7 | 22.4 | 24.3 | 26.2 | 29.1 | 29.1 |
| PS | 14.8 | 10.8 | 17.8 | 15.4 | 18.0 | 14.1 | 19.3 | — |
| PI | 0.6 | 3.5 | 1.8 | 2.2 | 0.3 | 2.9 | 3.7 | 5.5 |
| LPC | 1.0 | 3.8 | 0.9 | 1.8 | 1.7 | | | |
| PG | 1.1 | | | | | | | |
| SPH | 25.4 | 12.8 | 26.5 | 10.8 | 13.5 | 51.0 | 46.2 | 22.7 |
| UNIDENTIFIED | | | | | | | | |

[a]Dodge and Phillips (66).
[b]Nelson (122).
[c]Kleinig et al (123).

**TABLE 7-X.**
**Distribution of phospholipid classes in lung tissue of different animal species**

| Phospholipid Classes | Man[a] | Rat[a] | Rat[b] | Rat[c] | Pig[a] | Mouse[a] | Bovine[a] |
|---|---|---|---|---|---|---|---|
| | Animal Species | | | | | | |
| | % of total lipid P | | | | | | |
| PC | 47.5 | 44.9 | 52.3 | 57.5 | 49.0 | 43.7 | 39.5 |
| PE | 17.5 | 22.2 | 19.2 | 16.6 | | 20.1 | 21.2 |
| PS | 7.0 | 9.0 | 5.4 | 9.3 | 36.0 | 8.1 | 9.4 |
| PI | 3.2 | 3.9 | 5.4 | 2.6 | | 4.1 | 3.3 |
| PG | 2.5 | 2.2 | 2.2 | 3.3 | 0.8 | 2.4 | 2.0 |
| LbisPA | 1.5 | 0.3 | 2.3 | 1.0 | 0.8 | 0.3 | 0.5 |
| CL | 1.0 | 1.0 | 4.0 | 1.4 | 1.5 | 1.1 | 1.0 |
| PA | 0.5 | 0.3 | | 0.3 | | 0.4 | 1.4 |
| LPC | 2.0 | 1.3 | 1.2 | 0.5 | trace | 1.1 | 0.3 |
| LPE | 0.6 | 0.5 | | | | | |
| SPH | 11.1 | 10.0 | 7.6 | 7.5 | 12.0 | 9.3 | 16.1 |
| UNIDENTIFIED | | | | | | | |

[a]Baxter et al (124).
[b]Wang et al (125).
[c]Hallman and Gluck (126).

**TABLE 7-XI.**
**Distribution of phospholipid classes in bile of different animal species**

| Phospholipid Classes | Rat[a] | Sheep[b] | Steer[b] |
|---|---|---|---|
| | | % total lipid P | |
| PC | 90.0 | 53 | 89 |
| PE | 4.4 | trace | 4 |
| PS | 1.2 | | |
| PI | 1.4 | | |
| SPH | 2.7 | 8 | 8 |
| CL | — | | |
| LPC | 0.3 | 39 | trace |

[a] Kawamoto et al (127)
[b] Adams and Heath (128)

**TABLE 7-XII.**
**Distribution of phospholipid classes in adipose tissue of different animal species**

| Phospholipid Classes | Rat[a] | Rat[b] | Pig[b] | Beef[b] |
|---|---|---|---|---|
| | | % total lipid P | | |
| PC | 60.9 | 46.5 | 42.4 | 42.1 |
| PC ethers | | 4.2 | 3.7 | 3.7 |
| PE | 8.2 | 23.4 | 13.1 | 17.6 |
| PE ethers | | 11.4 | 18.0 | 20.0 |
| PS | 2.8 | | | |
| | | 2.1 | 3.4 | 3.0 |
| PI | 7.0 | | | |
| PA | | | | |
| | 7.7 | 3.6[c] | 8.1[c] | 4.1[c] |
| CL | | | | |
| LPC | 10.7 | 8.2 | 7.5 | 9.8 |
| UNIDENTIFIED | | | | |

[a] Abbreviated from Christee and Moore (68)
[b] Abbreviated from Grigor et al (129)
[c] Material at solvent front.

## Composition of Molecular Species

The phospholipids or lecithins isolated from different animal tissues vary greatly in the composition of the molecular species making up the individual lipid classes. The early work showed that the two positions of the glycerol moiety contained different fatty acids, with the sn-1-position being occupied mainly by saturated and the sn-2-position by unsaturated fatty acids, but the reverse isomers also exist. More recently preferential pairings of specific saturated and unsaturated fatty acids have been recognized. The exact composition of the molecular species varies with the tissue source of the phospholipid, but corresponding tissues or organs of different animal species possess closely similar patterns of molecular species (72, 73). The composition of molecular species of the individual phospholipid classes is little influenced by the extraction technique except in the case of plasmalogens, which are sensitive to acids, and of polyunsaturated acids, which are subject to losses upon bleaching. Since the physicochemical and biological properties of the phospholipid molecules are known to vary with the exact pairing of the component fatty acids, much effort has been expended in determining the composition of the molecular species of natural glycerophospholipids. The following section tabulates the molecular species of the major phospholipid classes of the commercially and nutritionaly more important tissues of selected animal species. Separate accounts of the fatty acid composition are not given. Approximate estimates of the total and positional distribution of fatty acids may be obtained by summing the estimates for the individual molecular species, which are recorded with the fatty acids in the sn-1-position on the left and those in the sn-2-position on the right.

**Phosphatidylcholine**

The phosphatidylcholines constitute the largest class of the glycerophospholipids of most animal tissues and frequently are referred to as the lecithins. The chemical composition and physico-chemical properties of phosphatidylcholine have been extensively investigated and generally have been equated with those of all phospholipids. In fact, the methods of determination of molecular species were first investigated and developed using natural phosphatidylcholines as model compounds.

### 1. Egg yolk

A complete determination of the molecular species of egg yolk phosphatidylcholines was first described by Kuksis and Marai (74). Prior to this time Tattrie (75) had established the differential fatty acid composition of the sn-1- and sn-2-positions of the phosphatide, while Renkonen (76, 77) and Nutter and Privett (78) had demonstrated that a rather detailed description of the major molecular species of egg yolk phosphatidylcholine may be obtained by an investigation of the diacylglycerol moieties of the parent phosphatide by argentation TLC in combination with a fatty acid analysis by GLC. The results of Kuksis and Marai (74), which were obtained by a combination of $AgNO_3$-TLC and high temperature GLC of the diacylglycerol moieties have been confirmed by Hasegawa and Suzuki (79) using GC/MS of the derived diacylglycerol acetates in combination with $AgNO_3$-TLC. More recently Myher and Kuksis (80) have reported a direct resolution of the molecular species of egg yolk phosphatidylcholines in the form of the TMS ethers of the derived diacylglycerols by GLC on polar capillary columns. This separation is similar to that commonly realized for the fatty acid methyl esters. Some 50 different molecular species were recognized and quantitated, but the estimates for the major components agreed closely with those reported previously. Others have resolved the molecular species of the egg yolk phosphatidylcholines in the form of the dimethyl esters of phosphatidic acid using normal phase HPLC with acetonitrile as eluant and refractive index detection (81).

The fatty acid composition and hence the molecular species make-up of egg phosphatidylcholines is subject to alterations by dietary fatty acids. Rhodes (82) has reported originally and Hanahan (83) has discussed extensively experiments in which chickens were maintained on diets containing varying degrees of unsaturation in the component fatty acids. The predominant feature of the results was that even though the proportion of the unsaturated fat in the diet was increased, only the sn-2-acyl groups reflected the change. Table XIII gives the principal molecular species of egg yolk phosphatidylcholines as obtained by different methods of analysis. The major molecular species contain the saturated fatty acids in the sn-1-position, and the unsaturated fatty acids in the sn-2-position, with palmitoyl oleate, palmitoyl linoleate, stearoyl oleate and stearoyl linoleate accounting for the bulk of the total mass. The occurrence of the reverse isomers is limited. Apparently only small amounts of the fully saturated and fully unsaturated phosphatidylcholines occur in the egg yolk.

## 2. Milk

Nutter and Privett (45) provided the first comprehensive analysis of milk phosphatidylcholine, although Morrison et al (37) had reported earlier detailed analyses of the positional distribution of fatty acids in the major milk phospholipids. Nutter and Privett (45) reported an essentially com-

**TABLE 7-XIII.**
**Molecular species of egg yolk phosphatidylcholines**

| Molecular Species | (a) | (b) | (c) | (d) |
|---|---|---|---|---|
| | | | Moles % | |
| 14:0 16:0 | 0.04 | | | 1.0 |
| 14:0 18:0 | 0.03 | | 0.07 | |
| 16:0 16:0 | 0.72 | | 1.0 | |
| 18:0 16:0 | 0.20 | | | |
| 18:0 18:0 | 0.01 | | | |
| 16:0 16:1 | 0.98 | 0.6 | 1.4 | 1.8 |
| 16:0 18:1 | 38.20 | 44.4 | 46.8 | 39.5 |
| 18:0 18:1 | 9.30 | 19.7 | 10.7 | 10.1 |
| 16:0 20:1 | 0.49 | | | |
| 16:1 16:1 | 0.04 | | | |
| 16:1 18:1 | 0.92 | | | |
| 18:1 18:1 | 2.96 | | 2.3 | |
| 18:1 20:1 | 0.08 | | | |
| 16:0 18:2 | 21.78 | 16.6 | 20.4 | 19.9 |
| 18:0 18:2 | 11.22 | 8.6 | 7.8 | 10.2 |
| 16:1 18:2 | 0.40 | | | |
| 18:1 18:2 | 1.50 | 1.8 | 1.1 | 2.3 |
| 20:1 18:2 | 0.10 | | | |
| 16:0 18:3 | 0.03 | | | |
| 16:0 20:3 | 0.45 | | | |
| 18:0 20:3 | 0.45 | | | 0.7 |
| 16:0 22:3 | 0.07 | | | |
| 16:0 20:4 | 2.52 | 3.1 | 1.6 | 1.9 |
| 18:0 20:4 | 3.36 | 4.9 | 2.1 | 5.3 |
| 20:4 20:4 | 0.12 | | | |
| 14:0 20:5 | 0.09 | | | |
| 16:0 20:5 | 0.08 | | | |
| 16:0 22:5 | 0.09 | | | |
| 18:1 20:4 | 0.48 | | | 1.5 |
| 18:1 22:4 | 0.26 | | | |
| 14:0 22:6 | 0.60 | | | |
| 16:0 22:6 | 1.84 | | 2.0 | 1.6 |
| 18:0 22:6 | 1.60 | | 0.7 | |
| 16:1 22:5 | 0.01 | | | |
| 18:1 22:5 | 0.04 | | | |
| 18:2 20:4 | 0.02 | | | |
| 18:2 22:4 | 0.03 | | | |

[a]Kuksis and Marai (74).
[b]Hasegawa and Suzuki (130).
[c]Myher and Kuksis (80).
[d]Kuksis et al (131).

plete account of the fatty acid pairings in the molecular species of the phosphatidylcholines of the milk serum, including all the species in amounts greater than 0.01% and accounting for some 113 of the theoretically possible 196 species. The technique consisted of a phospholipase C dephosphorylation of the phosphatides to the corresponding diacylglycerols, acetylation and a subsequent argentation TLC, with a further resolution of the individual fractions by HPLC. The final products were subjected to hydrolysis with pancreatic lipase to yield the proportions of the reverse isomers in each molecular species or small fractions thereof. Prior to this study, Blank et al (84) had obtained a less complete analysis by separation of the mercuric acetate adducts, a technique which does not involve destruction of the glycerophospholipid molecules. There was complete agreement between the estimates for the major molecular species obtained by the two different methods of analysis. Table XIV gives molecular species of the phosphatidylcholine of milk serum as reported by Nutter and Privett (45) and Blank et al (84).

### 3. Brain

Renkonen (76) was the first to resolve successfully the major molecular species of brain phosphatidylcholines, although complete analyses were not obtained. Renkonen (76) recognized the presence of diacyl, alkenylacyl and alkylacyl subclasses of brain phosphatidylcholines, which could be resolved following dephosphorylation, using adsorption TLC. Each subclass could be further separated according to degree of unsaturation by argentation TLC, and the recovered fractions further separated according to molecular weight or carbon number by high temperature GLC to obtain the exact pairing of the fatty chains. More than 40 molecular species or small groups of species were identified.

Montfoort et al (85) were the first to perform a quantitative determination of the molecular species composition of the phosphatidylcholines from different animal brains. In the rat brain, 1-palmitoyl 2-oleoyl and 1-stearoyl 2-oleoyl species were found to predominate but 1,2-dipalmitoyl species also were present in significant amounts, while palmitoyl and stearoyl arachidonates were present in only slightly lower amounts than the disaturates. Only negligible amounts of 1-saturated 2-linoleoyl species could be detected in the rat brains. The 1-palmitoyl 2-oleoyl species was also predominant in the rabbit and cow brain, while the dipalmitoyl species with much lower amounts of the stearoyl oleoyl species also were noted. Traces of palmitoyl arachidonate species also could be detected. It is interesting to note that the content of the dipalmitoyl species in the mammals is about one half of that in the birds. In the cow the level of this species in the brain approaches that in the lung. Dipalmitoylphosphatidylcholine occurs only in trace amounts in the liver and kidney of all animal species. In quantitative respect, 1-palmitoyl 2-oleoyl species was the most important component in the brains of all animal species tested. It occurs in this tissue in the somewhat increasing amounts in the order: rat, rabbit, pig, cow, chicken and duck. Linoleate species do not occur in significant amounts in the brain of the investigated animal species, except the birds, though appreciable quantities of arachidonic acid-containing molecules were detected.

Crawford and Wells (86) have determined the molecular species composition of rat brain phosphatidylcholine and have shown that the relative amounts of the molecular species remained nearly constant over the period 3–9 days with 35% saturated, 40% monoenoic, 6% dienoic, 11% tetraenoic, 2% pentaenoic and 5% hexaenoic species. Analyses of the fatty acid composition of the molecular species revealed that the polyunsaturated fatty acids 20:4 and 22:6 were predominantly paired with palmitic acid, while the same polyunsaturated acids in phosphatidylethanolamine were paired predominantly with stearic acid. Dipalmitoyl species accounted for 25% of the total.

Ogino et al (87) have determined the developmental changes in the molecular species of the choline glycerophospholipids in the brains of 17-, 19- and 21-day-old rat fetuses, and in 3-, 6-, 12-, 24- and 90-day-old rats. There were some significant changes. Thus, it was shown that the dipalmitoyl species increased to about 44% at around the 10th day and thereafter remained nearly constant. The palmitoyl oleate and palmitoyl stearate species decreased to about 17% and 6%, respectively, at that time, and then increased to about 30% and 14% of total, respectively. The stearoyl oleate species increased during the latter part of the development.

Kino et al (88) have determined the molecular species of choline glycerophospholipids in rat brain myelin and synaptosomes. The main species in myelin were palmitoyl oleate, stearoyl oleate and dioleate, whereas in synaptosomes they were mainly dipalmitoyl, palmitoyl stearoyl, palmitoyl oleaoyl and stearoyl oleoyl species. The myelin also contained 7% hexadecyl oleoyl species. There were no alkylacyl species detected in the synaptosomes.

Table XV lists the principal molecular species of phosphatidylcholines of brain of different animal species.

### 4. Liver

Detailed analyses of the molecular species of the hepatic phosphatidylcholine were first reported by Kuksis et al (89) and Van Golde et al (90). Some 26 molecular species were identified and quantitated. Van Golde et al (89)

**TABLE 7-XIV.**
**Molecular species of phosphatidylcholine of milk serum**

| Molecular Species | (a) | (b) | Molecular Species | (a) | (b) |
|---|---|---|---|---|---|
| | | Weight % | | | |
| 14:0 14:0 | 0.26 | | 18:0 14:0 | 0.70 | |
| 14:0 15:0 | 0.01 | | 18:0 15:0 | 0.08 | |
| 14:0 16:0 | 0.75 | | 18:0 16:0 | 1.41 | |
| 14:0 16:1 | 0.02 | | 18:0 16:1 | trace | |
| 14:0 17:0 | 0.01 | | 18:0 17:0 | 0.10 | |
| 14:0 18:0 | 0.15 | | 18:0 17:1 | 0.13 | |
| 14:0 18:1 | 1.88 | | 18:0 18:0 | 0.49 | |
| 14:0 18:2 | 0.04 | | 18:0 18:1 | 5.29 | 5 |
| 14:0 18:3 | 0.03 | | 18:0 18:2 | 2.25 | 2 |
| 14:0 20:3 | 0.12 | | 18:0 18:3 | 0.70 | |
| 14:0 20:4 | 0.02 | | 18:0 20:3 | 0.11 | |
| 15:0 14:0 | 0.07 | | | | |
| 15:0 15:0 | trace | | 18:1 14:0 | 1.10 | |
| 15:0 16:0 | 0.38 | | 18:1 15:0 | 0.20 | |
| 15:0 18:0 | 0.17 | | 18:1 16:0 | 10.28 | 10 |
| 15:0 18:1 | 0.52 | | 18:1 16:1 | 0.48 | |
| 15:0 18:2 | 0.05 | | 18:1 17:0 | 0.13 | |
| | | | 18:1 18:0 | 0.2 | |
| 16:0 14:0 | 1.56 | | 18:1 18:1 | 14.84 | 14 |
| 16:0 15:0 | 0.19 | | 18:1 18:2 | 3.81 | 3 |
| 16:0 16:0 | 2.94 | 3 | 18:1 18:3 | 1.22 | |
| 16:0 16:1 | 1.37 | | 18:1 20:3 | 0.21 | |
| 16:0 17:1 | 0.12 | | | | |
| 16:0 18:0 | 0.35 | | 18:2 14:0 | 0.12 | |
| 16:0 18:1 | 21.17 | 21 | 18:2 15:0 | trace | |
| 16:0 18:2 | 5.08 | 5 | 18:2 16:0 | 1.70 | |
| 16:0 18:3 | 0.85 | | 18:2 16:1 | 0.21 | |
| 16:0 20:3 | 0.24 | | 18:2 17:0 | 0.01 | |
| 16:0 20:4 | 1.25 | | 18:2 18:0 | 0.43 | |
| | | | 18:2 18:2 | 0.50 | |
| 16:1 16:0 | 0.91 | | | | |
| 16:1 16:1 | 0.12 | | 18:3 14:0 | 0.01 | |
| 16:1 17:1 | 0.02 | | 18:3 16:0 | 0.41 | |
| 16:1 18:1 | 1.51 | | 18:3 18:0 | 0.14 | |
| 16:1 18:2 | 0.23 | | 18:3 18:1 | 0.31 | |
| 16:1 20:3 | 0.16 | | | | |
| | | | 20:3 16:0 | 0.02 | |
| 17:0 14:0 | 0.08 | | 20:3 16:1 | 0.04 | |
| 17:0 15:0 | 0.08 | | 20:3 18:0 | 0.02 | |
| 17:0 16:0 | 0.41 | | 20:3 18:1 | 0.06 | |
| 17:0 16:1 | 0.02 | | | | |

**TABLE 7-XIV (Continued).**
**Molecular species of phosphatidylcholine of milk serum**

| Molecular Species | (a) | (b) | Molecular Species | (a) | (b) |
|---|---|---|---|---|---|
| | | Weight % | | | |
| 17:0 18:0 | 0.03 | | 20:4 16:0 | 0.06 | |
| 17:0 18:1 | 0.62 | | 20:4 18:0 | 0.05 | |
| 17:0 18:2 | 0.24 | | | | |
| 17:0 20:3 | 0.15 | | | | |
| 17:1 16:0 | 0.06 | | | | |
| 17:1 16:1 | 0.12 | | | | |
| 17:1 18:0 | 0.04 | | | | |

[a]Nutter and Privett (45).
[b]Blank et al (84).

**TABLE 7-XV.**
**Principal molecular species of phosphatidylcholine of brain of different animal species**

| Molecular Species | Rat[a] | Rabbit[a] | Pig[a] | Bovine[a] | Chicken[b,c] | Duck[b,c] |
|---|---|---|---|---|---|---|
| | | | Animal Species | | | |
| | | | | Mole % | | |
| 16:0 16:0 | 15 | 13 | | 14 | 26.8 | 29.1 |
| 18:0 16:0 | 6 | 3 | | 3 | 1.5 | 4.4 |
| 16:0 16:1 | | | | | | |
| 16:0 18:1 | 28 | 32 | 30 | 38 | 28.9 | 27.3 |
| 18:0 18:1 | 11 | 12 | 22 | 11 | 4.4 | 6.5 |
| 18:1 16:0 | 10 | 15 | 19 | 8 | 12.6 | 15.4 |
| 18:1 18:1 | | | | | 3.2 | 2.8 |
| 16:0 18:2 | | | | | 2.2 | 1.2 |
| 18:0 18:2 | | | | | | |
| 16:0 20:4 | 8 | 6 | 6 | 3 | 4.0 | 3.9 |
| 18:0 20:4 | 11 | 2 | 6 | 8 | 2.4 | 2.8 |
| 16:0 22:6 | | | | | 5.2 | 2.8 |
| 18:0 22:6 | | | | | 2.2 | 1.2 |
| UNIDENTIFIED | | | | | | |

[a]Moontfoort et al (85).
[b]Gonzalez-Ros and Ribera (132).
[c]Embryonic (21–28 days incubation).

showed that the composition of the species of the phosphatides could change dramatically with diet, the most profound alterations being observed on a diet deficient in essential fatty acids. Prior to this time Renkonen (76) and Van Golde and Van Deenen (91) had effected partial separations of the diacylglycerol moieties of the choline phosphatides, while Arvidson (55) had demonstrated the feasibility of $AgNO_3$-TLC resolution of the intact choline phosphatides of rat liver. More recently the molecular species of intact phosphatidylcholines of rat liver have been resolved by Patton et al (62) using HPLC on a reversed phase column along with methanol-water-acetonitrile 90.5:7:2.5 containing 20 mM choline chloride as the mobile phase. The peak elution profile was monitored by short wave-length UV (205 nm). The components were quantitated by collecting fractions and determining the organic phosphorus. The most complete analyses of the pairings of the fatty acids in the hepatic glycerophosphorylcholines have been reported by Myher and Kuksis (80), who succeeded in a resolution of all the species by GLC on polar capillary columns. The latter separations were done with the TMS or t-BDMS ethers of the derived diacylglycerol moieties. Table XVI gives the principal molecular species of phosphatidylcholines of liver of different animal species as determined by a variety of analytical methods. Separate values have not been given for the alkylacyl and alkenylacyl subclasses, which are present in very small amounts in hepatic phosphatidylcholines. Despite much variability in the completeness of the data, there is considerable similarity in the proportions and identity of the major molecular species of the choline phosphatides of the liver of different animal species.

### 5. Kidney

The phosphatidylcholines of the kidney also contain both diacyl and alkenylacyl species, although the latter usually do not exceed 10% of the total. In most instances the estimates for the corresponding diacyl and alkenylacyl species have been combined and reported together. Of special interest is the occurrence in very small amounts in the kidney of the sn-1-alkyl-2-acetyl-3-phosphoglycerocholine, which is identical in structure to the platelet activating factor and also lowers the blood pressure of artificially hypertensive animals (92). Table XVII lists the principal molecular species of the kidney of selected animal species.

**TABLE 7-XVI**
**Principal molecular species of phosphatidylcholine of liver of different animal species**

| Molecular Species | Rat[a] | Rat[b] | Rat[c] | Pig[d] | Rat[e,j] | Rat[e,k] | Rat[f] | Rat[g] | Cow[h] | Rabbit[i] |
|---|---|---|---|---|---|---|---|---|---|---|
| | | | | | Mole % | | | | | |
| 14:0 16:0 | 0.01 | | | 1.2 | 2.5 | | | | — | |
| 16:0 16:0 | 0.8 | | 0.9 | | | | 1.3 | 1.3 | 3 | |
| 18:0 16:0 | 0.2 | | | | 3.0 | | | | 3 | |
| 16:0 16:1 | 1.0 | 0.7 | 0.6 | | | | 0.1 | | | |
| 16:0 18:1 | 6.3 | 1.4 | 5.2 | 15.8 | 10.2 | 13.1 | 10.5 | 6.4 | 9 | 6 |
| 18:0 16:1 | 0.5 | | | | 2.3 | | | — | — | — |
| 18:0 18:1 | 1.2 | 3.4 | 1.6 | 7.8 | 6.5 | 2.7 | | 2.8 | 12 | 2 |
| 18:1 16:0 | | | | | | | | 2.0 | 2 | 2 |
| 16:0 18:2 | 15.7 | 14.4 | 15.2 | 12.8 | 1.6 | 11.8 | 21.6 | 12.6 | | 20 |
| 18:0 18:2 | 7.9 | 14.8 | 12.8 | 20.8 | 2.2 | 8.7 | 13.4 | 14.8 | | 24 |
| 18:1 18:1 | 2.0 | 6.4 | 0.3 | 0.6 | 7.6 | | | | | |
| 16:1 18:1 | 1.0 | 1.0 | 0.2 | | | | | 0.5 | | |
| 18:2 16:0 | | | | | | | | 1.6 | | |
| 16:1 18:2 | 0.4 | 1.1 | 0.6 | | | | 0.7 | | | |
| 18:1 18:2 | 1.4 | 3.8 | 1.4 | | | 3.0 | 0.3 | | | 3 |
| 20:1 18:1 | 0.2 | | | | | | | | | |
| 16:0 20:3 | 0.5 | 1.3 | | | 14.2 | 0.6 | | | 4 | |
| 18:0 20:3 | 0.5 | 1.6 | 0.4 | | 28.4 | 0.9 | | | 10 | |
| 18:2 18:0 | | | | | | | | 1.5 | | |
| 16:0 20:4 | 18.8 | 9.7 | 16.6 | 4.1 | 5.2 | 20.4 | 24.1 | 14.7 | 2 | 7 |
| 18:0 20:4 | 20.1 | 16.1 | 21.3 | 12.1 | 7.5 | 37.8 | 15.7 | 37.3 | 8 | 12 |
| 18:1 20:3 | 0.8 | | | | 5.4 | | | | | |
| 18:2 18:2 | 0.4 | 1.5 | 1.8 | | | | | | | 2 |
| 16:0 20:5 | 1.7 | 2.0 | 1.8 | | | | | | | |
| 18:0 20:5 | 0.4 | 2.6 | 1.5 | | | | | | | |
| 18:1 20:4 | 3.5 | | 2.2 | | | 5.3 | | 2.0 | | |
| 18:0 22:5 | 1.4 | 3.3 | 0.3 | | | | | | | |
| 16:0 22:5 | | | | | | | | | | |
| 16:0 22:6 | 4.8 | 4.0 | 4.6 | | | | 12.2 | | | |
| 18:0 22:6 | 5.6 | 3.6 | 2.7 | | | | | | | |
| 18:2 20:4 | 0.5 | 2.1 | 2.0 | | | | | | | |
| UNIDENTIFIED | | | | | | | | | 7 | 47 |

[a]Kuksis et al (89).
[b]Patton et al (62).
[c]Myher and Kuksis (80).
[d]Christie (133).
[e]Van Golde et al (90).
[f]Curstødt (134).
[g]Dyatlovitskaya et al (104).
[i]Montfoort et al (85).
[j]EFA deficient.
[k]after 216 hours on corn diet.

**TABLE 7-XVII**
**Principal molecular species of phosphatidylcholines of the kidney of different animal species**

| Molecular Species | Animal Species | | | | |
|---|---|---|---|---|---|
| | Rat[a] | Rabbit[b] | Pig[b] | Cow[b] | Rat[b] |
| | | | Moles % | | |
| 16:0 14:0 | 0.2 | | | | |
| 16:0 16:0 | 5.6 | | | | 7 |
| 18:0 16:0 | 0.7 | | | | |
| 18:0 14:0 | 0.4 | | | | |
| 14:0 18:1 | 0.3 | | | | |
| 16:0 16:1 | 1.6 | | | | |
| 18:0 16:1 | 1.6 | | | | 6 |
| 16:0 18:1 | 21.1 | 11 | 19 | 20 | 3 |
| 18:0 18:1 | 2.4 | 2 | 4 | 3 | |
| 18:1 18:0 | | 2 | 9 | 8 | |
| 18:1 16:1 | 0.3 | — | — | — | |
| 18:1 18:1 | 0.3 | 3 | — | — | |
| 16:0 18:2 | 21.0 | 24 | 24 | 18 | 10 |
| 18:0 18:2 | 6.2 | 23 | 16 | 6 | 11 |
| 18:2 16:0 | | 4 | 3 | | 8 |
| 16:1 18:2 | 0.5 | — | — | | |
| 18:1 18:2 | 3.0 | 9 | 4 | 4 | |
| 16:0 18:3 | 0.5 | — | — | 12 | — |
| 16:0 20:3 | 0.5 | — | — | 4 | |
| 18:0 20:3 | 0.5 | — | — | | |
| 18:2 18:1 | | 4 | — | | |
| 16:0 20:4 | 15.0 | 7 | 10 | 10 | 24 |
| 18:0 20:4 | 4.6 | | 8 | 2 | 28 |
| 18:1 20:3 | 1.2 | — | | | |
| 18:2 18:2 | 1.2 | — | | | |
| 16:0 20:5 | 1.0 | — | | | |
| 18:1 20:4 | 1.0 | 2 | | | |
| 16:0 20:5 | 1.0 | — | | | |
| 16:0 22:6 | 4.9 | — | | | |
| 18:0 22:6 | 0.9 | — | | | |
| 18:2 20:4 | 0.7 | — | | | |
| UNIDENTIFIED | | 9 | | 13 | |

[a]Kuksis et al (93).
[b]Montfoort et al (85).

### 6. Heart

The phosphatidylcholines of the heart are also made up of both diacyl and alkenylacyl species, with the latter accounting for about 10% of the total choline phosphatide in this tissue. The only complete analyses of molecular species are those of Kuksis et al (93). In Table XVIII the corresponding diacyl and alkenylacyl species have been combined and reported together according to their chain length and unsaturation of the acyl chain. Pugh et al (94) have characterized the minor amounts of the alkyl ether species recovered from bovine heart phosphatidylcholine following hydrogenation.

### 7. Skeletal muscle

The molecular species of the skeletal muscle phosphatidylcholines differ from those of the heart by the relative absence of the alkenylacyl species. Marai and Kuksis (63) have estimated that the plasmalogen content of the sarcotubular membranes of the skeletal muscle of rabbit, rat, chicken and man contained less than 1% plasmalogenic species. Likewise, Shamgar and Collins (95) failed to detect significant amounts of plasmalogens in the phosphatidylcholines isolated from the whole skeletal muscle of the rat. Table XIX gives the principal molecular species of the choline phosphatides of the sarcotubular membranes of the skeletal muscle of rabbit, rat, chicken and man and of the total skeletal muscle of the rat.

### 8. Erythrocytes

The molecular species of phosphatidylcholines of the erythrocytes also have been subjected to detailed analysis. Some 40 different components have been detected at levels exceeding 0.5% (89, 96). This estimate does not include the reverse isomers and the alkylacyl and alkenylacyl species, which are present in small amounts. The bulk of the species, however, is constituted by combinations of palmitic and stearic acids with the oleic, linoleic and arachidonic acids. The erythrocytes are relatively rich in the dipalmitoylphosphatidylcholines and there is considerable similarity among the species of the red blood cell phosphatidylcholines from different animal species. Table XX gives the principal molecular species of phosphatidylcholines of erythrocytes from different animal species. The values adopted from van Golde et al (97) were read from published bar graphs and are somewhat lacking in accuracy.

**TABLE 7-XVIII**
**Principal molecular species of phosphatidylcholines of the heart of different animal species**

| Molecular Species | | Rat[a] | Animal Species |
|---|---|---|---|
| | | | Moles % |
| 16:0 | 14:0 | 0.04 | |
| 16:0 | 16:0 | 1.4 | |
| 18:0 | 16:0 | 0.1 | |
| 18:0 | 14:0 | 0.4 | |
| 16:0 | 16:1 | 0.5 | |
| 18:0 | 16:1 | 0.5 | |
| 16:0 | 18:1 | 8.7 | |
| 18:0 | 18:1 | 3.0 | |
| 18:1 | 16:1 | 0.2 | |
| 18:1 | 18:1 | 1.9 | |
| 16:0 | 18:2 | 8.6 | |
| 18:0 | 18:2 | 12.7 | |
| 16:0 | 20:2 | 0.4 | |
| 16:1 | 18:2 | 0.3 | |
| 18:1 | 18:2 | 3.5 | |
| 20:1 | 18:2 | 0.2 | |
| 16:0 | 20:4 | 13.3 | |
| 18:0 | 20:4 | 21.7 | |
| 18:1 | 20:3 | 1.1 | |
| 18:2 | 18:2 | 1.9 | |
| 16:1 | 20:4 | 0.3 | |
| 18:1 | 20:4 | 1.3 | |
| 16:0 | 20:5 | 1.0 | |
| 14:0 | 20:5 | 0.3 | |
| 16:0 | 22:6 | 2.0 | |
| 18:0 | 22:6 | 5.0 | |
| 18:1 | 22:5 | 0.5 | |
| 18:2 | 20:4 | 2.5 | |

[a]Kuksis et al (93).

## 9. Lung

The lung secretes dipalmitoylphosphatidylcholine as a natural surfactant, which makes up 20–40% of the total lung phosphatidylcholine. The cow, however, appears to contain only about 10% dipalmitoylphosphatidylcholine

**TABLE 7-XIX**
**Principal molecular species of phosphatidylcholines of skeletal muscle of different animal species**

| Molecular Species | Rabbit[a] | Rat[a] | Chicken[a,c] | Man[a,c] | Rat[b,c] | Rat[b,d] |
|---|---|---|---|---|---|---|
| | | | Moles % | | | |
| 16:0 16:0 | 2.8 | 2.4 | 2.0 | 1.8 | — | — |
| 16:0 18:0 | 0.4 | 0.2 | trace | trace | — | — |
| 16:0 16:1 | | | | | | 7.4 |
| 16:0 18:1 | 26.2 | 9.2 | 30.2 | 15.0 | 8.8 | 34.6 |
| 18:0 18:1 | 0.9 | 0.8 | 1.1 | 1.0 | 2.6 | 5.2 |
| 16:0 18:2 | 40.4 | 29.8 | 20.0 | 40.0 | 31.3 | — |
| 18:0 18:2 | 3.9 | 6.8 | 2.0 | 4.0 | 5.2 | — |
| 18:1 16:1 | | | | | | 6.6 |
| 18:1 18:1 | 1.2 | 0.7 | 1.0 | 3.0 | — | 13.6 |
| 18:1 18:2 | 3.4 | 4.4 | 2.0 | 2.0 | 6.1 | — |
| 16:0 20:3 | 0.9 | 1.6 | 1.2 | 1.5 | — | 20.9 |
| 18:0 20:3 | | | | | | 3.5 |
| 16:0 20:4 | 8.7 | 19.2 | 20.0 | 12.0 | 35.3 | — |
| 18:0 20:4 | 2.7 | 5.2 | 4.0 | 3.0 | 6.6 | — |
| 18:1 20:3 | | | | | — | 8.2 |
| 18:1 20:4 | trace | 2.0 | 2.0 | 1.5 | 4.1 | — |
| 18:2 18:2 | 2.5 | 2.1 | 2.0 | 2.0 | — | — |
| 16:0 22:5 | 0.1 | 1.0 | 1.0 | 2.0 | — | — |
| 16:0 22:6 | 0.1 | 4.4 | 1.5 | 1.0 | — | — |
| 18:2 20:4 | 0.2 | 1.3 | 0.3 | 1.0 | — | — |
| 18:2 22:4 | trace | 0.8 | 0.1 | 0.5 | — | — |

[a]Marai and Kuksis (64, 65).
[b]Shamgar and Collins (95).
[c]16:0 includes 16:0A.
[d]EFA deficient.

species with the palmitoyloleoyl species contributing about 20% of the total. Rabbit and rat contain 10–20% of the total lung phosphatidylcholine in the form of palmitoyllinoleate. Table XXI lists the principal molecular species of the phosphatidylcholines of lung of different animal species.

## 10. Bile

The bile contains the surfactants secreted by the liver, which include diacylphosphatidylcholine as a prominent component. In most animal species the mono- and di-unsaturated species account for the bulk of the total,

**TABLE 7-XX.**
**Principal molecular species of phosphatidylcholines of erythrocytes of different animal species**

| Molecular Species | | Animal Species | | |
|---|---|---|---|---|
| | Rat[a] | Man[b] | Man[c] | Rabbit[c] |
| | | Moles % | | |
| 14:0 16:0 | 0.5 | 0.3 | | |
| 16:0 16:0 | 20.3 | 4.7 | 9.5 | 10.3 |
| 18:0 16:0 | 4.3 | 1.5 | 3.7 | 3.0 |
| 16:0 16:1 | 0.8 | 0.4 | 1.8 | 1.0 |
| 16:0 18:1 | 14.5 | 25.9 | 12.5 | 18.5 |
| 16:0 20:1 | 0.4 | | | |
| 18:0 16:1 | 0.8 | | 1.8 | 1.0 |
| 18:0 18:1 | 4.4 | 6.5 | 4.4 | 3.4 |
| 18:1 16:0 | | | 1.5 | 1.0 |
| 16:0 18:2 | 13.4 | 18.2 | 19.4 | 21.8 |
| 18:0 18:2 | 5.6 | 9.3 | 11.3 | 17.8 |
| 18:1 16:1 | 0.2 | | | |
| 18:1 18:1 | 0.8 | 2.7 | 4.8 | |
| 22:0 18:2 | 0.4 | | | |
| 18:2 16:0 | | | 1.0 | 1.2 |
| 16:0 20:3 | 0.4 | 1.5 | 3.5 | |
| 18:0 20:3 | 0.3 | 0.4 | 1.3 | 4.1 |
| 22:0 20:3 | 0.3 | | | |
| 18:1 18:2 | 0.7 | 1.4 | 7.0 | 7.5 |
| 18:1 20:2 | 0.3 | 0.3 | | |
| 18:2 18:1 | | | 2.3 | 0.7 |
| 14:0 20:4 | 0.5 | 0.3 | | |
| 16:0 20:4 | 5.4 | 8.4 | 6.6 | 2.0 |
| 18:0 20:4 | 7.5 | 4.2 | 3.4 | 1.4 |
| 18:1 20:3 | 0.8 | 0.7 | | |
| 18:2 18:2 | 2.7 | 0.6 | | |
| 16:0 20:5 | 0.5 | 0.9 | | |
| 16:0 22:5 | 0.7 | 0.5 | | |
| 18:0 20:5 | 0.3 | 0.5 | | |
| 18:0 22:5 | 0.6 | | | |
| 18:1 20:4 | 0.7 | 1.9 | | |
| 14:0 22:6 | 0.4 | 0.6 | | |
| 16:0 22:6 | 4.0 | 2.7 | | |
| 18:0 22:6 | 2.4 | 0.5 | | |
| 18:1 20:5 | 0.3 | 0.5 | | |
| 18:1 22:5 | 1.2 | | | |
| 18:2 20:4 | 3.0 | 1.0 | | |
| 18:2 22:4 | | 3.0 | | |

[a]Kuksis et al (135).
[b]Marai and Kuksis (96).
[c]Van Golde et al (97).

**TABLE 7-XXI.**
**Principal molecular species of phosphatidylcholine of lung of different animal species**

| Molecular Species | Animal Species | | | | | |
|---|---|---|---|---|---|---|
| | Rat[a] | Rat[b,c] | Rabbit[b] | Pig[b] | Cow[b] | Sheep[b] |
| | | | Mole % | | | |
| 14:0 16:0 | 3.7 | trace | | 2 | — | — |
| 16:0 16:0 | 41.4 | 28 | 22 | 29 | 12 | 25 |
| 18:0 16:0 | 0.9 | 3 | — | — | 3 | 10 |
| 18:1 18:1 | | | 3 | 2 | 2 | 3 |
| 14:0 18:1 | 6.4 | | — | — | — | — |
| 16:0 16:1 | 7.0 | 3 | 6 | — | — | — |
| 16:0 18:1 | 16.6 | 16 | 12 | 13 | 21 | 17 |
| 18:0 18:1 | 1.9 | | — | 3 | 8 | 3 |
| 18:1 16:0 | | 6 | 4 | 7 | 9 | 9 |
| 16:0 18:2 | 6.4 | 17 | 10 | 6 | 4 | 8 |
| 18:0 18:2 | 5.6 | 2 | 3 | 4 | 2 | 3 |
| 16:1 18:1 | 4.0 | | — | — | — | |
| 18:2 16:0 | | 2 | 2 | — | — | |
| 14:0 20:4 | 0.6 | | — | — | — | |
| 16:0 20:4 | 1.8 | 10 | 7 | 5 | 6 | 3 |
| 18:0 20:4 | 3.6 | 6 | 3 | 4 | 5 | — |
| 18:1 20:4 | | | 2 | 3 | 2 | — |
| 16:0 22:6 | trace | 6 | — | — | — | — |
| 18:1 18:2 | | | 3 | — | — | 4 |
| UNIDENTIFIED | | 1 | 27 | 22 | 14 | 15 |

[a]Kuksis et al (135).
[b]Montfoort et al (85).
[c]Toshima et al (136).

while numerous other species occur in smaller proportions. Table XXII lists the principal molecular species of bile of different animal species.

## 11. Adipose tissue

The phosphatidylcholine composition of the adipose tissue has received very little experimental attention, only one complete analysis having been published thus far. Table XXIII gives the principal molecular species of the phosphatidylcholines of the adipose tissue of the pig. The palmitoyl and stearoyl oleates and the stearoyl linoleate make up the bulk of the total phosphatide.

**TABLE 7-XXII.**
**Principal molecular species of phosphatidylcholine of animal bile**

| Molecular Species | | Animal Species | | | | |
|---|---|---|---|---|---|---|
| | | Pig[a] | Sheep[a] | Cow[a] | Rat[b] | Rat[c] |
| | | | Mole % | | | |
| 16:0 | 16:0 | | 2.8 | | 0.3 | trace |
| 16:0 | 18:0 | | 1.2 | | | trace |
| 16:0 | 16:1 | 0.9 | 1.0 | 3.0 | 0.5 | 0.6 |
| 16:0 | 18:1 | 22.6 | 39.7 | 37.6 | 17.9 | 8.2 |
| 18:0 | 18:1 | 3.6 | 7.3 | 9.9 | | 0.3 |
| 16:1 | 18:1 | 1.1 | 2.1 | 2.5 | | |
| 18:1 | 18:1 | 1.5 | 1.9 | 3.6 | | |
| 14:0 | 18:2 | | | | 0.1 | |
| 16:0 | 18:2(w6) | 26.9 | 9.6 | 11.3 | 55.9 | 53.7 |
| 18:0 | 18:2(w6) | 13.1 | 2.9 | 7.9 | 7.3 | 9.9 |
| 16:0 | 18:2 conj | | 9.0 | 1.4 | | |
| 18:0 | 18:2 conj | | 2.1 | trace | | |
| 16:0 | 18:3(w3) | trace | 6.7 | 3.5 | | |
| 18:0 | 18:3(w3) | trace | 2.2 | 1.6 | | |
| 16:0 | 18:3 conj | | 2.3 | | | |
| 18:0 | 18:3 conj | | 0.6 | | | |
| 16:0 | 18:3(w6) | 0.8 | | 1.3 | | |
| 18:0 | 18:3(w6) | 0.5 | | 1.4 | | |
| 18:0 | 20:3(w6) | | | | | |
| 16:1 | 18:2 | 0.8 | | 0.4 | 1.7 | 0.5 |
| 18:1 | 18:2 | 2.7 | | 1.4 | 1.0 | 2.2 |
| 16:0 | 20:4(w6) | 7.9 | 0.9 | 1.8 | 13.4 | 14.7 |
| 18:0 | 20:4(w6) | 3.6 | 0.7 | 0.9 | | 2.7 |
| 16:0 | 18:4(w3) | | | 1.2 | | |
| 16:0 | 20:4(w3) | | | | | |
| 18:0 | 18:4(w3) | | | 1.3 | | |
| 18:0 | 20:4(w3) | | | | | |
| 18:1 | 18:3 | | 0.7 | | | |
| 16:0 | 20:5/22:5 | 2.2 | 0.5 | 3.0 | | |
| 18:0 | 20:5/22:5 | 0.8 | 3.1 | 1.5 | | |
| 16:0 | 22:6 | 2.3 | 1.5 | 0.6 | 1.4 | |
| 18:0 | 22:6 | 0.5 | 0.5 | 0.2 | | |
| UNIDENTIFIED | | | | | | 7.0 |

[a]Christie (133).
[b]Curstedt (134).
[c]Kawamoto et al (127).

**TABLE 7-XXIII.**
**Principal molecular species of phosphatidylcholine in adipose tissue**

| Molecular Species | | Animal Species Pig |
|---|---|---|
| | | Mole % |
| 16:0 | 16:0 | 0.9 |
| 16:0 | 18:0 | 3.5 |
| 18:0 | 18:0 | 0.5 |
| 16:0 | 18:1 | 10.1 |
| 18:0 | 18:1 | 22.3 |
| 18:1 | 18:1 | 3.2 |
| 16:0 | 18:2 | 4.2 |
| 18:0 | 18:2 | 37.1 |
| 18:1 | 18:2 | 5.5 |
| 18:0 | 18:3 | 1.0 |
| 16:0 | 20:3 | 0.5 |
| 18:0 | 20:3 | 2.0 |
| 16:0 | 20:4 | 0.9 |
| 18:0 | 20:4 | 3.6 |
| 18:0 | 20:5 | 0.8 |

Christie and Moore (68).

### Phosphatidylethanolamine

Phosphatidylethanolamine is the second most abundant glycerophospholipid in most animal tissues and frequently accompanies phosphatidylcholine in commercial preparations of lecithin. Phosphatidylethanolamine also is known as cephalin in the older scientific literature. Phosphatidylethanolamine of most animal tissues contains a significant proportion of alkenylacyl and some alkylacyl species. The alkenylacyl species are notably acid-labile and give rise to lysophospholipids during extraction and chromatography in the presence of strong acids. Upon transmethylation with acidic catalysts, the alkenylacyl species yield dimethylacetals. Detailed determination of the molecular species of the phosphatidylethanolamines requires a preliminary resolution of the subclasses by adsorption chromatography. The alkylacyl and alkenylacyl species of phosphatidylethanolamine are hydrolyzed at a significantly slower rate by both phospholipase $A_2$ and C than the corresponding diacyl species. In a few instances the molecular species of the phosphatidylethanolamines have been resolved as the N-acetyl-O-methyl derivatives (98) and the N-trifluoro-acetyl derivatives (99,100).

### 1. Egg yolk

A complete analysis of the molecular species of egg yolk phosphatidylethanolamines was first reported by Holub and Kuksis (101). Prior to this time Renkonen (102), Noble and Moore (103) and Dyatlovitskaya et al (104) had reported partial analysis of egg yolk phosphatidylethanolamines based on a partial separation of dinitrophenyl derivatives. Table XXIV lists the molecular species of egg yolk phosphatidylethanolamine as reported by Holub and Kuksis (101). There are several preferential pairings of the saturated and unsaturated fatty acids. Thus, in the monoenes, palmitic and oleic acids are preferentially combined, while in the dienes, stearic acid is preferred over palmitic acid for pairing with linoleic acid. Likewise, there is a preferred association of arachidonic acid with stearic acid in the tetraenes. In contrast, palmitate was preferred over stearate in the combination with the docosatetraenoic and docosahexaenoic acids. Do and Ramachandran (105) have demonstrated the presence of small amounts of the 1-alkenyl and 1-alkyl-sn-glycero-3-phosphorylethanolamines in egg yolk phosphatidylethanolamines following mild alkaline hydrolysis.

### 2. Brain

Renkonen et al (76) has determined the composition of the molecular species of ox brain phosphatidylethanolamines, which may contaminate the lecithin preparations from this tissue. He has shown that this phospholipid class contains both alkylacyl and alkenylacyl derivatives in addition to the common diacylglycerol derivatives. The three subclasses of the ethanolamine glycerophospholipids can be separated by TLC by developing the plate first in an aliphatic and then in an aromatic solvent system, e.g. in hexane-diethyl ether 1:1 followed by toluene. However, the compounds must be first dephosphorylated with phospholipase C. The intact phosphatidylethanolamines may be resolved by argentation TLC following preparation of the methylated DNP-derivatives (106). Recently, Patton et al (62) have described the resolution of the molecular species of diacyl phosphatidylethanolamine by HPLC using methanol-water-acetonitrile 90.5:7:2.5 containing 20 mM choline chloride as the eluting agent and a variable wavelength UV detector at 205 nm as the means of monitoring the peak profile.

Crawford and Wells (86) have determined the molecular species composition of the phosphatidylethanolamine of rat brain from birth to weaning. The molecular species composition of phosphatidylethanolamine was deter-

**TABLE 7-XXIV.**
**Molecular species of egg yolk phosphatidylethanolamines**

| Molecular Species | Moles % |
|---|---|
| 16:0 18:1 | 11.47 |
| 18:0 18:1 | 10.03 |
| 18:0 16:1 | 0.24 |
| 18:1 16:0 | 0.48 |
| 18:1 18:0 | 0.48 |
| 16:0 20:1 | 0.14 |
| 20:1 16:0 | 0.19 |
| 20:1 16:0 | 0.14 |
| 18:1 16:1 | 0.21 |
| 18:1 18:1 | 2.40 |
| 16:0 18:2 | 6.24 |
| 18:0 18:2 | 11.98 |
| 16:0 20:2 | 0.20 |
| 18:0 20:2 | 0.26 |
| 16:1 18:2 | 0.06 |
| 18:1 18:2 | 1.84 |
| 18:1 20:2 | 0.05 |
| 20:1 20:2 | 0.05 |
| 16:0 20:4 | 5.32 |
| 18:0 20:4 | 28.46 |
| 18:1 20:3 | 1.09 |
| 16:0 22:4 | 2.32 |
| 18:0 22:4 | 1.50 |
| 16:1 22:4 | 0.12 |
| 18:1 20:4 | 1.39 |
| 18:2 20:3 | 0.68 |
| 18:0 22:5 | 0.33 |
| 18:1 22:4 | 0.47 |
| 20:1 20:4 | 0.06 |
| 16:0 22:6 | 6.30 |
| 18:0 22:6 | 4.29 |
| 18:1 22:5 | 0.10 |
| 18:2 22:4 | 0.01 |

Holub and Kuksis (101).

mined in brains of rats 3, 6 and 9 days old. The relative amounts of the molecular species remained relatively constant during this time period. Phosphatidylethanolamine contained 2% saturated, 8% monoenoic, 3% dienoic, 40% tetraenoic, 9% pentaenoic and 27% hexaenoic species. Table XXV gives the major molecular species of phosphatidylethanolamines of bovine brain.

It should be noted that in phosphatidylcholine the unsaturated fatty acids 18:1 and 18:2 are predominantly paired with 16:0, while in the phosphatidylethanolamines the 18:2 and 20:4 acids are paired with 18:0. Furthermore, dipalmitoylphosphatidylethanolamine accounted for a negligible amount of the total molecular species of phosphatidylethanolamine.

Kino et al (88) have determined the composition of the molecular species of the phosphatidylethanolamine of rat brain myelin and synaptosomes using GLC with mass spectrometry. The main molecular species of the ethanolamine glycerophospholipids in myelin were: 18:0 18:1 and vinyl 16:0 18:1, vinyl 18:0 18:1 and vinyl 18:1 18:1, whereas in synaptosomes they were 18:0 20:4 and 18:0 22:6, and vinyl 16:0 18:1, vinyl 18:0 18:1, and vinyl 18:1 18:1. Obviously the 1-alk-1-enyl 2-acyl type is present in greater abundance than the 1,2-diacyl type, accounting for about 80% in myelin. Since it is known that the lipid composition of brain shows organ specificity rather than animal species specificity, a similar composition would be expected for the brains of other animal species (107, 108). It should be noted that the ethanolamine glycerophospholipids contained the 1-alk-1-enyl 2-acyl type, the choline glycerophospholipids the 1-alkyl-2-acyl type of diradylglycerophospholipids as the predominant ether lipids.

### 3. Liver

The phosphatidylethanolamines of the liver are essentially free of the alkenylacyl and alkylacyl species. Those of the rat liver have been studied extensively and have provided much of the evidence for the metabolic heterogeneity of this class of glycerophospholipids of animal tissues. The phosphatidylethanolamines of the liver of other animal tissues have received only a cursory attention. Table XXVI gives the principal molecular species of phosphatidylethanolamines of liver in rat.

### 4. Kidney

The phosphatidylethanolamines of the kidney are characterized by one of the highest proportions of alkenylacyl species. Much of the vinyl ether is found in combination with the polyunsaturated fatty acids, but the exact diacyl/alkenylacyl proportion of any one chain length varies with the animal species. The physiological significance of the high proportion of the alkenylacyl species in the phosphatdylethanolamines of the kidney is not known. Table XXVII lists the principal molecular species of phosphatidyl-

**TABLE 7-XXV.**
**Principal molecular species of phosphatidylethanolamine of bovine brain.**

| Molecular Species | | Phosphatidylethanolamines | |
|---|---|---|---|
| | Alkenylacyl | Alkylacyl | Diacyl |
| | | Mole % | |
| 18:1 22:6 | 2.7 | 3.0 | 1.7 |
| 16:0 22:6 | 4.7 | 8.8 | 5.5 |
| 18:1 22:5 | 0.1 | 0.3 | 0.1 |
| 18:1 20:4 | 4.7 | 1.1 | 2.9 |
| 16:0 22:5 | 0.3 | 0.7 | 0.1 |
| 16:0 20:4 | 2.8 | 1.6 | 1.1 |
| 18:1 22:5 | 0.5 | 0.6 | 0.4 |
| 16:0 22:5 | 0.9 | 1.4 | 0.8 |
| 18:0 22:6 | 8.7 | 10.2 | 25.5 |
| 18:1 20:3 | 0.3 | 0.3 | 0.3 |
| 18:1 18:2 | | 0.3 | |
| 18:1 22:4 | 4.0 | 5.1 | 1.3 |
| 16:0 20:3 | 0.3 | 0.3 | 0.2 |
| 16:0 18:2 | | 0.5 | |
| 16:0 22:4 | 7.3 | 10.1 | 1.2 |
| 18:1 20:3 | 0.6 | 0.4 | 0.6 |
| 18:0 22:5 | 0.3 | 0.8 | 0.3 |
| 16:0 20:3 | 0.4 | 0.6 | 0.2 |
| 18:0 20:4 | 4.8 | 2.2 | 15.8 |
| 18:0 22:5 | 0.8 | 1.3 | 3.0 |
| 18:1 18:1 | 21.8 | 5.5 | 10.1 |
| 16:0 18:1 | 12.9 | 6.8 | |
| | | | 5.0 |
| 16:0 16:0 | 1.9 | 3.6 | |
| 18:1 X | | 1.6 | 0.3 |
| 18:0 18:2 | | 0.4 | |
| 18:0 20:3 | 0.3 | 0.3 | 0.9 |
| 16:0 X | 1.5 | 1.7 | 0.3 |
| 18:0 22:4 | 4.7 | 10.3 | 8.6 |
| 18:0 20:3 | 0.6 | 0.3 | 1.6 |
| 18:1 20:1 | 3.4 | 5.7 | 2.2 |
| 16:0 20:1 | 3.0 | 8.6 | 0.6 |
| 18:0 18:1 | 3.7 | 4.5 | 8.5 |
| 18:0 X | 0.7 | 1.0 | 0.8 |

Nakagawa and Horrocks (137).

ethanolamine of kidney of some animal species.

**TABLE 7-XXVI.**
**Principal molecular species of phosphatidylethanolamine of liver of different animal species**

| Molecular Species | | Animal Species | |
|---|---|---|---|
| | Rat[a] | Rat[b] | Rat[b,c] |
| | | Moles % | |
| 16:0 16:0 | trace | | |
| 18:0 16:0 | trace | | |
| 16:0 16:1 | 0.05 | 0.7 | 2.2 |
| 16:0 18:1 | 0.9 | 3.3 | 10.1 |
| 18:0 16:1 | 0.1 | — | — |
| 18:0 18:1 | 0.2 | 1.8 | 3.5 |
| 16:0 18:2 | 7.2 | 5.7 | |
| 18:0 18:2 | 5.3 | 3.7 | |
| 18:1 18:1 | 0.1 | | 3.1 |
| 16:1 18:1 | 0.1 | | 2.3 |
| 16:1 18:2 | 0.03 | | |
| 18:1 18:2 | 1.2 | 2.5 | |
| 16:0 20:3 | 0.3 | | 18.3 |
| 18:0 20:3 | 0.9 | | 24.7 |
| 16:1 20:3 | | | 4.2 |
| 16:0 20:4 | 14.2 | 19.6 | |
| 18:0 20:4 | 35.0 | 32.3 | 11.4 |
| 18:1 20:3 | 1.0 | | 13.4 |
| 18:2 18:2 | 0.5 | | |
| 16:1 20:4 | 0.1 | 7.3 | |
| 18:1 20:4 | 4.0 | 12.3 | |
| 16:0 22:5 | 1.1 | 6.6 | |
| 18:0 22:5 | 0.9 | | |
| 16:0 22:6 | 12.4 | 4.2 | 6.8 |
| 16:1 20:5 | 0.5 | | |
| 18:0 22:6 | 8.8 | | |
| 18:1 20:5 | 2.3 | | |
| 18:1 22:5 | 1.0 | | |
| 18:2 20:4 | 0.5 | | |

[a]Holub and Kuksis (72).
[b]Trewhella and Collins (138).
[c]EFA deficient.

## 5. Skeletal muscle

Like the phosphatidylethanolamine of the kidney, that of the skeletal muscle is characterized by a high proportion of the alkenylacyl and to a lesser extent the alkylacyl species. Table XXVIII gives the principal molecular species of the phosphatidylethanolamines of the skeletal muscle of rabbit, rat, chicken and man. There appears to be more variation between the values obtained in different laboratories than between the different animal species. The molecular species of composition of the phosphatidylethanolamines of the skeletal muscle appears to be extremely sensitive to the deprivation of the essential fatty acids. The significance of the high proportion of the alkenylacyl species in the phosphatidylethanolamine in the skeletal muscle of all the animal species is not known.

**Phosphatidylserines**

The phosphatidylserines are usually absent from animal lecithin preparations. However, phosphatidylserines are present in small amounts in most animal tissues. The composition of the molecular species of the phosphatidylserines varies more among the different animal tissues than that of any other class of glycerophospholipids, although within a specific tissue or organ type considerable uniformity is observed. Some tissues contain only a few species of phosphatidylserine, while others possess the entire complement of species generated from the common fatty acids. The phosphatidylserines are essentially free of the alkylacyl and the alkenylacyl species. The molecular species of phosphatidylserines have been resolved by $AgNO_3$-TLC as the N-trifluoroacetyl (99) and the N-trifluoroacetyl-0 methyl (100) derivatives. Patton et al (62) has resolved the molecular species of phosphatidylserine by HPLC on reversed phase columns using 30 mM choline chloride in methanol-25 mM $KH_2PO_4$-acetonitrile-acetic acid 90.5:7:2.5:0.8 as the eluting solution.

1. Brain.

The phosphatidylserine makes up a significant proportion of brain total phospholipid and most lecithin preparations from this tissue are contaminated with it. The brain contains only a few molecular species of phosphatidylserine, of which the stearoyloleate makes up more than 92% of the total. Table XXIX gives the principal species of phosphatidylserine of pig and ox brain.

**TABLE 7-XXVII.**
**Principal molecular species of phosphatidylethanolamine of the kidney of different animal species**

| Molecular Species | | Rat[a] | Dog[b] | Pig[b] |
|---|---|---|---|---|
| | | | Moles % | |
| 16:0 | 16:0 | 0.44 | | |
| 160A | 16:0 | | 0.2 | |
| 17:0A | 16:0 | | 0.2 | |
| 18:0A | 16:0 | | 0.1 | |
| 16:0 | 16:1 | | 0.3 | |
| 18:0 | 18:1 | 1.02 | 6.8 | 2.8 |
| 18:0 | 18:1 | | 1.1 | 2.1 |
| 16:0A | 18:1 | 0.75 | 1.7 | 2.9 |
| 17:0bA | | | 0.6 | 0.4 |
| 18:0A | 18:1 | 0.27 | 0.4 | 0.1 |
| 18:1 | 18:1 | 0.5 | 0.6 | 2.9 |
| 16:0 | 18:2 | 2.1 | 6.1 | 1.5 |
| 17:0 | 18:2 | | 0.2 | |
| 18:0 | 18:2 | 3.43 | 2.2 | 3.4 |
| 18:1A | 18:1 | 0.27 | 0.2 | |
| 16:0A | 18:2 | 0.37 | 1.0 | 2.2 |
| 17:0bA | 18:2 | | 0.4 | 0.3 |
| 17:0A | 18:2 | | 0.2 | |
| 18:0A | 18:2 | | 1.2 | 0.2 |
| 18:1 | 18:2 | 2.19 | 2.4 | 4.0 |
| 16:0 | 20:4 | 10.3 | 15.3 | 4.4 |
| 17:0 | 20:4 | 0.44 | 0.3 | 0.5 |
| 18:0 | 20:4 | 37.3 | 14.5 | 15.2 |
| 16:0A | 20:4 | 10.23* | 16.3 | 34.4 |
| 17:0bA | 20:4 | 0.48 | 3.9 | 1.5 |
| 17:0A | 20:4 | | 1.3 | |
| 18:0A | 20:4 | 5.86 | 9.3 | 3.8 |
| 16:0A | 20:5 | 0.75 | | |
| 16:1 | 20:4 | | 0.1 | 0.2 |
| 18:1 | 20:4 | 7.37 | 6.3 | 7.7 |
| 16:0 | 20:5 | 0.66 | | |
| 17:1A | 20:4 | | 0.1 | |
| 18:1A | 20:4 | 4.96 | 6.3 | 0.5 |
| 16:0 | 22:5 | 1.09 | | 0.2 |
| 18:0 | 22:5 | | | 0.3 |
| 16:0A | 22:6 | 1.89** | | |
| 18:2 | 20:4 | | | 0.5 |
| 16:0 | 22:6 | 1.89 | | 0.2 |
| 18:0 | 22:6 | | | 0.2 |
| 18:0A | 22:6 | 0.35*** | | |
| 18:1A | 22:6 | 0.29 | | |

**TABLE 7-XXVII (Continued).**
**Principal molecular species of phosphatidylethanolamine of the kidney of different animal species**

| Molecular Species | Rat[a] | Dog[b] | Pig[b] |
|---|---|---|---|
| | | Moles % | |
| Other ethers | 0.48 | | |
| Other esters | 2.55 | | |

[a]Myher and Kuksis (139).
[b]Yeung and Kuksis (140).

**TABLE 7-XXVIII.**
**Principal molecular species of phosphatidylethanolamine of skeletal muscle of different animal species**

| Molecular Species[a] | | Animal Species | | | | | |
|---|---|---|---|---|---|---|---|
| | | Rabbit[b] | Rat[b] | Chicken[b] | Man[b] | Rat[c,d] | Rat[c,e] |
| | | | | Moles % | | | |
| 16:0 | 16:1 | | | | | — | 3.4 |
| 16:0 | 18:1 | 13.8 | 3.4 | 11.4 | 5.0 | — | 4.5 |
| 18:0 | 18:1 | 6.2 | 1.4 | 3.5 | 3.0 | — | 8.3 |
| 18:1 | 16:1 | | | | | — | 6.0 |
| 16:0 | 18:2 | 8.3 | 6.6 | 6.0 | 10.0 | 2.9 | |
| 18:0 | 18:2 | 2.1 | 2.6 | 3.0 | 5.0 | 13.3 | |
| 18:1 | 18:1 | 3.4 | 2.0 | 0.5 | 3.0 | — | 10.8 |
| 16:0 | 22:3 | 5.2 | 1.5 | 4.5 | 4.5 | — | |
| 18:0 | 22:3 | 2.0 | 2.5 | 3.0 | 3.5 | — | |
| 16:0 | 20:4 | 18.3 | 13.0 | 20.0 | 22.0 | 5.7 | 0.9 |
| 18:0 | 20:4 | 10.4 | 8.5 | 15.0 | 18.0 | 46.7 | 6.4 |
| 16:0 | 22:4 | 2.5 | trace | 3.0 | 1.0 | — | |
| 16:0 | 22:5 | 5.2 | 3.0 | 3.5 | 3.0 | — | |
| 18:0 | 22:5 | 3.6 | 2.1 | 1.5 | 1.5 | — | |
| 16:0 | 22:6 | 1.8 | 15.5 | 5.0 | 3.0 | 2.9 | |
| 18:0 | 22:6 | 0.6 | 26.0 | 9.5 | 6.0 | 15.4 | |
| 18:1 | 20:4 | 3.9 | 1.9 | 1.5 | 1.0 | 4.3 | |
| 18:1 | 18:2 | | | | | 4.6 | 2.8 |
| 18:1 | 22:6 | | | | | 4.2 | |
| 18:0 | 16:1 | | | | | | 5.1 |
| 16:0 | 20:3 | | | | | | 8.7 |
| 18:1 | 20:3 | | | | | | 12.8 |
| 18:0 | 20:3 | | | | | | 30.0 |

[a]Fatty acids in the sn-1-position include both acyl and alkyl chains.
[b]Marai and Kuksis (64, 65).
[c]Shamgar and Collins (95).
[d]normal EFA.
[e]EFA deficient.

**TABLE 7-XXIX.**
**Principal molecular species of phosphatidylserines of brain of different animal species**

| Molecular Species | | Pig | Ox | Animal Species |
|---|---|---|---|---|
| | | | | Moles % |
| 16:0 | 18:1 | 4 | | |
| 18:0 | 18:1 | 93 | 92 | |
| 18:0 | 20:1 | 3 | 5 | |

Yeung et al (99).

### 2. Skeletal muscle

Phosphatidylserine makes up 3–6% of the total phospholipid of skeletal muscle and constitutes only a minor potential contaminant of lecithin prepared from this tissue. The phosphatidylserine of the skeletal muscle of different animal species contains most of the molecular species found in other glycerophospholipids, such as phosphatidylethanolamine, except for the absence of the alkylacyl and the alkenylacyl species. Table XXX gives the principal molecular species of the phosphatidylserines of the sarcotubular membranes of the skeletal muscle of the rabbit, rat, chicken and man.

## Phosphatidylinositols

Like the phosphatidylserines, the phosphatidylinositols also are usually absent from animal lecithin preparations. However, the phosphatidylinositols are present in small amounts in all animal tissues and may contaminate crude preparations of animal phospholipids. Like the phosphatidylserines, the phosphatidylinositols are made up of only a few major species. Alkylacyl and alkenylacyl species are not present among the molecular species of natural phosphatidylinositols.

### 1. Brain

The phosphatidylinositols make up 1–7% of brain total phospholipid. The phosphatidylinositols of brain contain about 60% stearoylarachidonate, with the rest of the mass being divided among most of the other combinations of the common fatty acids found in natural glycerophospholipids. Table XXXI lists the principal molecular species of bovine and guinea pig brain.

**TABLE 7-XXX.**
**Principal molecular species of phosphatidylserine of skeletal muscle of different animal species**

| Molecular Species | | Animal Species | | | | |
|---|---|---|---|---|---|---|
| | | Rabbit[a] | Rat[a] | Chicken[a] | Man[a] | Rabbit[b] |
| | | sarcotubular membranes | | | | |
| | | Moles % | | | | |
| 18:0 | 16:0 | | 2.0 | | | |
| 16:0 | 18:1 | | | | | 6 |
| 18:0 | 18:1 | 8.0 | 13.0 | 13.0 | 13.0 | 23 |
| 16:0 | 18:2 | | | | | 4 |
| 18:0 | 18:2 | 6.0 | 7.5 | 20.0 | 5.5 | 24 |
| 18:0 | 22:2 | 4.0 | 6.0 | 3.0 | 13.0 | — |
| 18:0 | 20:3 | 3.5 | 7.6 | 16.0 | 2.0 | 2 |
| 18:0 | 22:3 | 8.0 | 11.0 | 11.0 | 8.0 | 12 |
| 16:0 | 20:4 | | | | | 8 |
| 18:0 | 20:4 | 12.0 | 20.0 | 20.0 | 14.0 | 2 |
| 18:0 | 22:4 | 14.0 | 3.0 | 5.0 | 9.0 | |
| 18:0 | 22:5 | 5.0 | 8.0 | trace | 6.5 | 8 |
| 18:0 | 22:6 | 5.0 | 23.0 | 16.0 | 7.0 | |
| UNIDENTIFIED | | | | | | |

[a]Marai and Kuksis (64,65).
[b]Yeung et al (99).

## 2. Liver

The phosphatidylinositols of the liver contain about 77% of the total mass in the form of stearoylarachidonate and rank along with the phosphatidylserine of brain as one of the simplest glycerophospholipid classes of animal tissues. The physiological significance of this structural simplicity is not known, but the presence of the arachidonic acid in this phospholipid class has been noted as a possible source of fatty acids for the biosynthesis of prostaglandins and thromboxanes. Table XXXII lists the principal molecular species of phosphatidylinositols of rat, pig and sheep liver.

## 3. Skeletal muscle

The phosphatidylinositols make up a relatively high proportion of the total phospholipid of skeletal muscle (4–11%). The phosphatidylinositols of skeletal muscle have a composition of molecular species comparable to that of the brain and the liver, with about 60% of the total mass

**TABLE 7-XXXI.**
**Principal molecular species of phosphatidylinositol of the brain of different animal species**

| Molecular Species | | Bovine[a] | Guinea pig[b] |
|---|---|---|---|
| | | Moles % | |
| Saturates | | | 5.2 |
| 14:0 | 18:1 | 0.1 | |
| 14:0 | 16:1 | 0.2 | |
| 16:0 | 18:1 | 5.1 | 3.6 |
| 18:0 | 16:1 | 0.4 | |
| 18:0 | 18:1 | 4.0 | |
| 16:0 | 18:2 | 0.6 | |
| 18:1 | 16:1 | 0.1 | |
| 18:1 | 18:1 | 1.2 | 7.4 |
| 18:0 | 18:2 | 1.2 | |
| 16:0 | 20:2 | 0.5 | |
| 18:0 | 18:2 | 1.9 | |
| 18:1 | 20:1 | 0.4 | |
| 16:0 | 20:3 | 0.8 | |
| 18:1 | 18:2 | trace | 7.0 |
| 18:0 | 20:3 | 8.4 | |
| 16:0 | 20:4 | 5.7 | |
| 18:0 | 20:4 | 59.6 | |
| 18:1 | 20:3 | 0.7 | 63.3 |
| 18:1 | 22:3 | 5.7 | |
| 18:1 | 20:4 | 1.7 | |
| 18:2 | 22:3 | 0.8 | 13.5 |

[a]Holub et al (141).
[b]Luthra and Sheltawy (142).

being contributed by stearoyl arachidonate. Unusual is the relatively high content of the molecular species containing eicosatrienoic and docosatrienoic acids in combination with stearic acid. Table XXXIII gives the principal molecular species of the phosphatidylinositols of the sarcotubular membranes of the skeletal muscle of rabbit, rat, chicken and man.

## Sphingomyelins

Sphingomyelin is a choline-containing phospholipid and constitutes one of the most persistent contaminants of commercial preparations of lecithin. Since it is a minor component of tissue phospholipids it also remains a

**TABLE 7-XXXII.**
**Principal molecular species of the phosphatidylinositols of the liver of different animal species**

| Molecular Species | Rat[a] | Pig[b] | Sheep[c] |
|---|---|---|---|
| | | Moles % | |
| 16:0 16:1 | 0.05 | | |
| 16:0 18:1 | 0.20 | 0.6 | |
| 18:0 16:1 | 0.05 | | 60 |
| 18:0 18:1 | 0.20 | 1.4 | |
| 16:0 18:2 | 1.75 | 1.0 | |
| 18:0 18:2 | 2.20 | 4.5 | |
| 18:1 18:1 | 0.25 | 0.3 | 6 |
| 16:1 18:1 | 0.05 | | |
| 18:1 18:2 | 0.17 | 0.3 | |
| 16:0 20:3 | 0.43 | 0.4 | |
| 18:1 20:3 | 3.48 | | 11 |
| 18:0 20:3 | | 11.9 | |
| 16:0 20:4 | 4.14 | 2.4 | |
| 18:0 20:4 | 77.00 | 62.0 | |
| 18:2 18:2 | 0.83 | | 17 |
| 18:1 22:3 | 0.96 | | |
| 18:1 20:4 | 0.95 | 0.4 | |
| 16:0 22:5 | 0.08 | 0.4[d] | |
| 18:0 22:5 | 0.24 | 5.6[e] | 3 |
| 18:0 20:5 | | 0.6 | |
| 16:0 22:6 | 0.71 | | |
| 18:2 22:6 | 3.24 | | |
| 18:1 22:5 | 1.98 | | 4 |
| 18:2 20:4 | 0.24 | | |

[a]Holub and Kuksis (72).
[b]Hunter et al (143).
[c]Luthra and Sheltawy (144).
[d]Includes the 16:0 22:6 species.
[e]Includes the 18:0 22:6 species.

minor constituent of the lecithin preparation, and is usually included among the other phospholipids on the label. Natural sphingomyelins are heterogenous in the composition of their molecular species. In addition to species with both medium and long chain length fatty acids they contain hydroxylated fatty acids as a significant proportion of the total. The chain length distribution and the content of the hydroxylated fatty acid species varies with the tissue source of the phospholipid. Sphingomyelin composition is little affected by alterations in the dietary fat of the animal.

**TABLE 7-XXXIII.**
**Principal molecular species of phosphatidylinositol in skeletal muscle of different animal species**

| Molecular Species | | Animal Species | | |
|---|---|---|---|---|
| | Rabbit | Rat | Chicken | Man |
| | | Moles % | | |
| 18:0 16:0 | 1.0 | 1.0 | 3.0 | trace |
| 18:0 18:1 | 10.0 | 3.6 | 4.5 | 10.0 |
| 18:0 18:2 | 5.0 | 5.0 | 8.0 | 13.0 |
| 18:0 20:3 | 14.0 | 7.0 | 5.2 | 15.0 |
| 18:0 22:3 | 1.5 | 5.0 | 3.0 | 1.0 |
| 18:0 20:4 | 60.0 | 56.0 | 68.0 | 54.0 |
| 18:0 22:4 | 0.5 | 1.0 | trace | 0.5 |
| 18:0 20:5 | trace | 1.0 | trace | 0.5 |
| 18:0 22:5 | 2.0 | 5.5 | trace | 0.5 |
| 18:0 22:6 | trace | 7.6 | trace | trace |
| UNIDENTIFIED | | | | |

Marai and Kuksis (64,65).

### 1. Egg yolk

The sphingomyelins of the egg yolk are made up largely of medium chain length fatty acids, with the palmitoylsphingosines accounting for about 85% of the total. They are essentially free of hydroxylated fatty acids and sphingosine bases other than $C_{18}$. Table XXXIV lists the principal species of the sphingomyelins of the egg yolk of the hen.

### 2. Milk

The sphingomyelins of milk are much more complex than those of egg yolk and contain a wide range of fatty acid chain length, including small amounts of hydroxy acids, and a variety of nitrogenous bases. The major species, however, are made up of combinations of the $C_{16}$ and $C_{22}$–$C_{24}$ saturated acids and $C_{18}$ sphingosine. Table XXXV gives the principal molecular specie of sphingomyelins of human and bovine milk as indicated by their fatty acid composition (they contain only one major nitrogenous base). The phospholipid preparations of bovine milk fat also contain significant amounts of the ceramide mono- and di-hexosides (37).

**TABLE 7-XXXIV.**
**Fatty acid composition of various preparations of egg yolk sphingomyelins**

| Fatty Acid | a,c | b,d |
|---|---|---|
| | Mole % | |
| 14:0 | 1.4 | 1.0 |
| 15:0+18:1 | 0.5 | 0.3 |
| 16:0 | 82.3 | 84.3 |
| 17:0+20:1 | — | — |
| 18:0 | 5.9 | 5.2 |
| 19:0+22:1 | 1.7 | 0.7 |
| 20:0 | 1.0 | 1.0 |
| 21:0+24:1 | 4.5 | 4.7 |
| 22:0+25:1 | 1.7 | 1.5 |
| 23:0+25:1 | 0.3 | 0.6 |
| 24:0 | 0.7 | 0.7 |

[a]Do et al (145).
[b]Connelly and Kuksis (114).
[c]HPLC.
[d]GLC.

## 3. Brain

The brain contains the most complex mixture of sphingomyelins, a large proportion of which is made up of the hydroxy fatty acid containing species. Intact bovine brain sphingomyelin has been resolved into molecular species by HPLC by Jungalwala et al (109). A total of 10 major species containing the non-hydroxy fatty acids were obtained. The major species were the $C_{18}$ sphingosine amides of the $C_{18}$ and $C_{24}$ saturated and the $C_{24}$ monounsaturated fatty acids. Appreciable amounts of the $C_{18}$-sphinganine and the $C_{20}$-sphinganine base were also present. Other workers have analyzed the molecular species of sphingomyelins following preparation of the ceramides with comparable results (110). Table XXXVI lists the principal molecular species of sphingomyelins of bovine brain.

## 4. Plasma

Myher and Kuksis (80) have determined the molecular species of the sphingomyelins of the plasma using capillary GLC and mass spectrometry. For this purpose the sphingomyelins were first converted into the ceramides by hydrolysis with phospholipase C. The analyses were made with the TMS

**TABLE 7-XXXV.**
**Fatty acid composition of milk glycolipids and sphingomyelins of different animal species**

| Fatty Acids | Bovine[a] CMH | Bovine[a] CDH | Bovine[a] SPH | Human[b] SPH2 | Bovine[c] SPH | Bovine[d] SPH |
|---|---|---|---|---|---|---|
| | | | | Moles % | | |
| 12:0 | 0.8 | 0.4 | 0.3 | | | |
| 14:0 | 6.3 | 2.7 | 2.5 | 2.4 | 2.0 | 1.2– 1.9 |
| 15:0 br | 1.4 | 0.7 | trace | | | |
| 15:0 | 1.6 | 0.7 | 0.4 | 0.1 | 0.1 | |
| 16:0 | 29.8 | 25.0 | 22.1 | 16.1 | 12.8 | 10.5–17.1 |
| 16:1 | 2.6 | 1.2 | 0.8 | 0.8 | 0.6 | |
| 17:0 br | 0.9 | 1.3 | 0.2 | | | |
| 17:0 | 1.1 | 0.5 | 0.6 | 0.2 | 0.5 | |
| 18:0 | 11.1 | 16:2 | 4.5 | 12.1 | 11.8 | 1.7–10.8 |
| 18:1 | 16.7 | 13.9 | 5.0 | 0.4 | 1.0 | 1.4– 7.9 |
| 18:2 | 2.7 | 2.3 | 0.9 | — | 0.3 | 0.5– 1.8 |
| 19:0 | 0.7 | 0.8 | 0.2 | | | |
| 20:0 | 0.6 | 1.6 | 0.6 | 9.4 | 8.9 | |
| 20:1 | 0.2 | 1.1 | 0.1 | 0.4 | 0.5 | |
| 21:0 | 0.4 | 0.5 | 0.8 | 0.5 | 0.8 | |
| 22:0 | 5.4 | 9.3 | 14.7 | 16.4 | 19.5 | 15.5–20.7 |
| 23:0 | 8.6 | 12.5 | 27.0 | 3.6 | 4.0 | 27.5–40.0 |
| 24:0 | 5.1 | 7.1 | 14.8 | 15.4 | 19.5 | 12.8–22.3 |
| 24:1 | 1.1 | 1.3 | 1.9 | 19.3 | 15.4 | |
| Others | | | | | | |

[a]Morrison et al (37).
[b]Bouhours and Bouhours (146).
[c]Morrison and Smith (147).
[d]Huang and Kuksis (148).

and the t-BDMS ethers. Table XXXVII gives the principal molecular species of sphingomyelins of plasma of two major human plasma lipoproteins.

## Summary and Conclusions

A survey of the techniques of isolation of animal lecithins has shown that different tissues require different methods for effective extraction of phospholipids. Preparative routiness have generally given less complete recoveries than analytical procedures. Only egg yolk, milk and brain have served as major animal sources of commercial lecithins. In some instances the isolated lecithins have consisted of essentially pure phosphatidylcholine, while

**TABLE 7-XXXVI.**
**Principal molecular species of sphingomyelins of brain of different animal species**

| Fatty Acids | Bovine[a] | Animal Species |
|---|---|---|
| | | Mole % |
| 14:0 | — | |
| 15:0 + 18:1 | 0.5 | |
| 16:0 | 0.4 | |
| 17:0 + 20:1 | 1.2 | |
| 18:0 | 8.5 | |
| 19:0 + 22:1 | 1.1 | |
| 20:0 | 1.6 | |
| 21:0 + 24:1 | 24.2 | |
| 22:0 + 25:1 | 14.0 | |
| 23:0 + 26:1 | 11.9 | |
| 24:0 | 30.0 | |
| 25:0 | 5.0 | |
| 26:0 | 1.5 | |
| Total | 100.0 | |

Do et al (145).
[a]Derived from non-hydroxy cerebrosides.

in others the lecithin preparations have differed little from a crude total phospholipid extract of a tissue. In most cases the commercial lecithin preparations have been contaminated with other glycerophospholipids and sphingomyelin in proportions closely approximating those present in the original tissue. It may be pointed out that liver, kidney, heart and skeletal muscle are other major tissues which could also serve a sources of commercial lecithin. These phospholipids, however, are much more unsaturated than those of egg yolk and milk. Comparisons of the molecular species composition of the various glycerolipid classes in the original tissue and in various preparative extracts have revealed that little fractionation of species takes place, although the proportions of the different phospholipid classes may vary greatly. However, the polyunsaturated species may become peroxidized and removed during lecithin refining. Likewise, lecithins containing alkenylacyl species may become destroyed during refining, especially if acidic conditions are employed. The destruction of the latter components may lead to the formation of lysophospholipids, which can significantly alter the physico-chemical properties of the lecithin preparation. Since the alkenylacyl species are present in significant amounts only in the ethano-

**TABLE 7-XXXVII.**
**Principal molecular species of sphingomyelins of human plasma**

| Molecular Species | | Lipoprotein VLDL | HDL |
|---|---|---|---|
| | | Moles % | |
| 16:0 | 18:1′ | 30 | 18 |
| 18:0 | 16:1′ | 2 | 2 |
| 16:0 | 18:2′ | 3 | 3 |
| 18:1 | 16:1′ | trace | trace |
| 18:0 | 18:1′ | 5 | 3 |
| 20:0 | 16:1′ | 1 | 1 |
| 18:0 | 18:2′ | 2 | 2 |
| 18:1 | 18:1′ | 1 | 1 |
| 18:1 | 18:2′ | trace | trace |
| 22:0 | 18:1′ | 5 | 8 |
| 23:0 | 17:1 | trace | trace |
| 24:0 | 16:1 | trace | 1 |
| 22:0 | 18:2 | 2 | 5 |
| 22:1 | 18:1′ | 1 | trace |
| 24:1 | 16:1′ | 2 | 3 |
| 22:1 | 18:2′ | trace | trace |
| 24:0 | 18:1′ | 5 | 6 |
| 24:1 | 18:1′ | 13 | 15 |
| 24:0 | 18:2′ | 3 | 5 |
| 24:1 | 18:2′ | 3 | 5 |
| 24:2 | 18:1′ | 1 | 2 |
| 24:2 | 18:2′ | trace | trace |
| UNIDENTIFIED | | | |

Myher et al (110).

lamine phosphatides, their degradation is not a serious problem unless the lecithins have been prepared from such tissues as the kidney, heart, skeletal muscle and brain, which contain high proportions of phosphatidylethanolamine rich in alkenylacyl species.

The tabulations of the lipid classes and molecular species are limited to the major phospholipids and the major tissues. Complete analyses of molecular species are available only in a few instances, most studies not distinguishing even between the fatty acid composition of the sn-1- and the sn-2-position. The present review does not provide separate listing of the fatty acid composition as it can be readily calculated from the composition of the molecular species. Comparisons among different animal species have established that there exists a remarkable similarity in the lipid class and molecular species composition of the organs, although all tissues have not been compared. Furthermore, the degree of similarity increases with increasing degree of subfractionation of the tissue and membranes. The composition of the molecular species is influenced by the fatty acid composition of the diet, but not all tissues or lipid classes are equally affected. Since animal diets tend to be rather uniform, the fatty acid composition of the molecular species is not subject to extreme variations. It may therefore be concluded that the chemical composition of a commercial preparation of lecithin will depend more on the tissue and method of preparation of the extract and purification of the lipid than on the diet of the animal. The final product may vary in composition from a crude total phospholipid extract to pure phosphatidylcholine.

**Acknowledgment**

The studies by the author and his collaborators were supported by funds from the Medical Research Council of Canada, and The Ontario Heart Foundation, Toronto, Ontario

## References

1. Levene, P. A., and I. P. Rolf, J. Biol. Chem. 72:587 (1937).
2. Pangborn, M. C., J. Biol. Chem. 188:471 (1951).
3. Hanahan, D. J., M. B. Turner and M. E. Jayko, J. Biol. Chem. 192:623 (1951).
4. Saunders, L., Pharm. Pharmacol. 9:834 (1957).
5. Singleton, W. S., M. S. Gray, M. L. Brown and J. L. White, JAOCS 38:565 (1961).
6. Hanahan, D. J., Biochem. Prep. 9:55 (1962).
7. Small, D. M., M. C. Bourges and D. G. Dervichian, Biochim. Biophys. Acta 125:563 (1966).
8. Rhodes, D. N., and C. A. Lea, Biochem. J. 65:526 (1957).
9. Singleton, W. S., M. S. Gray, M. L. Brown and J. L. White, JAOCS 42:53 (1965).
10. Radin, N. S., J. Lipid Res. 19:922 (1978).
11. Nielsen, J. R., Lipids 15:481 (1980).
12. Folch, J., I. Ascoli, M. Lees, J. A. Meath and F. N. Le Baron, J. Biol. Chem. 191:833 (1951).
13. Gornall, D. A., and A. Kuksis, J. Lipid Res. 14:197 (1973).
14. Ramesh, B., A. V. Prabhadesai and C. V. Viswanathan, JAOCS 56:585 (1979).
15. Borgstrom, B., Acta Physiol. Scand. 25:101 (1952).

16. Sen Gupta, A. K., Fette, Seifen, Anstrichm. 78:111 (1976).
17. Hanahan, D. J., J. C. Dittmer and E. Warashina, J. Biol. Chem. 228:685 (1965).
18. Renkonen, O., J. Lipid Res. 3:181 (1962).
19. Marsh, M., and R. T. Holzbach, Clin. Chim. Acta 43:87 (1973).
20. Rouser, G., J. O'Brien and D. Heller, JAOCS 38:14 (1961).
21. Lundberg, B., Acta Chem. Scand. 27:3545 (1973).
22. Collins, F. D., Nature 188:297 (1966).
23. Davenport, J. B., Lipids 4:308 (1969).
24. Tipton, C. L., J. W. Paulis and M. D. Pierson, J. Chromatogr. 14:486 (1964).
25. Berry, J. F., and B. Kaye, Lipids 3:386 (1968).
26. Fager, R. S., S. Shapiro and B. J. Lifman, J. Lipid Res. 18:704 (1977).
27. Geurts van Kessel, W. S. M., W. M. A. Hax, R. A. Demel and J. de Gier, Biochim. Biophys. Acta 486:524 (1977).
28. Porter, N. A., R. A. Wolf and J. R. Nixon, Lipids 14:20 (1978).
29. Geurts van Kessel, W. S. M., M. Tieman and R. A. Demel, Lipids 16:58 (1981).
30. Hanson, V. L., J. Y. Park, T. W. Osborn and R. M. Kiral, J. Chromatogr. 205:393 (1981).
31. Gross, R. W., and B. E. Sobel, J. Chromatogr. 197:79 (1980).
32. Majonnier, T., and H. C. Troy, The Technical Control of Dairy Products, Majonnier Bros. Co., Chicago, IL, 1922, pp. 106.
33. Crane, J. C., B. E. Horrall, J. Dairy Sci. 26:935 (1943).
34. Walstra, P., and J. J. De Graaf, Netherlands Milk Dairy J. 16:283 (1962).
35. Duthie, A. H., and S. Patton, J. Dairy Sci. 48:170 (1965).
36. Cerbulis, J., J. Agric. Food Chem. 15:784 (1967).
37. Morrison, W. R., E. L. Jack and L. M. Smith, JAOCS 42:1142 (1965).
38. Privett, O. S., L. J. Nutter and R. A. Gross, Dairy Lipids and Lipid Metabolism, edited by H. F. Brink and D. Kritchevsky, Avi Publishing Co., Westport, CT, 1968, pp. 99–115.
39. Patton, S. S., and T. W. Keenan, Lipids 6:58 (1971).
40. Galanos, D. S., and V. M. Kapolas, Biochim. Biophys. Acta 98:278 (1965).
41. Bligh, E. G., and W. J. Dyer, Can. J. Biochem. Physiol. 37:911 (1959).
42. Morrison, W. R., and L. M. Smith, Biochim. Biophys. Acta 84:759 (1964).
43. Morrison, W. R., and L. M. Smith, J. Lipid Res. 5:600 (1964).
44. Santha, I. M., and K. M. Narayanan, J. Food Sci. Technol. 15:24 (1978).
45. Nutter, L. J., and O. S. Privett, J. Dairy, Sci. 50:298 (1967).

46. Rouser, G., G. Galli, E. Lieber, M. L. Blank and O. S. Privett, JAOCS, 41:836 (1964).
47. Cho, B. H. S., D. M. Irvine and J. B. M. Rattray, Lipids 12:983 (1977).
48. Rouser, G., G. Simon and G. Kritchevsky, Lipids 4:599 (1969).
49. Wittcoff, H. L., The Phosphatides, Reinhold Publ. Co., New York, NY, 1951.
50. Folch, J., M. Lees and G. H. Stanley, J. Biol. Chem. 226:497 (1957).
51. Rouser, G., G. Kritchevsky and N. Yamamoto, Lipid Chromatographic Analysis, Vol. 1, edited by G. V. Marinetti, Marcel Dekker, New York, NY 1967, pp. 99–162.
52. Nelson, G. J., in Analysis of Lipids and Lipoproteins, edited by E. G. Perkins, American Oil Chemists' Society, Champaign, IL, 1975, p. 1.
53. Phillips, F., and O. S. Privett, Lipids 14:590 (1979).
54. Skipski, V. P., and M. Barclay, Methods Enzymol 14:530 (1969).
55. Arvidson, G. A. E., J. Chromatogr. 103:201 (1975).
56. Jungalwala, F. B., J. E. Evans and R. H. McCluer, Biochem. J. 155:55 (1976).
57. Kaltaranta, J. K., and S. P. Bessman, Anal. Chem. 53:1232 (1981).
58. Chen, S. S., and A. Y. Kou, J. Chromatogr. Biomed. Applic. 227:25 (1982).
59. Siakotos, A. N., and G. Rouser, JAOCS 42:913 (1965).
60. Rouser, G., A. N. Siakotos and S. S. Fleischer, Lipids 1:85 (1966).
61. James, J. L., G. A. Clawson, C. H. Chan and E. A. Schmukler, Lipids 16:541 (1981).
62. Patton, G. M., J. M. Fasulo and S. J. Robins, J. Lipid Res. 23:190 (1982).
63. Gray, G. M., and M. G. Macfarlane, Biochem. J. 81:480 (1961).
64. Marai, L., and A. Kuksis, Can. J. Biochem. 51:1248 (1973).
65. Marai, L., and A. Kuksis, Can. J. Biochem. 51:1365 (1973).
66. Dodge, J. T., and G. B. Phillips, J. Lipid Res. 8:667 (1967).
67. Rose, H. G., and M. Oaklander, J. Lipid Res. 6:428 (1965).
68. Christie, W. W., and J. H. Moore, Comp. Biochem. Physiol. 47B:409 (1974).
69. Galanos, D. S., and V. M. Kapolas, J. Lipid Res. 3:134 (1962).
70. Drozdowski, B., and R. G. Ackman, JAOCS 46:371 (1969).
71. Weihrauch, J. L., and Y-S. Son, JAOCS 60:1971 (1983).
72. Holub, B. J., and A. Kuksis, Adv. Lipid Res. 11:1 (1978).
73. Ohno, K., T. Akino and T. Fujiwara, Revs. Perinatal Med. 2:227 (1978).
74. Kuksis, A., and L. Marai, Lipids 2:217 (1967).
75. Tattrie, N. H., J. Lipid Res. 1:60 (1959).
76. Renkonen, O., JAOCS 42:298 (1965).

77. Renkonen, O., Lipids 1:160 (1966).
78. Nutter, L. J., and O. S. Privett, Lipids 1:234 (1966).
79. Hasegawa, K., and T. Suzuki, Lipids 10:667 (1975).
80. Myher, J. J., and A. Kuksis, Can. J. Biochem. 60:638 (1982).
81. Hsieh, J. Y-K., D. K. Welch and J. G. Turcotte, Lipids 16:761 (1981).
82. Rhodes, D. N., Biochem. J. 68:380 (1958).
83. Hanahan, D. J., Lipid Chemistry, John Wiley and Sons, New York, NY, 1960, Ch. 3, pp. 42–105.
84. Blank, M. L., L. J. Nutter and O. S. Privett, Lipids 1:132 (1966).
85. Montfoort, A., L. M. G. Van Golde and L. L. M. Van Deenen, Biochim. Biophys. Acta 231:335 (1971).
86. Crawford, C. G., and M. A. Wells, Lipids 14:757 (1979).
87. Ogino, H., T. Matsumura, K. Satouchi and K. Saito, Biochim. Biophys. Acta 574:57 (1979).
88. Kino, M., T. Matsumura, M. Gamo and K. Saito, Biomed. Mass Spectrom., 9:363 (1982).
89. Kuksis, A., W. C. Breckenridge, L. Marai and O. Stachnyk, JAOCS 45:537 (1968).
90. Van Golde, L. M. G., W. A. Pieterson and L. L. M. Van Deenen, Biochim. Biophys. Acta 152:84 (1968).
91. Van Golde, L. M. G., and L. M. G. Van Deenen, Biochim. Biophys. Acta 125:496 (1966).
92. Blank, M. L., F. Snyder, L. W. Byers, B. Brooks and E. E. Muirhead, Biochem. Biophys. Res. Commun. 90:1194 (1979).
93. Kuksis, A., W. C. Breckenridge, L. Marai and O. Stachnyk, J. Lipid Res. 10:25 (1969).
94. Pugh, E. L., M. Kates and D. J. Hanahan, J. Lipid Res. 18:710 (1977).
95. Shamgar, F. A., and F. D. Collins, Biochim. Biophys. Acta 409:116 (1975).
96. Marai, L., and A. Kuksis, J. Lipid Res. 10:141 (1969).
97. Van Golde, L. M. G., V. Tomasi and L. L. M. Van Deenen, Chem. Phys. Lipids 1:282 (1967).
98. Sundler, R., and B. Akesson, J. Chromatogr. 80:233 (1973).
99. Yeung, S. K. F., A. Kuksis, L. Marai and J. J. Myher, Lipids 12:529 (1977).
100. Bjerve, K. S., J. Chromatogr. Biomed. Applic. 232:39 (1982).
101. Holub, B. J., and A. Kuksis, Lipids 4:466 (1969).
102. Renkonen, O., Acta Chem. Scand. 21:1108 (1967).
103. Noble, R. C., and J. H. Moore, Can. J. Biochem. 43:1677 (1965).
104. Dyatlovitskaya, E. V., V. I. Volkova and L. D. Bergelson, Biokhymiya 32:1227 (1967).

105. Do, U. H., and S. Ramachandran, J. Lipid Res. 21:888 (1980).
106. Renkonen, O., J. Lipid Res. 9:34 (1968).
107. O'Brien, J. S., and E. L. Sampson, J. Lipid Res. 6:537 (1965).
108. Sun, G. Y., and L. A. Horrocks, Lipids 5:1006 (1970).
109. Jungalwala, F. B., V. Hayssen, J. M. Pasquini and R. H. McCluer, J. Lipid Res. 20:579 (1979).
110. Myher, J. J., A. Kuksis, W. C. Breckenridge and J. A. Little, Can. J. Biochem. 59:626 (1981).
111. Privett, O. S., M. L. Blank and J. A. Schmit, J. Food Sci. 27:463 (1962).
112. Cook, W. H., and W. G. Martin, in Structural and Functional Aspects of Lipoproteins in Living Systems, edited by E. Tria and A. Scanu, Academic Press, NY, 1969, pp. 579–615.
113. Gornall, D. A., and A. Kuksis, Can. J. Biochem. 49:51 (1971).
114. Connely, P. W., and A. Kuksis, Can. J. Biochem. 61:63 (1983).
115. Siakotos, A. N., G. Rouser and S. Fleischer, Lipids 4:239 (1969).
116. Siakotos, A. N., and S. Fleischer, Lipids 4:234 (1969).
117. Scott, T. W., B. P. Setchall and J. M. Bassett, Biochem. J. 104:1040 (1967).
118. Wuthier, R. E., J. Lipid Res. 7:544 (1966).
119. Simon, G., and G. Rouser, Lipids 4:607 (1969).
120. Osanai, A., and T. Sakagami, J. Biochem. 85:1453 (1979).
121. Getz, G. S., W. Bartley, D. Lurie and B. M. Notton, Biochim. Biophys. Acta 152:325 (1968).
122. Nelson, G. J., Biochim. Biophys. Acta 144:221 (1967).
123. Kleinig, H., H. Zentgraf, P. Comes and J. Stadler, J. Biol. Chem. 246:2996 (1971).
124. Baxter, C. F., G. Rouser and G. Simon, Lipids 4:243 (1969).
125. Wang, N. S., R. V. Kotas, M. E. Avery and W. M. Thurlbeck, J. Appl. Physiol. 30:362 (1971).
126. Hallman, M., and L. Gluck, Biochim. Biophys. Acta 409:172 (1975).
127. Kawamoto, T., T. Akino, M. Nakamura and M. Mori, Biochim. Biophys. Acta 619:35 (1980).
128. Adams, E. P., and T. J. Heath, Biochim. Biophys. Acta 70:688 (1963).
129. Grigor, M. R., A. Moehl and F. Snuder, Lipids 7:766 (1972)
130. Hasegawa, K., and T. Suzuki, Lipids 8:631 (1973).
131. Kuksis, A., W. C. Breckenridge, J. J. Myher and G. Kakis, Can. J. Biochem. 56:630 (1978).
132. Gonzalez, J. M., and A. Ribera, Lipids 15:279 (1981).
133. Christie, W. W., Biochim. Biophys. Acta 316:204 (1973).
134. Curstedt, T., Biochim. Biophys. Acta 369:196 (1974).
135. Kuksis, A., L. Marai, W. C. Breckenridge, D. A. Gornall and O.

Stachnyk, Can. J. Physiol. Pharmacol. 46:511 (1968).
136. Toshima, N., T. Akino and K. Ohno, Tohoky J. Exp. Med. 108:265 (1972).
137. Nakagawa, Y., and L. A. Horrocks, J. Lipid Res. 24:1268 (1983).
138. Trewhella, M. A., and F. D. Collins, Biochim. Biophys. Acta 296:34 (1973).
139. Myher, J. J., and A. Kuksis, Can. J. Biochem. Cell Biol. 62:352. (1984).
140. Yeung, S. K. F., and A. Kuksis, Can. J. Biochem. 52:830 (1974).
141. Holub, B. J., A. Kuksis and W. Thompson, J. Lipid Res. 11:558 (1970).
142. Luthra, M. G., and A. Sheltawy, J. Neurochem. 27:1503 (1976).
143. Hunter, H. L., W. W. Christee, and J. H. Moore, Lipids 8:68. (1973).
144. Luthra, M. G., and A. Sheltawy, Biochem J. 126:1231. (1972).
145. Do, U. H., P. T. Pei and R. D. Minard, Lipids 16:855 (1981).
146. Bouhours, J–F., and D. Bouhours, Lipids 16:726 (1981).
147. Morrison, W. R., and L. M. Smith, Lipids 2:178 (1967).
148. Huang, T. C., and A. Kuksis, Lipids 2:453 (1967).

# Chapter Eight

## Commercial Lecithin Products: Food Use of Soybean Lecithin

**W. E. Prosise**
Thomas J. Lipton Inc.
Englewood Cliffs, NJ 07632

### Historical Perspective

The production date of the first commercial lecithin may well be lost in antiquity. Exploration of lecithin products, however, can be traced back as early as 1793, when Fourcray presented evidence for the existence of complex fatty components (1). Nineteen years later, Vanquelin isolated phosphorus-containing fatty materials from animal brains (2) and in 1846–47, Gobley (3) isolated an impure phosphatide mixture from egg yolk, which he named "lecithin" from "lekithos," the Greek word for egg yolk. In his research, Dr. Gobley found that, for centuries, ancient Greeks had used dried egg yolk to reduce the surface tension of fats and oils applied to soften leather.

In 1891, fats containing phosphorus were first isolated from plant seeds (4). The complexity and expense of the isolation process, however, limited any commercial production.

The first readily available, commercial soybean lecithin was manufactured from expeller soybean foots in the late 1920s. The foots were composed of phosphatides, phytins, sterols, glycerides, carbohydrates, gums, water and some meal. It was necessary to process this "sludge" through several extractions, evaporations and other steps to produce good, stable lecithin (5). This expensive processing did not help to promote lecithin's commercial availability. The price of crude lecithin was ca. $1.40/lb—very high, even by today's standards (6).

The 1930s brought widespread use of commercial solvent extraction techniques for vegetable oil production. Because "degumming" of crude vegetable oil became necessary for shipping stability, a large supply of crude lecithin "gums" was produced. These "gums" were obtained in sufficient quantity to necessitate their becoming an item of commerce (7).

The 1930s and 40s brought extensive and diverse research into developing new lecithin applications, as well as product refinements and modifications. Justification for this work rested on disposing of the lecithin surplus. This 20-year period also brought the large-scale commercial marketing of lecithin products. The period's literature is profuse with articles and patents on soybean lecithin by such large companies as Glidden Company, Ross & Rowe and American Lecithin Company.

Commercial lecithin products were being sold for such diverse applications as chocolate and confectionery products (8), margarines (9), baking (10, 11), textiles (12), insecticides (13, 14), paints (15), pasta products (16) and many other applications still active today.

Although soybean oil processing technology has greatly improved over the last 40 years, the basic processes for obtaining crude lecithin have proven themselves by withstanding the test of time. Virtually the same recovery techniques as those developed in the 1930s are employed today (17). Modern commercial lecithin products are based on crude lecithins obtained through updated versions of the degumming operations developed over 50 years ago.

From now on, the word "lecithin" will be used to mean soybean lecithin, since it is the only important commercial lecithin available today.

The following review of crude lecithin composition and general physical/chemical properties of lecithin products is fundamental to understanding the diversity of modern commercial lecithin products, their properties and their applications.

## Composition of Crude Lecithin

For the last 50 years, crude lecithin has had a soybean oil content of 30–50%. Phosphatidylcholine, one of the major active components, is found in most crude lecithins at levels of ca. 16%. Phosphatidylethanolamine, occasionally called cephalin, is also a major component of soybean lecithin. Crude lecithin contains ca. 14% phosphatidylethanolamine. The third principal phospholipid observed in soybean lecithin is a complex mixture of inositol phosphatides, which are present at levels of ca. 12%. Miscellaneous low level components include phosphatidic acid, various glycolipids, phosphatidylserine, carbohydrates, sterols and tocopherols. Moisture is typically present at levels of ca. 1% or less.

Most commercial lecithins are formulated from crude lecithins having this composition. In practice, commercial lecithin products are not marketed by phosphatide content, but are marketed under a set of unique chemical/physical properties. These properties, as indicated by product specifications, must be understood, since they are used to characterize specific lecithin types.

## Chemical/Physical Specifications and Properties Used to Characterize Commercial Lecithins

The following is a discussion of the major chemical/physical measurements applicable to soybean lecithin. This background is necessary to obtain an understanding of the lecithin types and their applications.

### Acetone-Insolubles (AI)

In general, lecithin products are defined by their phosphatide content. The amount of acetone-insolubles matter (% AI) is approximately representative of these phosphatides. The phosphatides are insoluble in acetone, whereas the triglycerides are soluble. The composition of the AI portion is taken to be roughly constant. This is why the lecithin AI can be used to approximate phosphatide content. In crude lecithins, AI is usually synonymous with activity. However, upgraded products are often formulated for total activity. In addition to the AI portion, acetone-soluble components in upgraded products can be functionally active with respect to viscosity control, emulsification, oxidative stability and end-use compatability (18).

The acetone-insoluble index is AOCS Official Method Ja 4-46 (19).

### Acid Value (AV)

The acid value (AV) is the titratable acidity of lecithin in mg KOH/g of sample. A product's AV is representative of acidity contributed by the phospholipids as well as acidity from free fatty acids.

The AV is usually not indicative of pH, as the chemical nature of the phosphatides impart buffering qualities to most systems. Lecithins typically exhibit a neutral pH value in aqueous media. An AV above 36 may indicate degradation of the lecithin due to improper processing or substandard soybeans. AV should not be confused with free fatty acid content, pH or mineral acid. The correct method to assay for free fatty acids is to titrate only the acetone-soluble portion, thereby leaving any contribution from the phospholipids in the AI portion. One cannot simply titrate whole lecithin (AI plus AS fraction) to determine free fatty acid content.

AV is determined by AOCS Official Method Ja 6-55 (19).

### Moisture

The water content of lecithin products is usually less than 1.5%. As a consequence of lecithin's essentially moisture-free state, lecithin products have very low water activity and do not adversely contribute to the microbiological profile of most food systems. Most lecithin products are preserved by this essentially moisture-free state. Higher moisture levels in lecithin usually indicate a greater potential for spoilage or chemical degradation.

Moisture level is determined by Karl Fisher aquametric titration (AOCS Official Method Tb 2-64 (19) and FCC III Official Method (20)). A less accurate moisture level can be determined by azeotropic toluene distillation (AOCS Official Method Ja 2-46 (19)). One cannot determine lecithin moisture by vacuum oven methods. These methods are known to degrade lecithin products and give false moisture levels.

### Hexane-Insolubles (HI)

The level of hexane-insoluble (HI) matter is one measure of the purity of lecithin products. HI matter usually consists of residual fiber, but is often particulate material introduced during processing (e.g., filter aid). The level of HI in crude lecithin should never exceed 0.3% and rarely exceeds 0.1%. This is an official Food Chemicals Codex test method (20). HI matter in lecithin is detrimental to clarity and usability in specific applications. Some of these will be discussed later in this chapter. In modern lecithins, the presence of any HI matter is usually indicative of poor lecithin processing. Lower quality lecithin grades, often containing up to 0.3% HI materials, are often used in less demanding applications including chocolates, margarines and other food applications.

### Color

The color of lecithin is fundamentally an aesthetic quality standard. Many formulators instinctively dislike dark-colored ingredients. Although it is demonstrably true that many lecithin applications require lightly colored ingredients, most lecithin applications can be satisfied by the amber color grades currently available. When added to aqueous systems, even the darkest lecithins become creamy white.

Historically, lecithins have been color graded as unbleached, single-bleached and double-bleached. An unbleached product was not treated with bleach, a single-bleached product was treated with only one type of bleach (usually hydrogen peroxide) and a double-bleached product was treated with 2 types of bleach (usually hydrogen peroxide and benzoyl peroxide).

Although these grades continue to exist by name, the bleaching methods used to manufacture them are no longer fixed. High quality, relatively light colored, "unbleached" lecithins are achieved through modern manufacturing methods. Additionally, today's double-bleached product may have been treated with only a small quantity of one type of bleach. Products are presently bleached to a color specification only, regardless of bleaching technique or quantity. By convention, the amber color tones of lecithin are measured on the Gardner color scale. The color range covered by lecithin is generally Gardner 9–17 in an undiluted form.

I have found no evidence that bleaching with either hydrogen peroxide or benzoyl peroxide functionally modifies lecithins. Bleaching seems only to affect the pigments, which are usually not functionally active.

The color of most lecithin products will darken on prolonged heating. Color stability can be achieved, however, by avoiding exposure of lecithin to temperatures over 140 F. Recently, one manufacturer released a "heat-resistant" lecithin. This lecithin maintains its light color for extended periods, even at elevated temperatures.

## Other Physical/Chemical Properties and Quality Criteria

### Consistency

Lecithins are available in both fluid and plastic (solid) forms. Fluid lecithins generally follow Newtonian flow characteristics. The viscosity profile of lecithins is a complex function of acetone-insoluble content, moisture, mineral content, acid value and the combined effects of assorted additives such as vegetable oils and surfactants. Generally, higher AI and/or moisture content yields higher viscosity lecithins, whereas an increased AV often decreases viscosity. Certain minerals, such as calcium and other additives, can also adjust lecithin's viscosity.

Plastic lecithins are available in several forms. Oil-free lecithins are plastic, due to the removal of their nascent oils, i.e., their residual soybean oils. They are generally powdered or granular in form. Plasticized crude lecithins are typified by high AI, low AV, high moisture or their content of certain minerals. One, or any combination of these, can produce plastic lecithins. It is often difficult to differentiate between a very high viscosity lecithin and a plastic one. In practice, perhaps no differentiation is needed since handling is often similar. The viscosity of crude lecithin is decreased (or plastic lecithins are fluidized) by warming to a maximum of 140 F.

### Clarity

In the early years of its production, lecithin was almost always isolated from unfiltered crude oil. As a consequence of this lack of oil filtration, high levels of hexane-insoluble matter were invariably present in the lecithin. The heavier HI partitioned with the lecithin "gums" on separation from the oil. This lipid-insoluble material causes fluid lecithins to appear hazy. With modern miscella and oil filtration techniques, lecithins with very low, or even no HI content are produced. Consequently, modern lecithins are usually quite clear. Additionally, moisture can contribute to lecithin's lack of clarity. Higher moisture levels (generally over 1%) can cause haziness even in 0% HI lecithins. This low level moisture influence is not surprising, since even mildly hydrated lecithin tends to be milky-opaque in appearance. Haziness, besides causing aesthetic problems, can cause real application-oriented problems. If caused by HI material, the haziness can settle out over time, causing solid particulate material to appear on the bottom of an otherwise clear food product containing lecithin. Many applications cannot tolerate fibrous material.

HI material in a sprayable lecithin product can plug aerosol spray nozzles causing erratic operation. This is of particular importance in lecithins for pan release applications. The use of lecithin as an iron oxide dispersant for magnetic recording tapes is another example. HI material in lecithin that becomes impregnated on a tape surface can cause signal drop-out and undue wear on a tape player pickup head.

Generally, clarity is quantified by modified HI tests. However, these tests are only marginally useful, due to their failure to recognize moisture as a contributing factor. Light transmittance methods can be meaningful in specific cases, but are usually found inadequate, again due to the presence of a variety of factors contributing to haziness. I have found a combination of tests, including moistures and a modified HI test, to be the most useful for quantifying the clarity of lecithin products.

## Commercial Lecithin Types

For many years, the National Soybean Processors Association Trading Rules (21) have divided lecithin products into 6 grades based on viscosity and color (Table I—NSPA Specifications). In most cases, this is an oversimplification. A more adequate regulatory definition for lecithin (and hydroxylated lecithin) is presented in Food Chemicals Codex III (Tables II and III). These use sets of some of the unique physical/chemical tests that were discussed previously to define lecithin products. In the USA, lecithin manufacturers must make all lecithin products (and hydroxylated lecithin products) within these specifications. The FCC, however, allows for slight variations in lecithin product types. These variations permit lecithin manufacturers to produce through minor modifications a variety of functionally specific lecithin products.

**TABLE 8-I.**
**Soybean Lecithin Specifications**[a]

| Grade: | Fluid natural lecithin | Fluid bleached lecithin | Fluid double-bleached lecithin |
|---|---|---|---|
| Analysis | | | |
| Acetone insoluble, min | 62% | 62% | 62% |
| Moisture, max | 1% | 1% | 1% |
| Benzene insoluble, max | 0.3% | 0.3% | 0.3% |
| Acid value, max | 32 | 32 | 32 |
| Color, Gardner, max | 10 | 7 | 4 |
| Viscosity, poises, @ 77 F, max | 150 | 150 | 150 |
| **Grade:** | **Plastic natural lecithin** | **Plastic bleached lecithin** | **Plastic double-bleached lecithin** |
| Analysis | | | |
| Acetone insoluble, min | 65% | 65% | 65% |
| Moisture, max | 1% | 1% | 1% |
| Benzene insoluble, max | 0.3% | 0.3% | 0.3% |
| Acid value, max | 30 | 30 | 30 |
| Color, Gardner, max | 10 | 7 | 4 |
| Penetration, max | 22 MM | 22 MM | 22 MM |

[a]Year book and trading rules, National Soybean Processors Association (1982–83).

The following is a brief summary of some of the lecithin products available today.

### Crude Lecithins

Crude, or commodity grade, lecithin is the highly variable viscous amber fluid that is probably familiar to most users. It meets the specifications in the Food Chemicals Codex and the NSPA Trading Rules and is generally

**TABLE 8-II.**
**Food Grade Soybean Lecithin Specifications[a]**

| | |
|---|---|
| Acetone-insoluble matter | not less than 50.0% |
| Acid value | not more than 36 |
| Hexane-insoluble matter | not more than 0.3% |
| Water | not more than 1.5% |
| Peroxide value | not more than 100 |
| Limits of impurities | |
| Arsenic (as As) | not more than 3 parts per million (0.0003%) |
| Heavy metals (as Pb) | not more than 40 parts per million (0.004%) |
| Lead | not more than 10 parts per million (0.001%) |

[a]Food Chemicals Codex, 3rd edition, 1981.

**TABLE 8-III.**
**Food Grade Hydroxylated Lecithin Specifications[a]**

| | |
|---|---|
| Acetone-insoluble matter | not less than 50.0% |
| Acid value | not more than 70 |
| Benzene-insoluble matter | not more than 0.3% |
| Iodine value | between 85 and 95 |
| Peroxide value | not more than 100 |
| Water | not more than 1.5% |
| Limits of impurities | |
| Arsenic (as AS) | not more than 3 parts per million (0.0003%) |
| Heavy metals (as PB) | not more than 40 parts per million (0.004%) |
| Lead | not more than 10 parts per million (0.001%) |

[a]Food Chemicals Codex, 3rd edition, 1981.

traded as a commodity (i.e., large volumes with supply-specific pricing).

Crude lecithin usually is adjusted to tighter specifications within the NSPA trading rules by bleaching and by additions of soybean oil, soybean fatty acids and often by mineral and/or moisture adjustments.

The general characteristics of crude lecithins include the following: (a) most variable in composition and function—least quality controlled, (b) intense flavor profile, (c) least expensive, (d) most difficult to hydrate into aqueous systems, (e) among the most difficult lecithins to handle due to their thick consistencies, (f) among the most lipophilic lecithins.

### Fluidized Lecithins

Crude, as well as upgraded lecithin products, are sold in fluidized form for easier handling, pumpability, ease of application or even sprayability. These can be easily coated onto powders to effect instantizing, or sprayed onto surfaces as cooking releases and depanning agents (22).

Lecithin products are fluidized by a variety of techniques. Additions of fatty acids, special mineral salts or special diluents can aid fluidization. Most highly fluid lecithins (under ca. 2,000 centipoise) contain lower percentages of AI matter. Some, however, can be made highly fluid by the addition of special diluents such as mineral acids, or by special processing methods.

### Highly Filtered Lecithin Products

Upgraded, as well as crude, lecithin products can be highly filtered for utmost purity and clarity. Filtration simply removes HI material. Highly filtered lecithin products are marketed as encapsulated nutritional supplements, for high purity pharmaceutical grades and for high technology applications requiring the ultimate in purity.

Several filtration techniques have proved effective in producing highly filtered lecithins. The most common methods involve the efficient filtration of crude oil prior to degumming. The oil can be filtered either before or after desolventizing. The solvent-laden oil is the most commonly filtered, due to its reduced viscosity. The rate of oil filtration can be greatly expedited using this technique.

Although it is possible to filter lecithin directly, it is rarely done, due to the expense involved. Lecithin marketing companies who manufacture their own lecithin generally produce better quality grades of filtered lecithins. Distributors are totally under the control of the degumming or oil extraction plants in obtaining well filtered lecithin. Since many of these plants do not have lecithin as a main business, they produce "by-product grade" lecithin which is often inadequately filtered.

In addition to product quality benefits, filtration facilitates lecithin processing. HI material normally has an abrasive, and therefore detrimental, effect on lecithin processing equipment. Its removal prior to lecithin degumming appears to reduce equipment maintenance.

### Compounded Lecithin Products

Compounded lecithins are perhaps the largest category of lecithin products. This is probably a result of the ease with which they are manufactured. They are simply blended. Lecithins blended with selected additives can exhibit modified properties and functionalities, providing specific activities for a given system. Lecithins exhibit synergistic actions with some additives. Often, compounded lecithin products are designed simply for compatibility to a particular system.

Common lecithin additives include special oils, polysorbates, monoglycerides and modified monoglycerides, lanolin and lanolin derivatives, solvents, plasticizers and other surfactants. These are either added to the wet gums prior to drying, or are blended with the dry fluid lecithin at elevated temperatures. The latter is the most common practice.

### Chemically Modified Lecithins

Crude lecithin has been successfully hydrogenated, hydroxylated, ethoxylated, halogenated, sulfonated, acylated, ozonized and phosphorylated, to name but a few operations (5). Only rarely have viable commercial products been developed through these efforts. The only chemically modified lecithin that is presently food-approved is hydroxylated lecithin (23) (Table III). This is a product that is more hydrophilic and, therefore, exhibits better oil-in-water emulsifying properties than crude lecithin. Hydroxylated lecithins are prepared by treating crude lecithin with hydrogen peroxide in the presence of lactic or acetic acid and water. The reaction is followed by neutralization and drying to a fluid consistency.

### Lecithin Products Produced by Fractionating Crude Lecithin

Due to the differential solubility of its components, crude lecithin can be fractionated by many solvents or solvent blends. Some of the commercially available fractionated lecithins and the fractionating solvents are outlined below.

*Oil-free lecithin.* Fractionation with acetone results in isolating mixed phosphatides, the active portion of crude lecithin essentially free of triglycerides. These products, marketed as oil-free lecithin and oil-free phosphatides, are of plastic consistency and are usually available in either granular or powdered forms. The acetone-soluble portion, or that which is removed, consists of triglycerides, flavor components, pigments and dramatically smaller percentages of free fatty acids, phospholipids and sterols. This acetone-soluble portion is of no known practical importance. Of significance, however, is the removal of these components from the AI fraction. When compared to crude lecithin, oil-free lecithin is more hydrophilic and seems to have better emulsifying activity than its AI alone might suggest. Perhaps the more lipophilic properties of crude lecithins are due to an inhibitory effect of the triglycerides on the more hydrophilic AI portion. The removal of odor/flavor components with the oil, of course, produces blander lecithins. Since it is produced by simple fractionation, oil-free lecithin might be considered a natural extract.

Since these oil-free lecithins are among the lecithin products more costly to process, most manufacturers closely control their quality. Guaranteed specifications often include tight microbiological profiles and other specifications more stringent than those guaranteed with other lecithin product types. Consequently, the price of oil-free lecithin can be up to 10 times that of some of the standard fluid grades.

*Alcohol-extracted fractions.* Commercial extractions of oil-free or crude lecithin with alcohols result in lecithin products which are enriched in specific phosphatide components. Commercial products which are alcohol-soluble contain concentrated phosphatidylcholine and only low levels of inositol phosphatides. The alcohol-insoluble products are enriched in the inositol phosphatides, whereas their phosphatidylcholine content is reduced significantly. The phosphatidylethanolamine component is approximately equally partitioned between the alcohol fractions. Extensive alcohol and mixed solvent fractionation, combined with other techniques, produces alcohol-fractionated lecithin products that can be greatly enriched in particular phosphatides. Solvent concentration, temperature and mixing efficiency can be used to control the enrichment of these phosphatides.

The alcohol fractions of lecithin are truly different in their functional properties. The alcohol-soluble grades tend to be better oil-in-water emulsifiers, whereas the insoluble grades are more effective water-in-oil emulsifiers. When compared to either crude or oil-free lecithin, alcohol fractions are more functionally specific.

*Other products from fractionation.* Adsorption and distribution chromatography are utilized for lecithin fractionation. Lecithin production employing these methods is very limited. Products consisting of up to 95% phosphatidylcholine, however, are commercially available (24). Uses for phosphatidylcholine have been limited to clinical studies, higher priced pharmaceuticals and some high technology applications, due to the high commercial price of these products.

## Functions and Applications

Lecithin products are employed primarily for their functional benefits to food products. They are rarely employed for their nutritional benefits and are never used as "fillers". These functional characteristics are simplified in terms of the chemical structure of lecithin's major phospholipids (Fig. 1). Phosphatide molecules contain 2 long-chain fatty acids esterified to glycerol, as well as a phospho-diester bonding a choline, inositol or ethanolamine group. A phosphatide's fatty acid end is nonpolar and thereby lipophilic (or oil-like). Conversely, its phospho-diester of the previously mentioned substituents is zwitterionic (or bipolar) which explains the hydrophilic (or water-like) properties of this portion of the molecule (7).

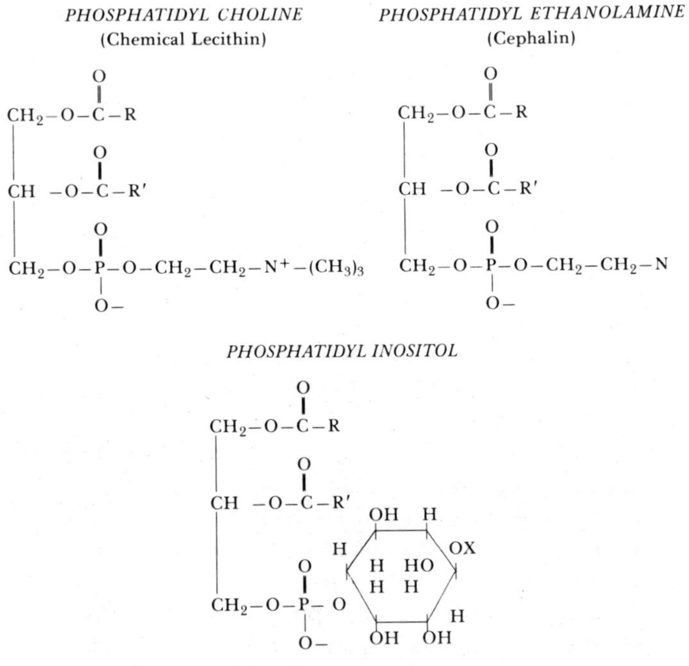


Fig. 8-1. Three principal components of soybean lecithin.

The combined hydrophilic and lipophilic properties of phospholipid molecules give them surface-active effects. Surface-active agents, or surfactants, are defined by the existence of both lipophilic and hydrophilic portions within one compound. This explains a surfactant's ability to orient itself at the interface between immiscible substances where both portions of the molecule are chemically satisfied. This creates a reduction in surface tension between the immiscible substances, allowing easier mixing.

All lecithin products are surfactants. As surfactants they can exhibit a variety of functions common to other surfactants. They also have some unique surfactant functionalities.

**Colloidal Dispersants**

A colloidal dispersant is a substance that aids in forming, or stabilizing, a system of either large molecules or small particles. Lecithin products function in the following types of foods which can be classified as colloidal dispersions.

*Emulsions.* Emulsions are very finely divided dispersions of one liquid within another immiscible liquid. Various lecithin products function as emulsifiers in foods, i.e., they break up one immiscible phase within another and allow intimate mixing.

Examples of food emulsions include milk, butter, margarine, puddings, chocolate, bakery items, cheeses, milk and cream replacers and ice cream.

Crude lecithins are excellent water-in-oil (dispersed phase is water, continuous phase is oil) emulsifiers. However, modified lecithins can function to emulsify either water-in-oil or oil-in-water emulsions, depending on the type of lecithin modification and the specific parameters of a system. These system parameters can include pH, types of components, component ratios, solids content and others. Unlike crude lecithins, hydroxylated lecithins are stable in acid systems (pH $\geq$ 3.5). Fractionated lecithins can be manufactured for specific emulsion types. Since lecithin emulsifying activity is dependent on its phosphatide ratios, changing the ratios can alter its emulsifying capabilities. When working with lecithins, one might readily conclude that lecithin products seem to emulsify on a very system-dependent basis. This is due to amphoteric (+,−) polar ends on the individual phosphatides. This makes it impossible to assign a general emulsifier rating (such as HLB) to lecithin products.

Most processed food emulsions are not stabilized by emulsification alone. The particle size of the dispersed phase has to be much smaller for dynamic stability than is practically possible with foods. Emulsifier/stabilizer systems, however, are normally used to make stable food emulsions. Lecithin will break up (emulsify) the particles, while a stabilizer (water-soluble polymer, etc.) will hold the particles in a dispersed orientation giving a stable emulsion.

*Solubilizations.* Solubilization is often used as another term for the production of stable microemulsions using surfactants. Microemulsions are stable simply by having dispersed phases with very small particle sizes (i.e., less than 500 Å). Generally no stabilizers are required. Most lecithins can aid in the production of microemulsions, an example being water-soluble dyes in oil systems. Modified or fractionated lecithins might also produce oil in water microemulsions; however, the best microemulsion can be obtained with synergistic lecithin-surfactant mixtures.

*Solid particle dispersions (sols).* Sols are finely divided solids dispersed in a continuous liquid phase. Most lecithin products can function effectively in forming (or reforming) sols. Lecithin products are still some of the best, most effective, surfactants for dispersing sols. This seems to be due to lecithin's affinity for solids/liquid surface interfaces. Phosphatides seem particularly attracted to particles containing metals and metal salts. Examples of food sols include some liquid chocolates, instant drinks, some frosting mixes, pigmented foods and others. Nonfood applications include paints, inks and other pigmented coatings. Lecithin products can function in paints to facilitate pigment dispersion and redispersion, and also to regulate (generally reduce) viscosity. Upgraded lecithins are generally superior Sol dispersants for paints and coating.

*Foams.* Foams are dispersions of a gas in a liquid media. In most aqueous systems, de-oiled lecithin products function as good foaming agents. Examples include whipped toppings, ice creams, and many types of candies. Upgraded lecithin products have also been employed as effective defoaming agents in foams caused by powdered proteins in water. This is an excellent example of the system specificity of lecithin products.

### Wetting Agents

Wetting refers to the ability of a soluble particle to be readily dispersed or solubilized. Wetting agents can be identical to Sol dispersants, if the particles to be wetted are solid and remain so. More often, however, wetting agents are employed to control the rate of moisture uptake to a hydratable, or potentially soluble particle.

Lecithin products are effective wetting agents for a wide variety of powdered or granular products. *Instantizing* effects can be obtained by including the proper lecithin product in a food product's formula. Generally, a very hydrophilic powder (such as a pregellatinized starch) will tend to hydrate so quickly that it will simply "ball up" on the surface of water. A very lipophilic lecithin product will work best if coated onto the powder, thereby protecting the powder from too rapid hydration, yet still hydrating. Thus, lipophilic lecithin facilitates wetting. One lecithin company produces a special lecithin product mixed with monoglycerides for this application.

If a more lipophilic powder needs instantizing, a hydrophilic lecithin is recommended. Acetylated monoglycerides are more easily wetted by first coating with hydroxylated lecithin. Examples of other foods that can be instantized with lecithins include cocoa powders, instant drink mixes, powdered coffee whiteners, milk replacers, cake mixes, powdered instant puddings, and instant soups and sauce mixes. Several manufacturing techniques are employed to produce instant products. Specific lecithin products that are compatible with each of these manufacturing techniques are commercially available.[26] Dry food powders can be spray coated with highly fluid lecithin products. Additionally, powders can be co-spray dried with more hydrophilic lecithins such as oil-free forms, or hydroxylated products. Lecithin products are also available that will assist the wetting of agglomerated powders. Agglomeration is the primary instantizing operation, lecithin simply "fine tunes" instantizing. Other lecithin wetting agents are applicable to more product-specific manufacturing techniques.

### Release Agents

Cooking surfaces often require lubrication in order to keep foods from sticking. Food processors have traditionally used fats for this purpose. However, lecithin products have been found to function more effectively than animal or vegetable fats. It is generally assumed that lecithin functions by modifying the boundary between the fat, or oil, and the solid release surface. Therefore, oil must be included with phosphatides in order to be effective release agents. Phosphatides will not function alone, however, phosphatides with oil in water dispersions also effect release. Hydroxylated lecithins are preferred for these aqueous dispersions.

Very low viscosity lecithin products are used as food processing release agents in frozen waffle manufacture, processed cheese manufacture (slice release), and vegetable dehydration processes. These are easily sprayable onto release surfaces. Lecithin products are included in almost all griddle frying fats, as well as bakery depanning agents. Food processors spray low viscosity lecithin products onto machinery including the product contact surfaces of oven conveyers, flame broiling lines, and other equipment to reduce product sticking and breakup, and to facilitate clean-up. Heat resistant, low viscosity lecithin products have superior properties for many of these applications.

### Crystallization Control

Lecithin products can control crystallization in various food systems. In foods containing sugars or fats, the presence of as little as 0.5% lecithin produces altered crystal sizes or structures that can have positive effects on product texture and viscosity. This is important in cookie fillings, butter containing maple syrups, ice cream toppings, and similar products.

**TABLE 8-IV.**

| Application | Function(s) |
|---|---|
| Food Processing | |
| Ingredient in: | |
| Margarine | Emulsifier, antispattering agent |
| Confections and snack foods | |
| Chocolate | Crystallization control, viscosity control, antisticking |
| Caramels | |
| Coatings | |
| Instant foods | |
| Cocoa powders | Wetting and dispersing agent, emulsifier |
| Instant drinks | |
| Instant cocoa | |
| Instant coffee | |
| Protein drinks | |
| Dietetic drinks | |
| Coffee whiteners | |
| Milk replacers | |
| Cake mixes | |
| Puddings | |
| Instant toppings | |
| Commercial bakery products | |
| Breads | Crystallization control, emulsifier, wetting agent, release agent (internal & external) |
| Rolls | |
| Donuts | |
| Cookies | |
| Cakes | |
| Pasta products | |
| Pies | |
| Cheese products | |
| Pasteurized processed cheese and cheese food | Emulsifier, release agent |
| Imitation cheese | |
| Meat and poultry processing | |
| Meat and poultry glazes and basting compounds | Browning agent, phosphate disperant |
| Pet foods | Dietary supplement, release agent, emulsifier |
| Bacon | |
| Dairy and imitation dairy products | |
| Infant, milk formulas | Emulsifier, wetting and dispersing agent, antispattering agent, release agent |
| Milk and cream replacers | |
| Egg replacers | |
| Imitation eggs | |
| Whipped toppings | |
| Ice cream | |

**TABLE 8-IV (Continued).**

| Application | Function(s) |
|---|---|
| Dairy products (continued) | |
| Flavored milks | |
| Flavored butters (garlic, etc.) | |
| Basting butters | |
| Miscellaneous products | |
| Peanut spreads | Crystallization control, emulsifier |
| Salad products | |
| Flavor and color solubilization | |
| Packaging aid | |
| Polymer package—interior coating | Release agent, sealant |
| can interior coating | |
| sausage casing coating | |
| stocking net | |
| Processing equipment | |
| Frying Surfaces | Internal (in product) and/or external release agent, lubricant |
| Extruders | |
| Conveyors | |
| Broilers | |
| Dryers | |
| Blenders | |
| Evaporators | |

### Applications

General lecithin application areas include confections and snack foods, instant foods, commercial bakery products, dairy and imitation dairy products, and meat and poultry processing. The various companies marketing lecithin products often have specific upgraded products available for each of these applications.

Use levels for most lecithin products are generally 0.2–1.0%, based on finished product weight. If no other information is available, a general "rule of thumb" is to start with 0.5% lecithin on a finished product basis. These use levels are generally consistent with those of chemical surfactants.

## Specific Food Applications (Table IV)

The most widespread uses of crude lecithin products are in confections and margarine. Crude lecithin has some interesting functions in both of these applications. However, this author shall not dwell on these more traditional and widely understood applications.

Many bakery applications have traditionally used crude lecithins. However, the only use in baking that is relatively recent, is lecithin's function as a release agent. Lecithin products that are highly fluidized with oils and/or fatty acids have found widespread applicability as release agents, due to their sprayability, surface spreadability, handleability, and improved release functionality. The improved properties of heat resistant lecithin products are just beginning to be recognized by companies producing baking release agents.

Lecithin release agents are employed industrially in applications such as frozen waffle manufacture; bread, roll, and cookie baking; pizza baking, pie baking; and in pasta products. Most institutional griddle frying fats are formulated with lecithin, solely for its release functionalities. In retail markets, all spray pan release products are based on lecithin. Included here are retail baking pan releases containing flour for effective cake depanning, as well as the more traditional frying releases.

Lecithin has been used extensively as a wetting agent and emulsifier in instant foods. Foods including cocoa-powders, instant drinks, instant cocoa, instant flavored coffees, powdered protein drinks, coffee whiteners, instant puddings, cake mixes, and instant toppings have widely employed application specific lecithins. A multitude of different lecithin products are used in these applications.

The most widespread method to incorporate lecithin here, is as an external coating on the powder particles. The particular type of lecithin to be employed largely depends on the hydrophilicity, or lipophilicity of the powder system. In most cases, a more lipophilic, or oil-like powder requires a more hydrophilic lecithin, and vice-versa.

In recent years, instant food manufacturers have become increasingly interested in oil-free lecithins. These are powdered or granular, but can be fluidized by addition to water or fats. Additionally, fluid products are now available from some lecithin manufacturers, that are *based on* oil-free lecithin. Some of these are very highly fluid (a few hundred centipoise) and are the most bland products available today. These have obvious advantages for instant food manufacturers.

Another major application for lecithin products, is in dairy and imitation dairy products. Almost all infant formulas contain either hydrophilic or oil-free lecithins as fat emulsifiers. Egg replacers employ oil-free lecithin as a very bland emulsifier, while whipped toppings utilize hydrophilic lecithins as emulsifiers, and sometimes to assist foaming.

Lecithin products have recently proven themselves in several new application areas. Among these are the application of highly fluid lecithin to promote slice separation in high moisture processed cheeses. Specialty fluid lecithin products are sprayed, or wiped onto sheets of processed cheese prior to slicing and stacking. A second newer application employs oil-free lecithin as a key ingredient to emulsify and assist in immobilizing animal fat in canned, or frozen meat containing products. When properly formulated, this dramatically reduces, or eliminates "fat-capping" in products such as canned chili, sloppy joes, gravies, and others containing high levels of animal fat.[27]

There are, of course, numerous other applications for lecithin products in the food industry. It is important to realize that lecithin is one of today's most versatile and useful food ingredients. The future of lecithin products is assured by the development of new, highly functional lecithin products targeted at specific applications.

## References

1. Fourcray, Ann. Chim. 16:282 (1793).
2. Vanquelin, Ann. Chim. 81:37 (1812).
3. Gobley, M. J., Pharm. (3), 8, 81, 161 (1846)–op. cit. 11, 409 (1847); 12, 1 (1847)-op. cit. 17, 401 (1850).
4. Loew, O., Biol. Zenk. 11:269 (1891).
5. Stanley, J., Soybean and Soybean Products, edited by K. S. Markley, Vol. II, Interscience Publishers, Inc., NY, p. 601.
6. Eichberg, J., Soybean Digest 2, No. 3, 3, 12 (1942).
7. Sullivan, D. R., and B. F. Szuhaj, AOCS National Meeting Paper No. 39 (1975).
8. Arent, A. G., Mfg. Confectioner 10:36 (1930)–op. cit., Rewald, B., pp. 47–83 (May 1930).
9. Reynolds, M. C., and A. K. Epstein, U.S. Patent 1,815,739 (1931).
10. Glabau, Baker's Weekly 68:58 (1930).
11. Anon., Siebel Tech. Rev., Vol. 5, No. 4, pp. 103, 131 (1931).
12. Mechoels, O., Melliand Textilber 12:123 (1931), Rev. Chem. Abstr. 26:3673 (1932).
13. Rewald, B., U.S. Patent 1,938,854 (1933).
14. Bousquet, E. W., U.S. Patent 2,006,227 (1935).

15. Morg, O., and F. Wagner, Farbe and Lack, p. 535, Rev. Chem. Abstr. 27:197 (1933).
16. Jesser, H., Chem. Zrg. 58:532 (1934).
17. Prosise, W. E., AOCS Short Course Presentation NY, Processing and Quality Control of Fats and Oils (1980).
18. Prosise, W. E., and F. Flider, IFT Basic Symposium Presentation, St. Louis, MO (1979).
19. Link, W. E., Official and Tentative Methods of the American Oil Chemists' Society, 3rd edn., 1977.
20. Food Chemicals Codex, National Academy of Sciences, Washington, DC, 3rd edn., 1981, pp. 166–167 and 147–148.
21. Year Book and Trading Rules 1982–83, National Soybean Processors Association, Washington, DC, p. 79 (1982).
22. Anon., Lecithin Release Agents, Brochure, Central Soya Company, Inc., Fort Wayne, IN, 1980.
23. Code of Federal Regs., 21, 172.814, General Services Administration, Washington, DC, 1981.
24. Anon., Soluthin 100 Technical Data Sheet No. 120, Byron Chemical Company, Long Island City, NY, 1982.
25. Anon., Instantizing with Metarine, Brochure, Lucas Meyer Company, Hamburg, Germany, 1982.
26. Anon., Lecithin Instantizing Agents, Brochure, Central Soya Company, Inc., Fort Wayne, IN, 1979.
27. Anon., The Solution to Fat Problems is The System, Brochure, Central Soya Company, Inc., Fort Wayne, IN, 1980.

# Chapter Nine

## Nonfood Uses of Lecithin

**J. C. Schmidt and F. T. Orthoefer,**
A. E. Staley Manufacturing Company,
Decatur, IL 62525

### Introduction

Lecithin has widely accepted use as a multifunctional additive in a variety of foods (1). However, many current food applications (e.g. baking, confectionary and margarine) are in mature market areas. The most significant growth potential for lecithin appears to be among the large domain of other products grouped together as "nonfoods".

There is no industry-wide reporting of lecithin sales and so the exact percentage breakdown of uses is difficult to determine. This is complicated by the fact that as much as 30–40% of the domestic market involves hundreds of small volume users who are serviced by distributors. Additionally, some customers may have both food and nonfood applications for the lecithin. It may be used as an emulsifier in candy, for example, and may also be used as a conveyor-belt coating for its release properties. For the most part, the greatest percentage of lecithin is used in foods. The second highest use is in animal feeds. It seems likely that no more than ca. 30% of the more than 80 million pounds of domestic lecithin currently produced involves nonfood and nonfeed uses. Paints and other coatings consume a major portion of the nonfood lecithin. Pharmaceuticals, including diet supplements, and inks and cosmetics also consume significant portions. All other nonfood uses currently consume only small volumes of lecithin.

Much patent activity involving lecithin occurred from the late 1920s through the 1940s. These early developments provided the foundation for today's markets. A number of review articles provide a perspective of this early growth (2–4). More recent reviews and papers on lecithin provide updated information (5–12). A monograph on the biological functions of phospholipids is also available (13).

There are literally thousands of citations yearly in the chemical literature involving lecithin or phospholipids in general. These citations again include many patents. From January 1967 to October 1983, there were over 800 foreign and domestic patents on lecithin (including food patents) listed by Chemical Abstracts. The nonfood patents among these have been the major focal point for new material for this review. Animal feed uses are also included. Successful commercialization of some of the cited patents, particularly in agricultural, pharmaceutical, or cosmetic areas, may be subject to USDA or FDA approval. Some of the patented ideas and formulations may never be successfully commercialized; however, such cases may perhaps serve as seeds for successful future ideas.

We begin with a brief overview of the functions of lecithin that make it useful as an additive in nonfoods and feeds. The applications are then discussed and some prognostications are made on future growth areas based on patent activity.

## Lecithin Functionality

Lecithin has a broad range of functions which make it amenable to many different formulations. It is surface active and can modify the boundary layers between many types of substances. Between solid phases lecithin can function as a lubricant and slip or release agent. Between solid and liquid phases lecithin can act as a wetting and dispersing agent. By coating the surface of solids, lecithin reduces surface tension and reduces the attractive forces between the particles. Thus, the solid is more easily dispersed and suspended in a liquid phase. Between immiscible liquid phases lecithin can again reduce the surface tension and act as an emulsifying agent. Because of its HLB of ca. 4 (14), lecithin functions primarily as a water-in-oil emulsifier (as in margarine). It can also function as an oil-in-water emulsifier however (as in salad dressing) and the oil-in-water emulsifying properties can be improved by using additives with the lecithin or by modifying it chemically.

On the molecular level lecithin can form complexes with a variety of inorganic salts and with organic substances such as carbohydrates and proteins. The lecithin modifies the solubility and crystallization behavior of these substances. Small amounts of inorganic salts may be dispersed in organic media using lecithin, for example. Tocopherols present in lecithin contribute antioxidant properties.

The functions that have been cited for lecithin in nonfood uses are summarized in Table I. Some of the cited functions appear to be closely related. For example, addition of lecithin to a mixture of solids may "condition" the mixture, making it grind or machine easier, thus making the lecithin a "machining aid" or "grinding aid." These functionalities may occur because the lecithin coats and lubricates the particles. Some of the functions appear to be opposites, such as "adhesion aid" or "coupling agent" versus "antistick agent". These properties are consistent with the polar and nonpolar chemistry of phosphatides, however, and the predominant property will depend on the nature of the phases with which the lecithin interacts.

**TABLE 9-I.**
**Miscellaneous Functions of Lecithin**

a. Adhesion aid
b. Antibleed agent (as in fat bloom)
c. Anticorrosive
d. Antidusting agent
e. Antioxidant
f. Antispatter agent
g. Biodegradable additive
h. Catalyst
i. Color intensifier
j. Conditioning agent
k. Coupling agent
l. Dispersing agent, mixing aid
m. Emollient, softening agent
n. Emulsifier or surfactant
o. Flocculant
p. Grinding aid
q. Lubricant
r. Liposomal encapsulating agent
s. Machining aid
t. Modifier
u. Moisturizer
v. Nutritional supplement, vitamin source
w. Penetrating agent
x. Photosensitive agent
y. Plasticizer
z. Promoter
aa. Release agent, antistick agent
bb. Spreading aid
cc. Stabilizer
dd. Strengthening agent
ee. Suspending agent
ff. Synergist
gg. Viscosity modifier
hh. Water repellant
ii. Wetting agent
jj. Biologically active agent

## Nonfood Applications

Nonfood and feed applications for lecithin are discussed below. Many similar patents may be filed in a given application area. Generally, only representative references, usually the most recent, will be cited as examples. Commercial soybean lecithin is generally used in these applications. In some cases, especially where indicated, use of modified lecithin may be desirable to achieve enhanced functionality.

The applications are summarized in Table II along with functions from Table I that are applicable. Levels of use are in the range 0.05 to 1.5% or as indicated in parenthesis.

### Adhesives

It is interesting that lecithin, which often functions as a release agent, can also be useful in adhesives. It may function as a plasticizer in some of these formulations and as a dispersing aid. Tape (15) and ceiling tile (16) are two applications. An adhesive layer for attaching magnetic tape coatings can also contain lecithin (17).

Lecithin is especially useful in adhesive formulations for laminated materials. For example, it can be formulated with phthalate esters and a copolymer to make an adhesive for laminated paper useful for high speed passage through computer equipment (18).

### Adsorbents and Flocculants

Aluminum oxide coated with phosphatidylcholine will adsorb petroleum from seawater and the mixture will sink without releasing the oil (19). Lecithin has also been patented as a stationary phase adsorbed on polyester or nylon fibers for the chromatographic separation of granulocytes (20).

Lecithin used in conjunction with mineral spirits can serve as a flocculant for the removal of solid residues of coal and coke from acidified aqueous slurries (21).

### Agricultural and Agriproduct Processing Uses

The principal agricultural use of lecithin is in animal feeds for calves, cattle, pigs, poultry, fur-bearing animals, fish and pets. A portion of lecithin not isolated for commercial use is added to animal feeds. Additionally, perhaps 10–15% of commercial lecithin is sold for feed.

**TABLE 9-II.**
**Nonfood Applications for Lecithin, Levels of Use and Cited Functions**

| Application | Functions |
|---|---|
| Adhesives[a] | l, y |
| Adsorbents | a, k, o |
| Animal feeds (1–3%) | d, e, g, l, n, s, v, ii, aa |
| Catalysts | h, n, t, ii |
| Ceramics and glass | l, aa, hh |
| Cosmetics and soaps (1–5%) | e, l, j, m, n, r, u, v, w, cc, ii |
| Detergents | c, n |
| Dust control | a, d |
| Dyes (0.5–2%) | l, k |
| Explosives | d, n, cc |
| Fertilizers | d, j, bb |
| Inks (0.5–3%) | i, l, n, p, cc, ee, ii |
| Leather (1–2% of oil) | j, m, n, q, w |
| Magnetic tapes (0.5–1.5%) | e, l, n, q, ii |
| Masonry and asphalt products | c, l, n, y, aa, dd, ii |
| Metal processing | c, f, o, q, aa |
| Paints and other coatings (0.5–5% of pigment) | e, i, l, n, p, z, bb, gg, cc, ee, ii |
| Paper | l, m |
| Pesticides (0.1–5%) | a, e, g, l, n, w, bb, cc, ff, jj |
| Petroleum and other fuel products (0.005–2%) | a, c, e, l, n, q, cc |
| Pharmaceuticals | e, n, r, s, v, cc, ii, jj |
| Polymers, including rubber (0.5–1–5%) | b, e, n, l, t, y, aa, cc, dd |
| Printing, photocopying and photography | i, l, p, x, ii |
| Release agents (2–10%) | n, q, aa |
| Textiles (0.2–0.5%) | d, j, l, m, n, q, aa, hh, ii |
| Waste treatment | a, l, o |

[a]The usage level is 0.05–1.5% or as indicated in parenthesis.

Lecithin serves principally to emulsify fats in the feeds, improving their digestion and thus increasing weight-gain efficiency. This is especially important in calf milk replacer formulations and fish foods. In fish, liver dysfunction can result from fat-supplemented diets that are too low in emulsifiers (22). Some compounded and chemically modified lecithins function better as fat emulsifiers than does the crude commercial lecithin. This is due to enhancement of the oil-in-water emulsifying function of the lecithin. Lecithin also serves as an appetite stimulant, antioxidant and vitamin and mineral source in these feeds and promotes healthy coats in the animals. Recent patent activity in the feeds areas includes bloat control (23) and particulate lipid-protein (24) formulations.

A combination of lecithin, a food grade acid, and water is recommended as a preconditioning spray in the grain-rolling, feed-production industry (25). The spray can be applied to the grain to allow rapid moisture penetration. This softens the grain, reduces roller resistance and results in production increases of 25–50% per hour according to the patent. Incorporating lecithin into feeds may also reduce dust that can lead to losses and to explosion hazards (26).

Lecithin can be used to inhibit the hardening of hygroscopic salts such as potassium fertilizers (27). It has been recommended as a nutrient for mushroom growth (28) and as an additive for the growth medium of predaceous insects (29).

Lecithin can be used as an additive in agricultural pesticide formulations. It can function as an emulsifier and improve adhesion and penetration of insecticides, according to Japanese workers (30). These same researchers report that lecithin can combine with the fatty substances of vegetable leaves forming a coating which aids in the rejection of plant viruses. Lecithin may improve plant tolerance to certain insecticides. In one study the phytotoxic effect of a pesticide formulation containing oil was eliminated using lecithin as an additive (31). Sodium bicarbonate containing 0.1% lecithin inhibited citrus common green mold, cucumber powdery mildew and rice blast better than did the bicarbonate alone (32).

Phospholipids formulated with alcohols, glycols and/or glycol ethers are reported to have antibacterial effects in agrichemical applications (33). Lecithin is a component of coating compositions recommended for a variety of fruits and vegetables (34). Postharvest treatment of apples with aqueous emulsions containing lecithin and various fats and oils can reduce the incidence of soft scald that occurs during cold (−1 C) storage (35). Fungicidal preparations reported to be useful for stored oranges contain organic acids and lecithin (36). Lecithin is exempt from minimal residue requirements when used as an emulsifier ingredient of coatings or pesticide formulations for growing crops and commodities after harvest (37).

### Antioxidants and Corrosion Inhibitors

Lecithin functions as an antioxidant in a number of vitamin and other pharmaceutical formulations and in cosmetics. The phosphatides themselves may scavenge oxygen but lecithin also contains tocopherols, which are effective antioxidants. The combination of these substances can have a synergist effect. For example, in a vitamin formulation, it was shown that ascorbic acid, alpha-tocopherol, a selenium salt and lecithin all acted synergistically to improve the oxidative stability of the product (38). The combination of lecithin and alpha-tocopherol will also inhibit degradation of vegetable oils (39). Synergistic effects are also reported between phosphatides and the antioxidants quercetin (40) and erythorate (41). Powdered forms of lecithin have been shown to have antioxidant properties (42).

Corrosion is a form of oxidation. Soybean lecithin can be added to turbine or spindle oils to form an anticorrosive lubricant (43). Oil-free lecithin as well as certain derivatized and hydrolyzed lecithins were found to be useful as coatings to inhibit the rusting of steel in salt spray tests (44).

## Catalyst Applications

It is not unreasonable to presume that lecithin could have some catalytic or cocatalytic effect in multiphase systems, due to its surface active properties if only via its action as an emulsifier. Phosphatides occur naturally in rubber latex and are known to function as accelerators in vulcanization (2,3). Lecithin is reported to function as an initiator in the production of poly(oxymethylene) (45). No studies on the general applicability of lecithin as a surfactant in emulsion polymerization were found. Small amounts of lecithin, such as could occur in soaps used as emulsifiers, were found to have no inhibitory effects on polymer formation in the standard GR–S styrene-butadiene polymerization system (46). Linolenate and linoleate soaps were found to have a retarding effect on the rate of polymerization, however. Lecithin is reported to be useful as an emulsifier in the curing of aqueous dispersions of unsaturated polyesters (47). The products are more easily removed from their molds and have improved mechanical properties when the lecithin is used.

Phosphatidylcholine and its lyso derivative are reported to exhibit a cocatalytic effect in the reduction of $N_2$ to $N_2H_4$ and $NH_3$ using a Mo(III) -Mg(II) catalyst (48). Yields of both nitrogen products were markedly higher using the cocatalyst system. Other phosphatides had no enhancing effect. In a fermentation application, 1.5% soybean lecithin acts as an inducer in the preparation of cholesterol esterase using a strain of Pseudomonas bacteria (49).

Aside from its role as a catalyst, initiator or modifier, lecithin may have ancillary use in catalyst systems as part of a protective coating. For example, lecithin can be used as part of a coating composition to make Raney nickel nonpyrophoric (50).

### Ceramics and Glass

In the manufacture of ceramics and glass, lecithin is reported to be useful as a release agent and for dispersion of pigments (9). A dilute aqueous dispersion of lecithin applied to glass will make the glass hydrophobic (51).

### Coatings

As previously indicated, coatings make up a major portion of the lecithin nonfoods market. These include paints, waxes and polishes, and wood preservatives. Lecithin is also widely used as a release agent and applications are currently being found in magnetic tape coatings. These latter two areas are covered separately below.

Lecithin serves as an emulsifier and wetting, dispersing and stabilizing agent in both oil- and latex-based paints. It shortens mixing time, aids brushing and increases the covering power of the paint (52). Lecithin coats the particles of metal oxide pigments rendering them readily dispersible. The minimum lecithin required to coat red iron oxide pigments is reported to be 0.01%, based on pigment weight (53), although 5% or more by weight may be used normally.

If more than one pigment is used, the lecithin coating helps maintain a uniform mixture (54). Various formulations and special lecithin blends have been patented for improved functionality in specific areas. Examples of these applications include an aerosol scratch filler and primer (55), multicolored coatings (56), an antifouling composition (57), thick paint coatings (58), magnetic paint (59) and anti-pickup paint for steel foundry molds and cores (60).

Metal salts of lecithin have been patented as paint additives (61). These salts are reported to improve pigment dispersion, act as drying promoters and inhibit yellowing of the product. The same patent promotes the use of phosphorylated lecithin as a rust inhibitor in primer paints.

An interesting coatings application for lecithin is in producing nonpyrophoric Raney nickel (50). The lecithin creates a lipophilic surface on the metal facilitated with a wax, fat or organic polymer.

Polyamide coatings containing lecithin are recommended as a finish coat in nylon leather substitutes (62). Lecithin is also used for its release and sealant properties in polymer coatings for food containers including can interior coatings, sausage casing coatings and stocking nets for hams and other meats (9). It can be used as a foam stabilizer in foam fire extinguishing agents (63).

**Cosmetics and Soaps**

Phosphatides occur naturally in skin and in other biological membranes (13, 22). Lecithin, then, is a natural choice for cosmetic formulations. These preparations include skin creams and lotions, foundations and cleansing creams, sunscreens, soaps, bath oils, shampoos and hair conditioners, shaving creams, preshave and after-shave lotions, nail enamels, face powder, eye color creams, lipstick and hair sprays. In these formulations, lecithin serves one or more of the following functions: emulsifier, wetting agent, foam stabilizer, antioxidant, conditioning and softening agent, moisturizer, penetrating agent, emollient, vitamin source and pigment dispersant.

Examples of recent patent activity include a formulation for face powder containing lecithin and an organic acid which is reported to be easy to apply and gives a water-repellent coating free of glossiness (64). A cosmetic cream for hair and skin is composed of guar extract, lecithin and an oily plant extract (65). Skin and nail compositions containing a phosphatide-trialkanolamine complex have been reported (66). Foamy permanent wave compositions containing lecithin have been patented (67). Even perfume compositions may contain small amounts of lecithin (68).

A number of the patents involving cosmetics concern the nature of the emulsions used in the formulations. Stable water-in-oil emulsions with a high water content (75%) can be made using lecithin as emulsifier by tempering the emulsion (69) or using additives such as fatty acid derivatives of amino acids (70). Polyalcohols such as sucrose monooleate can be used along with lecithin to prepare emulsions (71).

Some of the cosmetic patents involve liposomes. These multicompartment membranes have been used to encapsulate vitamins to be incorporated into skin whitening lotions or creams, for example (72). Liposomes are discussed separately below.

The relative high market value of cosmetics make them a fertile area for use of modified and specially refined lecithins. These modified forms often impart added functionality not available from the lower priced commercial grades. Liposomes made of pure phosphatidylcholine were more effective than those made of soybean or egg lecithin in softening and restoring dry, aged skin, for example (73). The ethanol-soluble extract of granular lecithin is recommended as a spray for conditioning hair (74). The spray eliminated electrostatic charge, made the hair easy to comb and held curls at high humidity. In nail enamels, settling of suspended materials and streaking effects are prevented by adding lecithin, particularly hydroxylated lecithin, and a cosmetically acceptable acid (75).

### Detergents

A patent assigned to a major detergent manufacturer describes the use of soybean lecithin to improve the detergency of anionic detergent compounds in the cleaning of dishes containing dried or baked soils (76). Compositions for cleaning aluminum and zinc and aluminum-zinc alloy surfaces are improved by adding lecithin and lignin to the mixture (77). The cleaning effectiveness is increased and the surface corrosion is decreased. Lecithin is a component of detergent compositions in halogenated hydrocarbon solvents for washing precision-detailed items (78). Lecithin has been added to the solvents and surfactants of a dry cleaning formulation to increase the cleaning power (79). Japanese workers describe the use of soybean lecithin to improve fish cleaning compositions containing alkali metal salts and soap (80). Fractionated lecithins improve the detergency of N-acylglutamate salts (81). Phosphatidylcholine is used as a solubilizing agent for low molecular weight basic proteins in an electrophoretic analysis technique (82).

### Dust Control

Lecithin is reported to control cotton dust in textile manufacturing (6). It can be used for aluminum dust explosion prevention in bomb manufacturing by coating the aluminum particles with compositions containing wax, lecithin and nitrocellulose (83). Dustless agrochemical powders have been formulated using inorganic carrier powders, petroleum resins and lecithin (84).

### Explosives

Lecithin can be used as an emulsifer and stabilizer in explosive formulations involving ammonium and sodium nitrate (85), hydrazine mononitrate (86) and aluminum with TNT and other organic explosives (87).

### Leather

Leather tanning is a collagen protein purification and stabilization process that initially involves pH extremes as the animal hide is first limed to loosen hair and swell the collagen and then pickled with sulfuric acid and treated with chromium sulfate (88). In the next stage of the process, called fat liquoring, the hide is treated with an oil-in-water emulsion to reduce the amount of fiber cohesion during the drying process and provide lubrication. It is in this fat liquoring stage that lecithin, particularly hydroxylated lecithin, finds use as an emulsifer and penetrating agent (10).

### Liposomes

Liposomes are closed lipid bilayer membranes (vesicles) which are spherical in shape and usually 25 nanometers to 2 microns in diameter (89, 90). The membrane may enclose a single aqueous compartment or may be in concentric layers resembling tree rings with alternating aqueous layers. Liposomes are formed by allowing phospholipids to swell in aqueous solution. Alternately, methanol, ethanol or ether solutions of phosphatides may be injected into Tris- and phosphate-buffered saline (91). Size can be controlled by various means including sonication (92) and extrusion through polycarbonate membranes (93). Giant unilamellar liposomes up to 100 microns in diameter can be prepared under certain conditions in the presence of methyl glucoside (94). The aqueous compartments between the bilayers can be used to entrap therapeutic molecules. The permeability of the liposome can be reduced by incorporating cholesterol into the bilayers or by modifying the lipid composition (by using dipalmitoyl phosphatidylcholine, for example). Liposomes are being studied as delivery and sustained release vehicles for drugs, vaccines, enzymes and hormones to be used orally or intravenously.

Some progress with pharmaceuticals has been noted. A decrease in blood glucose was found in diabetic rats fed liposomes containing insulin, for example (95). This could lead to oral forms of insulin for man. Liposomes have been found to enhance antibody production against entrapped protein (96) which could make them useful antiviral agents. A water-insoluble drug has been transported to leukemia cells in vitro using liposomes (97). Liposomes which incorporate antitumor agents have been patented (98, 99) and patents have been issued on a number of other therapeutic and diagnostic uses (100, 101). The incorporation of the cell moderater cAMP into liposomes has been patented (102). Still needed are liposomes that will transport drugs more selectively to target organs. Surface modification of the liposomes should achieve this effect (89).

Liposomes can also be used to encapsulate cosmetics and foods (103) and chelating agents (104). Magnetically-localizable liposomes have been prepared (105).

### Magnetic Tape

Magnetic tape coatings are a relatively new area for lecithin use and much recent patent activity has been observed, especially among Japanese workers.

Some of these tapes are reported to be suitable for video applications (106). Lecithin functions as a dispersing agent, lubricant, wetting agent and antioxidant in these formulations. Deoiled, granular products with low hexane insolubles are being promoted for this application.

A typical coating contains iron oxides, sometimes doped with cobalt or chromium; a polymer system which acts as a binder; various additives, including up to 1.5% lecithin (dsb) and solvents such as methyl ethyl ketone and cyclohexanone (107–111). This material is milled into a dispersion and deposited onto a tape and dried.

Lecithin can also be used in a tape coating containing abrasive cleaning compounds for cleaning tape recorder heads (112). A magnetooptical imaging film can be made by coating a support with a photosensitive resin containing magnetic material (113). The coating is prepared using lecithin and the film is especially useful for chinese character pattern memory.

### Masonry and Asphalt Products

Lecithin is used as an emulsifier, wetting and dispersing agent, plasticizer and/or release agent in concrete, asphalt and tar shingling, linoleum tiles and surface sealants and caulking compounds (2, 3, 6, 9).

The compressive strength of concrete is reported to be increased by 32–115% without causing significant reduction in handling properties by adding 0.1–3.0% lecithin or a mixture of lecithin with glycerol or other similar additives (114). A rust-preventive release oil for steel concrete forms contains naphthenic base mineral oil, dimerized linoleic acid and 4–8% lecithin (115).

### Metal Processing

Lecithin is reported to function as a lubricant, anticorrosive, flushing aid, and/or release agent in a variety of metal processing applications (9). These include aluminum sheet and foil rolling, wire drawing and casting, cutting, welding (spatter release) and polishing areas. Lecithin is also used in a polyurethane foam sealant for the rolled sheet metal joints of barrels, drums and cans (116).

Low levels of lecithin (0.05–0.1%) are reported to improve the properties of water glass-containing coatings for foundry cores and molds (60). Metal ions are reported to be extractable from acid solution with a solution of phosphatides (117). Flotation of the tin oxide mineral cassiterite using lecithin has been reported (118).

## Paper, Printing, Photocopying and Photography

Lecithin is reported to be useful as a dispersing and softening agent in paper manufacturing (9). Partially saponified lecithin has been recommended for this use (119). Lecithin has functionality in printing inks similar to those mentioned for paints. It also increases the intensity of colors and serves as a grinding and remixing aid. Some patented products include multidetectable ink compositions (120) and magnetic ink (121). Lecithin is also useful in formulations for removing offset and printing inks (122).

The most patent activity in this area recently has involved toner formulations for photocopying (more properly called electrophotography or transfer xerography). This process involves electrostatic imaging to ultimately transfer an image to a plain piece of paper (123). Powdered toners are used in the transfer of the image from an electrostatically charged drum to the paper. There are a number of possible variations in this process especially in the transfer step. These variations have resulted in recent patents involving a variety of toner developer formulations including mixed toners (124) encapsulated dry toners (125) magnetic toners (126) and liquid formulations of toners (127). Carbon black or iron oxide are the toners dispersed by lecithin in these particular formulations. Copiers that print in color, make hard copy from computer display terminals, and print microfiche are currently being perfected. The latter application requires very high resolution toners.

Oil-free phosphatides can be used as photosensitive agents for making positive or negative photographic images for lithographic prints, printed circuits or similar uses (128, 129). An electrosensitive film useful for duplicating designs uses lecithin (130). Lecithin can be used as a dispersing aid for oil-soluble photographic additives (131).

## Pesticides

Lecithin is reported to be an active agent in mosquito control (10, 132). The exact mechanism of action is unknown but the lecithin forms stable films on water surfaces which may interfere with the breathing mechanism of the mosquito puppae.

Lecithin can be used as an emulsifier for various pesticide formulations. Machine oil and phospholipid mixtures have been patented as fungicides (133). Lecithin functions to improve adhesion and penetration of the pesticide and may improve plant tolerances to these substances. Lecithin-containing formulations are active against mites (134) and against worm infestations on apples (135). Other pesticide applications are discussed above under Agricultural Uses.

### Petroleum and Fuel Products and Lubricants

Lecithin can serve as an antioxidant, anticorrosive, antiwear and detergent in petroleum applications. These include gasoline, fuel oils, greases, lubricating oils and cutting oils. A very small amount of lecithin in gasoline, 1–10 lbs per 1000 barrels, inhibits gumming and corrosion due to moisture and also stabilizes tetraethyllead if present (3,6). Many of the patents on these applications date back to the early 1940s but some of the newer formulations also include lecithin. These again include fuel additives with lecithin compounded with imidazolines (136) and with various other additives (137, 138). Combinations of various metals or metal salts, such as magnesium and barium salts, with lecithin, are reported to be useful as fuel oil additives for the reduction of $NO_x$ emissions by up to 25% (139, 140). Lecithin can also be used as part of an emulsifier formulation for dispersing water in fuel oil. This fuel mixture burned with 31% less $NO_x$ emissions than the fuel oil alone (141). Lecithin is also used in lubricating oils (142–144), as a water-in-oil emulsifier for hydraulic fluids (145) and in drilling muds (6). Lecithin is a component of an aqueous siloxane lubricant used in tire shaping (146). Lithium stearate grease is reported to be more heat stable when it contains a small amount of lecithin when used as a spray-on lubricant for non-ferrous metals (147).

Stable dispersions of powdered coal and fuel oil can be prepared using lecithin as a dispersing agent (148). Lecithin can also be used as a binder for bituminous coal (149). It functions as a dispersant in solid propellants containing beryllium hydride (150).

### Pharmaceuticals

Phospholipids are components of biological membranes in all living systems. They play important roles in biological processes such as cell permeability, the regulation of membrane-bound enzymes, and lipid and cholesterol transport (6, 13). The neurological importance of phospholipids is emphasized by the phospholipid composition of dry brain tissue which is about 25% (151). Because of this biological activity, it is no surprise that a great variety of applications for lecithin occurs in the pharmaceutical field.

Phospholipids, or specific phosphatides such as phosphatidylcholine (PC), are being investigated as active pharmaceutical agents. Studies have been underway for sometime to determine how phospholipids might be useful in cardiovascular disease, liver disorders and brain disorders such as Huntington's chorea, spinocerebellar hereditary ataxia, late dyskinesia and Alzheimer's dementia.

A recent review was prepared for the FDA on the effects of consumption of lecithin and choline on cardiovascular and neurological systems (152). There are conflicting results in the literature. Administration of PC to human subjects lowered serum cholesterol only in some of the studies. Some limited success has been observed in the amelioration of symptoms of tardive dyskinesia and Alzheimer's disease. The authors of the review conclude that further clinical trials are worthwhile but will require sufficient amounts of 90–95% purity PC.

Phosphatide derivatives with antilipemic and antiarteriosclerosis activity have been patented (153), as well as compositions which reduce platelet adhesiveness in flowing blood (154).

Several patents have been granted for the use of phosphatide-containing pharmaceuticals to combat mental disorders (155, 156). An active lipid fraction has been fractionated from egg yolk which is reported to be more useful than crude or high PC lecithin (157). This active lipid fraction reportedly gives brain cells from old mice the permeability observed with young cells. Although the active fraction contains cholesterol which is not present in soybean lecithin, soybean lecithin is also claimed to yield an active fraction.

Synthetic whole blood containing phosphatides has been patented (158) as well as lecithin derived clotting factors (159). Compositions containing phosphatides have been patented for the treatment of hyaline membrane disease (160). Lecithin is also being studied for treatment of skin disorders such as psoriasis (6). It has been patented for the treatment of nervous tension from prolonged physical fatigue (161) and even as a component of a formulation for breaking the smoking habit (162). Animal studies indicate that side effects of nonsteroidal antiinflammatory agents can be controlled by the simultaneous administration of phospholipids (163).

Lecithin can be used in an antigen formulation for the diagnosis of syphilis (164). Lysolecithins can be used as hemolysis agents in a test to detect and differentiate between normal and abnormal erythrocytes such as would occur in hepatitis or cirrhosis patients (165).

A patent has been issued for cancinostatic and immunostimulating phospholipid—lysophospholipid mixtures which show anticancer activity in animals (166). By administration of the mixture along with a fat or oil in the form of an emulsion, the strong hemolysis and serum-binding properties of the lysophospholipids are reported to be overcome. Antitumor effects of lysophosphatides have also been noted by other workers (167).

The use of liposomes as carriers of pharmaceutical preparations was previously discussed. Liposomes have also been used to encapsulate hemoglobin cells (168). Liposomes may be administered orally and act as time release agents. Intact liposomes are apparently not absorbed from the gastrointestinal tract, but injectable lecithin preparations are also being developed (169).

Aside from being an active pharmaceutical agent or a liposome-type transport agent, lecithin also has more traditional functions as emulsifier, wetting, dispersing and stabilizing agent in pharmaceutical applications. Representative formulations incorporating lecithin include beta-lactam antibiotics (170), steroids such as diethyl stilbesterol (171), prostaglandins (172), vitamins (173) and an antidiarrhea absorbent (174). Lecithin has been used as an aid to facilitate microencapsulation of pharmaceuticals with ethyl cellulose (175).

### Polymers

Lecithin functions as a pigment-dispersing agent and mold-release agent in the preparation of various polymer products. It facilitates dispersion of polymer resins in plasticizers to form plastisols (6).

Polymerization of vinyl chloride in the presence of poly(vinyl alcohol) and lecithin gave a poly(vinyl chloride) which was more thermally stable than polymer prepared without the lecithin (176). A polyurethane foam prepared in the presence of lecithin had improved nonshrink character (177). Rubber gloves made from a latex containing lecithin and soybean protein developed tackiness much less rapidly than the same formulation without the additives (178).

A curing agent for polyurethane contains lecithin as a dispersing agent (179). Lecithin can be incorporated into expandable polystyrene beads as an antilumping agent (180). It can be used as an initiator in the production of poly(oxymethylene) (45).

Phosphatides are naturally occurring emulsifiers in rubber latex. Additional lecithin may be added to serve other functions. Phosphatides have plasticizing effects which facilitate working of the rubber and they have been shown to be have positive effects on the vulcanization (2,3,181). Lecithin can be used in a solvent-based rubber for forming rubber gaskets (182) and in a formulation for a bag lubricant for tire molding (183).

### Release Agent

Lecithin can function as an antistick agent on a wide variety of processing equipment such as frying surfaces, broilers, blenders, conveyors, extruders, molds, tables and utensils. As such it can be blended into the product prepared on the equipment, in the case of foods or plastics, for example, or it can be coated onto the surface of the equipment by spraying or dipping. Lecithin can be diluted with hydrocarbon propellants and dispensed from aerosol spray cans (184) or formulated with soy oil, mineral oil and alcohols to be dispensed from finger pump bottles or squeeze bottles (185, 186). A solid stick lubricant has also been formulated (187).

Lecithin was mentioned above as a release agent for laminated papers, plastics and foils. It can also be used for pressure-sensitive sheets for transferring colors (188–190) or for correcting typewriter errors (191).

### Textiles

Saponified lecithin is recommended for improving softness and hydrophilicity of cotton (119). Lecithin reduces stickiness of cellulose fiber modified for use as absorbents (192). It is useful as a spinning lubricant for cotton and wool (2). In general, lecithin is recommended for its softening and conditioning effects especially in sizing and finishing (6). It is also useful in a spray to reduce cotton dust (6).

### Waste Treatment and Pollution Control

Compositions containing lecithin can be added to wastewater in treatment plants to prevent obstruction by fats and oils and to increase the biodegradability of the oils (193). Lecithin can also be utilized in the removal of crude oil from fresh or salt water (194, 195). Organic impurities may be removed from air by mixing the air with an aerosol of particles coated with lecithin (196).

## Future Prospects

Just as in some food uses, some of the nonfood applications of lecithin are in mature market areas such as paints and inks. Judging from patent activity, however, growth in some areas seems reasonably assured. Among coatings, the most active patent area has been in magnetic tapes. Japanese workers, in particular, have obtained quite a number of patents.

In recent years, pharmaceutical patents accounted for almost 25% of the nonfood patent activity. Pharmaceutical applications, particularly those involving liposomes, should require increasing quantities of refined lecithins. An increased demand for lecithin as a dietary supplement is also anticipated. Cosmetic uses of lecithin should increase particularly for hair care products. Lecithin has the virtue of being derived from a natural product as well as having good functionality for the cosmetics.

Lecithin sales as magnetic tape, xerographic toner, explosive and detergent additives could also be on the increase according to the patent activity. Overall, the future looks good for lecithin in nonfood uses if good contact between commercial processors and patent developers can be maintained.

## References


1. See, for example, earlier chapters in this monograph.
2. Wittcoff, H., "The Phosphatides", Reinhold Publishing Corp., New York, 1951.
3. Stanley, J. in "Soybeans and Soybean Products," K. S. Markley, ed., Vol. 2, Interscience Publishers, Inc., New York, 1950, pp. 593–647.
4. Stanley, J. in Kirk—Othmer "Encyclopedia of Chemical Technology", 1st edn., Vol. 8, pp. 309–326 (1952).
5. Sartoretto, P., Ibid, 2nd edn., Vol. 11, pp. 343–361 (1967).
6. Eichberg, J., Ibid, 3rd edn., Vol. 14, pp. 250–269 (1981).
7. Graf, T., Flavors, 1976, pp. 218–221.
8. Wolf, W. J. and D. J. Sessa in "Encyclopedia of Food Science," M. S. Peterson and A. H. Johnson, eds., AV1 Publishing Co. Inc., Westport, CT, pp. 461–467 (1978).
9. Szuhaj, B. in "Proc.-World Soybean Res. Conf. II," 1979, C. T. Frederick, ed., Westview Press, Boulder, Co. (pub 1980), pp. 681–91.
10. Van Nieuwenhuyzen, W., JAOCS 58:886–888 (1981).
11. Cherry, J. P., et al, JAOCS 58:903–913 (1981).
12. Sietz, F. G., Seifen, Oele, Fette, Wachse, 107:629–631 (1981), CA 96:87380.
13. Ansell, G. B., et al, "Form and Function of Phospholipids" Elsevier, New York, 1973.
14. Becher and Birkmeier, JAOCS, *41,* p. 169 (1964) (cited in Eichberg).
15. Klepetar, M. and C. R. Kucera, FR 2011241 (1970).
16. Lennox, P. F. and J. B. Kirkland, U.S. Patent 3293054 (1966).
17. Falko, A. T., et al, USSR SU 815755 (1981).
18. Semko, L., et al, USSR SU 861459 (1981).
19. Etschenberg, E., et al, Ger. Offen 2930544, (1981).
20. Horikoski, K. and T. Abe, Ger. Offen. 3109493 (1981).
21. Crosby, A. C. and D. C. Campen, U.S. Patent 4,323,365 (1982).
22. Altfeld, J., et al, eds; "50 Years Lucas Meyer—25 Years Lecithin," Lucas Meyer Co., Hamburg, 1973, pp. 74–76 and references cited therein.

23. Holmes, R. F. and C. K. Kandel, South Africa, ZA 7405866 (1975).
24. Cloosterman, A. B. M. and S. Van Kranenburg, European Patent 57989 A2, Aug 18, 1982, CA 97:180608.
25. Mommer, R. P., U.S. Patent 3,682,653 (1972).
26. Burgi, F. Arbeiten DLG, p 125, 1969 (cited in Altfeld et al).
27. Koehler, H. et al, E. Ger. DD 138649 (1979).
28. Holtz, R. B., U.S. Patent 4,370,159 (1983).
29. Yazlovetskii, I. G., et al, USSR SU 688161 (1979).
30. Jpn. Chem. Weeky, p 6, (1974).
31. Kalocsai, G., et al, Hung. Teljes 18,326 (1980).
32. Homma, Y., et al, Nippon Noyaku Gakkaishi, *6,* 145–53 (1981), CA 95:127244.
33. Okauchi, M., et al, Japan Kokai Tokkyo Koho JP76104033 (1976).
34. Shillington, W. L. and J. J. Liggett, U.S. Patent 3533810 (1970).
35. Wills, R. B. H., et al, J. Sci Food Agric, *31,* 663–6, 1980; CA94:28976.
36. Misato, T., et al, Japan Kokai Tokkyo Koho 7622825 (1976).
37. Fed. Regis., *46,* 18315–16, Mar 24, 1981.
38. Yaeger, L. L. and J. Bjorksten in Autoxid. Food Biol. Syst. [Proc. Int. Workshop] 1979, M. G. Simic and M. Karel, eds., Plenum, NY pub. 1980, pp. 405–11, CA94:119735.
39. Nipon Oils and Fats KK, Jpn 55089383 (1980).
40. Hudson, B. J. F. and S. E. O. Mahgoub, J. Sci Food Agric, *32,* 208–10 (1981).
41. Asaki Denka Kogyo K. K., Jpn 82 143398 (1982).
42. Nisshin Flour Milling Co., Jpn 74004480 (1974).
43. Ajinomoto, Co., Inc., Jpn 81 41383 (1981).
44. Pardun, H., Seifen, Oele, Fette, Wasche, *107,* 409–11 (1981).
45. Setterquist, R. A., U.S. Patent 3394092, (1968).
46. Blackley, D. C., "Emulsion Polymerization: Theory and Practice," Wiley, New York, 1975, pp. 282–286.
47. Otsuki, T. and K. Nagashima, Japan Kokai Tokkyo Koho 7545086 (1975).
48. Didenko, L. P., et al, Dokl. Akad, Nauk SSSR, 254 (3), p 643–5 (1980), CA94:37034.
49. Beaucamp, K., et al, Ger. Offen. 2,933,648 (1981).
50. Nayler, P., European Patent Application 20,123, 1980, CA94:128176.
51. Wegner, J., Ger. Offen 2519187 (1976).
52. Kronstein, M., Mod. Paint Coatings *68,* p 59 (1978) cited in Eichberg (6).
53. Renko, D., et al, Kem. Ind., *30,* p 9–14 (1981), CA 94:210352.

54. Kronstein, M. and J. Eichberg, U.S. Patent 4056494, (1977).
55. Kendall, S. E., U.S. Patent 4372991 (1983).
56. Mann, J. D. British Patent Application GB 2078243A (1982).
57. Sakai, S., Jpn Kohai Tokkyo Koho, 79 83936, (1979).
58. Kosinska, M., et al, Pol. 107,705, Nov 3, 1980, CA94:105037.
59. Victor Co. of Japan, Ltd., Jpn. Kokai Tokkyo Koho 80 151,069 (1980).
60. Bulshtein, R. I., et al, USSR SU 624697 (1978).
61. Kronstein, M., and J. Eichberg, U.S. Patent 2,997,398 (1961).
62. Kuriyama, K. and M. Toyama, Japan Tokkyo Koho, 7448167 (1974).
63. Tsujimito, H., et al, U.S. Patent 4049556 (1977).
64. Shono, Y., Japan Kokai Tokkyo Koho 79110335 (1979).
65. Caster, S. A. V., Fr Demande FR 2437829 (1980).
66. Keith, A. D. and P. Frest, PCT International Patent, WO 8102673, Oct 1, 1981, CA 96:011692.
67. Mizumaki, K., Japan Kokai Tokkyo Koho 7973135 (1979).
68. Takei, Y., Japan Kokai Tokkyo Koho 7941338 (1979).
69. List, P. H. and R. Von Kleinsorgen, Ger. Offen 2,928,040 (1981).
70. Oreal Cosmetics K.K., Jpn. Kokai Tokkyo Koho JP 82 35931 (1982).
71. Nippon Surfactant Kogyo KK, Jpn. Kokai Tokkyo Koho JP 81 70,826 (1981).
72. Kanebo Cosmetics Inc., Jpn. Kokai Tokkyo Koho 81 120,612 (1981).
73. Rovesti, P., and S. B. Curri, Riv. Ital EPPOS, 63, 23–6, 1981, Chem. Abstr.94:162588.
74. Spitzer, J., et al., U.S. Patent 4,314,573 (1982).
75. Socci, R., et al., U.S. Patent 4302442 (1981).
76. Bernardino, L. W., U.S. Patent 4,297,251 (1981).
77. Duntsan, A. Y., USSR SU 924,176 (1982).
78. Car Ind. Exp. Cons, USSR SU 495,345 (1976).
79. Ushakovia, M. I., et al., USSR SU225845 (1968).
80. Kao Soap Co. Ltd, Jpn. Kokai Tokkyo Koho 81 28,298 (1981).
81. Takisawa, et al., Japan Kokai Tokkyo Koho JP 75 70410 (1975).
82. Willard, K. E., et al., Anal Biochem, 100, 289–98 (1979).
83. Plauson, R. A., and W. McBride, U.S. Patent 3998676 (1976).
84. Nikon Nohyaku Co., Ltd., Jpn. Kokai Tokkyo Koho 81 59701 (1981).
85. Seto, P.F.L., and R. Binet, Braz. Pedido PI BR 79 07,969, (1980).
86. Cooper, J., and V. Sujansky, European Patent Application EP44664 (1982).
87. Ulsteen, K., European Patent Application EP 35,376 (1981).
88. Bailey, D. G., et al., in Kirk-Othmer's Encyclopedia of Chemical Technology, 3rd edn., V 14, 200–231, Wiley-Interscience, NY, 1981.
89. Colley, C. M., and B. E. Ryman, Trends in Biochemical Science, 1976, pp. 203–205.

90. Paapahadjopoulos, D., Liposomes and their Use in Biology and Medicine, NY Acad. Sci., NY, 1978.
91. Lasic, D. D., et al., Farm Vestro (Llubljana), 31, 187–95 (1980), Chem. Abstr.94:117354.
92. Pick, U., Arch. Biochem. Biophys., 212, 186–94 (1981).
93. Papahadjopoulos, D., et al. in Liposomes Immunobiol. Proc. Natl Symp., 1980, 151–64, Chem. Abstr.95:49331.
94. Oku, N., et al., Biochem. Biophy. Acta, 692, 384–8 (1982).
95. Patel, H. M., and B. E. Ryman, FEBS Lett., 62, 60–63 (1976) (cited in Colley (89)).
96. Allison, A. C., and G. Gregoriadis, Nature 252, 252 (1974) (cited in Colley (89)).
97. Fruehling, J., et al., Eur. J. Cancer, 16, 1409–16 (1980).
98. Rahman, Y–E, and E. A. Cerny, U.S. Patent 3993754 (1976).
99. Asahi Chemical Industry Co. Ltd, Jpn. Kokai Tokkyo Koho 81 22,724 (1981).
100. Baldeschwieler, J. D. et al., European Patent Application EP 28917 (1981), U.S. Patent Application 92342 (1979).
101. Papahadjopoulos, D. P., et al., Gen Offen 2907303 (1979).
102. Watanabe, K., Ger. Offen. 3,016,976 (1980).
103. Handjani, R. M., and G. Vanlerberghe, U.S. Patent 4,217,344 (1978).
104. Rahman, Y. E., U.S. Patent 3932657 (1976), U.S. Patent 4016290 (1977).
105. Morris, R. M., European Patent Application EP 42249 (1981).
106. Fujiki, K., Gen. Offen. DE 3,100,686 (1981).
107. Kawakara, H. Gen. Offen. DE 3,114,646 (1982).
108. Hosaka, A. et al., Ger. Offen. DE 3104702 (1982).
109. Intermagnetics Corp, Fr. Demande 2,448,560 (1979).
110. Tamura, H., Ger. Offen. 3,017651 (1980).
111. Ohlinger, M., et al., Ger. Offen, 2942646 (1981).
112. Buzniak, M., et al., Poland Patent PL 117092 (1982).
113. Toray Industries, Inc., Jpn. Kokai Tokkyo Koho 80 83,041 (1980).
114. Proell, W. A., U.S. Patent 3537869 (1970).
115. Rue, H. M., U.S. 3775340 (1973).
116. Silckerodt, W., Ger. Offen. DE 1957827 (1971).
117. Aslanov, N. N. et al., USSR SU 629239 (1978).
118. Moody, G. L., Canada CA 895594 (1972).
119. Foa, M, Fr Demande 2475053 (1981).
120. McDonough, W. F., et al., U.S. Patent 3928226 (1975).
121. Friedman, H. et al., France FR 1548700 (1968).
122. Hajduk, G. et al., Poland PL 104287 (1979), Chem. Abstr.92:112454.

123. Wolff, N. E., and J. W. Weigh in Kirk-Othmer's Encyclopedia of Chemical Technology, 3rd edn., Vol 8, 794–826 (1979).
124. Canon, K. K., Japan Tokkyo Koho, JP8129267 (1981).
125. Matkan, J., U.S. Patent 4307169 (1981).
126. Gruber, R. J., and S. B. Bolte, U.S. Patent 4303749 (1981).
127. Ricoh Co. Ltd, Jpn. Kokai Tokkyo Koho JP 81 78,844 (1981).
128. Hayes, L. P., Canada Chem. Abstr. 945800 (1974).
129. Hayes, L. P., et al., South Africa ZA 6906125 (1970).
130. Shimomukai, S., et al., Gen. Offen. DE 2916494 (1979).
131. Fuji Photo Film Co., Ltd, Jpn. Kokai Tokkyo Koho, 80 88045 (1980).
132. McMullen, A. I., British Patent 1357952 (1974).
133. Kumiai Chemical Industry Co. Ltd, Japan Tokkyo Koho JP 8108804 (1981).
134. Nisshin Oil Mills, Ltd., Jpn. Kokai Tokkyo Koho JP 81140910 (1981).
135. Ritzmann, G., et al., Ger. Offen DE 3010041 (1981).
136. Thayer, H. I., U.S. Patent 3644393 (1972), U.S. Patent 3615291 (1971), U.S. Patent 3527583 (1970).
137. Honnen, L. R., U.S. Patent 4128403 (1978).
138. Cornelius, T. E., III, and A. E. Bischof, U.S. Patent 3547605 (1970).
139. Yamauchi, T., Jpn. Kokai Tokkyo Koho JP 7843706 (1978).
140. Katsuta, K. et al., Jpn. Kokai Tokkyo Koho JP 7814706 (1978).
141. Nakajima, N., Jpn. Kokai Tokkyo Koho JP 7844167 (1978).
142. Dorer, C. J., U.S. Patent 3284409 (1966).
143. Gaston, G. E., U.S. Patent 3313607 (1967).
144. Salenko, V. I. et al., USSR SU 699007 (1979).
145. Wiese, H. F., U.S. Patent 3269946 (1966).
146. Cekada, J. Jr., Ger. Offen. DE 2125948 (1971).
147. Tokai Rika Co. Ltd., Jpn. Kokai Tokkyo Koho JP 8208296 (1982).
148. Nisshin Oil Mills, Ltd, Jpn. Kokai Tokkyo Koho, JP 8231993 (1982).
149. Seidel, H. et al., East Ger, DD159549 (1983).
150. Thompson, W. W., U.S. Patent 4111729, (1978).
151. Devel, H. J., Jr., The Lipids, Their Chemistry and Biochemistry, Vol. 2, 1955, p. 743.
152. Wood, J. L., and R. G. Allison, Fed. Proc. 41:3015 (1982).
153. Davidson, H., and H. Betzing, Ger. Offen. 2756866 (1979).
154. Martin, W., U.S. Patent 4061738 (1977).
155. Nakajima, T., et al., Gen. Offen. DE 2907778 (1979).
156. Growdon, J. H., and R. J. Wurtman, U.S. Patent 4346085 (1982).
157. Shinitzky, M., et al., European Patent Application EP 74251 (1983).
158. Ecanow, C. S., and B. Ecanow, U.S. Patent 4343797 (1982).
159. Horikoski, I., et al., U.S. Patent 4348384 (1982).

160. Tetsuro, F., et al., Ger. Offen. DE 3021006 (1981).
161. Ernst, E., and H. Mies, Munchenes Med. Woch. 15,713 (1960), cited in Eichberg (6).
162. Gerritsen, F. B. J., Belgium Patent BE 890671 (1982).
163. Nattermann, A., and Co., Jpn. Kokai Tokkyo Koho 80 89,225, 1980.
164. Sennikov, G. A., et al., USSR SU 629927 (1978).
165. Nunogaki, Y., Jpn. Kokai Tokkyo Koho JP 78104296 (1978).
166. Kodama Y., et al., U.S. Patent 4,372,949 (1983).
167. Munder, P. G., and O. Westphal, Ger. Offen. DE 2619686 (1977).
168. Miller, I. F., and L. Djordjevich, U.S. Patent 4133874 (1979).
169. Altman, R. F. A., U.S. Patent 4252793 (1981).
170. Sherico Ltd., Ger. Offen. DE 2756079 (1978).
171. Ludwig, N. H., and W. A. White, U.S. Patent 3808338 (1974).
172. Gordon, M., and J. A. Weisbach, Canada Chem. Abstr. 932324 (1973).
173. Yehara, R., et al., Jpn. Kokai Tokkyo Koho JP 8049313 (1980).
174. Bullous, G. H., British Patent 1182463 (1970).
175. Samejima, M., et al., Ger. Offen. 3,039,908 (1981).
176. Heijkoop, A., Ger. Offen. DE 2213927 (1972).
177. Dahm, M., Ger. Offen DE 2015159 (1971).
178. Terajima, E., Jpn. Tokkyo Koho JP 7205415 (1972).
179. Caruso, P. P., U.S. Patent 4282344 (1981).
180. Wright, H. A., U.S. Patent 3520833 (1970).
181. Meshandin, A. G., et al., USSR SU 905,236 (1982).
182. Cahill, G. F., and I. J. Arons, U.S. Patent 3344099 (1967).
183. Cekada, J., Jr., U.S. Patent 3532624 (1970).
184. Follmer, D. W., U.S. Patent 3896975 (1975).
185. Doumani, C., U.S. Patent 41557700 (1970).
186. Sejpal, V. O., U.S. Patent 4142003 (1979).
187. Mahler, D. G., and C. Doumani, U.S. Patent 4023912 (1977).
188. Maniar, P. N., Ger. Offen. DE 2117075 (1971).
189. P. E. Kinzli, et al., Ger. Offen. DE 2117214 (1971).
190. Newman, D. A., Ger. Offen. DE 2116902 (1971).
191. Dabisch, W., U.S. Patent 3515572 (1970).
192. Fuji Spinning Co. Ltd., Jpn. Kokai Tokkyo Koho 8100306 (1981).
193. Fusey, P., Fr. Demande 2392941 (1978).
194. Jatzkowski, K., et al., East Ger. DD 149056 (1981).
195. Olivieri, R., et al., Ger. Offen. DE 2739428 (1978).
196. Merrill, E. W., Ger. Offen. DE 2044685 (1971).

# Chapter Ten
## Modified Lecithins

**J. C. Schmidt and F. T. Orthoefer**
A. E. Staley Manufacturing Co.
Decatur, IL

Crude commercial lecithin is a mixture of oil and phosphatides with minor amounts of other substituents. A change in the relative concentration of any one of these components, or an alteration of its chemical structure, may cause some change in the physical or chemical properties of the lecithin. Lecithin can exist as a liquid, plastic or free-flowing solid. Its color, solvent solubility, surfactancy and chemical reactivity all can be modified. These modifications, in turn, will alter the functional properties of the lecithin in a given application. In this chapter the various methods used for modifying commercial lecithin will be discussed.

A method for classifying lecithin to include the modified forms has been proposed (1). One version of a classification scheme is given in Table I. Bleached and fluid varieties of commercial lecithin have been defined by the National Soybean Processors' Association (2). The simplest method for modifying lecithin is by the addition of a nonreactive substance. Crude lecithin to which additives other than the usual fluidizing or bleaching agents have been added have been called compounded lecithins. If the additives react with the lecithin to alter the chemical structure of one or more of the lecithin components, the lecithin is said to be chemically modified. Much of the chemistry for making compounded and chemically modified lecithin has been known for some time (3–5).

Crude lecithin can be further refined by various fractionation methods to selectively separate its components. Acetone and ethanol are the most common solvents associated with these methods. Much of the recent patent activity on modified lecithins involves these fractionation procedures. A classification for purified phosphatides has been included in Table I to reflect the state of the art. Fractionation procedures yielding phosphatidylcholine of greater than 90% purity have been reported.

Most modified lecithins currently marketed are prepared for use in either food, feed or pharmaceutical applications. A few products, mainly those for use in paints and similar coatings, contain non-edible ingredients.

**TABLE 10-I**
**Classification of Soybean Lecithins**

I. Crude commercial lecithin
  A. Plastic
    1. Unbleached
    2. Single bleached
    3. Double bleached
  B. Fluid
    1. Unbleached
    2. Single bleached
    3. Double bleached
II. Compounded
III. Chemically modified
IV. Refined
  A. Deoiled
    1. As is
    2. Custom blended
  B. Fractionated
    1. Alcohol soluble
      a. As is
      b. Custom blended
    2. Alcohol insoluble
      a. As is
      b. Custom blended
  C. Purified phosphatides

## Crude Commercial Lecithin

Crude lecithin obtained from water degumming of soybean oil at room temperature forms a liquid crystalline phase that imparts a plastic consistency to the product. Plastic lecithin is functionally equivalent to the fluid grades but is seldom preferred because it cannot be pumped or poured without being heated. Crude lecithin then is usually modified at least to the extent of being fluidized. This is accomplished by dilution of the phosphatides to the 62–64% range using soybean oil and an additional additive, most commonly 2–4% fatty acids (6). Mineral acids may also be added. The acids are added to the point where the total acidity of the lecithin, as measured by the acid value (7), is in the range of about 20 to 32. A fluid lecithin may also be obtained by addition of small amounts of divalent metal ions such as calcium and magnesium (8). These ions occur naturally in the lecithin and can be supplemented to attain a concentration of approximately 0.4%. At this concentration fluid lecithins of 70 AI or higher can be prepared.

Crude lecithin may also be bleached with peroxides to form lighter colored products. Bleaching involves oxidation of the carotenes and other color bodies within the lecithin; it does not significantly alter the functionality of lecithin. Bleached products are traditionally grouped with the unbleached forms of crude lecithin. No distinction is made between bleached and unbleached forms as far as Generally Recognized as Safe (GRAS) status is concerned (9).

Lecithin destined for certain applications may require more rigorous than usual initial refining conditions. Refined lecithin used for magnetic tape should be derived from crude lecithin with very low peroxide value and low hexane insolubles, for example.

## Compounded Lecithins

Crude lecithin can be combined with surfactants or other additives or blended with solid carriers to form products with special properties. Such blends have been called compounded lecithins.

Crude lecithin does not disperse well in water. Blending in surfactants such as polysorbate or ethoxylated monoglycerides facilitates rapid water dispersibility of the lecithin (10). Other blends may contain mineral oil, xylene or other organic solvent. Compounded lecithins are particularly useful in pigment and paint applications. Low-cost powdered products can be prepared by spray drying lecithin onto proteins such as skim milk and whey or soy protein (11). Flaked or powdered products can be made by spray cooling lecithin with high melting mono- and diglycerides (11).

## Chemically Modified Lecithin

Lecithin contains a number of chemical functional groups that make it susceptible to a variety of chemical reactions. These groups include carbon-carbon double bonds, esters, phosphonate esters, amines and hydroxyl groups. Lecithin has been acetylated, hydroxylated, hydrolyzed, hydrogenated, halogenated, phosphorylated and sulfonated, to name the most widely quoted chemistries. Of these, only acetylated and hydroxylated, and perhaps some hydrolyzed, lecithins are produced in significant commercial quantities at the present time.

**Acetylated Lecithin**

Lecithin is acetylated using acetic anhydride. The anhydride can be added in small amounts to crude soybean oil just prior to the degumming step or it can be added to the wet gums (12). Acetylation occurs primarily on the amino group of phosphatidylethanolamine. This effect, combined with the acetic acid produced as a byproduct, can be used to improve the fluidity of crude lecithin. Anhydride processing also improves the water dispersibility of the lecithin and enhances its oil-in-water emulsifying properties (13). This would make it advantageous for use in the manufacture of coffee whiteners or for oil-in-water cosmetic emulsions, for example.

**Hydroxylated Lecithin**

This product is a light colored lecithin with increased water dispersibility and enhanced oil-in-water emulsifying properties. It is produced by reacting crude lecithin with hydrogen peroxide and lactic acid or a peracid (14). Normally, hydroxylation until a 10% reduction in iodine value is achieved will give the desired properties. A linear effect in oil-in-water emulsification enhancement is reported up to about a 25% IV reduction. The IV reduction and isolation of dihydroxystearic acid indicate that the fatty acid double bonds are reactive sites for the hydroxylation. The ethanolamine group is also modified (11). Hydroxylated lecithin is useful in applications where a water dispersible lecithin is desirable. It is especially useful in baking applications where it can improve the dispersion of fats and retard staling.

**Hydrolyzed Lecithins**

Crude lecithin is readily hydrolyzed in the presence of strong acids or bases. Enzymes can be used for very selective hydrolysis. Prolonged treatment with acids or bases leads to fatty acids, glycerophosphoric acid and inositolphosphoric acid or their salts with mixtures of amino compounds and carbohydrates (4). Still further degradation to glycerol, phosphoric acid and inositol eventually occurs. Acid or basic hydrolysis must be stopped at an early stage, preferably with just one fatty acid cleaved, to obtain useful products. Hydrolysis can be carried out at pH 3.5 and 63 C with hydrochloric acid in the presence of moisture to a prescribed acid value followed by neutralization with sodium hydroxide (15). Aqueous hydrolysis at 95 C following citric acid treatment has also been reported (16). Methanolysis with a magnesium oxide catalyst yields mixtures of alpha and beta lysolecithins and glycerolphosphatides (17). Brief treatment with aqueous strong bases will give similar results.

Much better control of hydrolysis products is obtained using enzymes. A number of phospholipase enzymes are available. Selective cleavage of the alpha or beta fatty acid can be achieved or the polar end groups can be hydrolyzed to give phosphatidic acid. The use of phospholipase $A_2$ at reaction temperatures of 50–70 C to cleave the beta fatty acid is the most industrially promising enzymatic reaction (11).

Partially hydrolyzed lecithins have greatly enhanced oil-in-water emulsifying properties. They do not lose their emulsifying action in the presence of calcium and magnesium ion as rapidly as do the unmodified types. These have been produced commercially, especially in Europe, and have found application in calf milk replacement formulations to improve the emulsification and digestibility of fats.

### Hydrogenated Lecithins

Lecithin can be hydrogenated to a stearin-like solid which has greater oxidative stability and is less hygroscopic but has reduced solubility in the usual solvents. The reaction can be carried out at 75–80 C in at least 70 atmospheres pressure using a flaked nickel catalyst (18). The iodine value is usually reduced to the 10–30 range. Nickel or palladium catalysts are preferred over platinum. Phosphatides are not hydrogenated as readily as soybean oil and as the hydrogenation pressure and temperature is lowered, the oil can be selectively hydrogenated. Ethyl actate, benzene or mineral oil can be used as solvents for the hydrogenation. In chlorinated solvents or in mixtures of these solvents with alcohol, much lower temperatures and pressures can be used for hydrogenation particularly when a palladium catalyst is used (19). For example, using a 10% palladium on carbon catalyst to hydrogenate corn lecithin in dichloromethane resulted in a reduction from 88 to 31.7 IV in 20 min at 22 psi and temperatures not exceeding 80 C. Hydrogenated lecithin has been recommended as a lubricating oil additive (20). It has been reported to function well as an emulsifier and as an inhibitor of fat bloom in chocolate (14). Hydrogenated lecithin has been used to form emulsions for intravenous injections (21).

### Other Modified Products

Lecithin will add chlorine or bromine across the double bonds to make halogenated products useful in lubricant applications (22). Iodine can be added by warming granular lecithin dissolved in acetic acid in the presence of iodine and using aluminum or magnesium metal as a catalyst (23, 24). Lecithin reacts with sulfur dioxide and an aldehyde, ketone or phenol to make derivatives which readily disperse in water and resist precipitation by acids and salts (25). These products are reported to have application in the textile and tanning industries (4). A patent has been issued on polyoxyethylene derivatives of lecithin prepared by reaction of lecithin with ethylene oxide (26) but a later reference disclaims the existence of such products (5). Phosphoric acid and metal salt derivatives useful in coating applications also have been reported (27).

Lecithin rapidly darkens at temperatures in excess of 70 C due to browning reactions involving its sugars and amines. Additives which would inhibit this reaction or refining processes which would remove the sugars could result in a more heat stable lecithin. Small amounts of lecithin in oils can be stabilized by using arginine-glutamic acid salt, sodium glutamate or sodium bicarbonate (28–30). Corn oil containing 3% lecithin and 2.3% of one of these additives could be heated to 180 C without discoloration while controls turned dark brown. A heat stable lecithin, prepared by an undisclosed process, was recently introduced into the US market (31).

Lecithin can be sterilized chemically by heating lecithin solutions 30–60 min in an autoclave at 121 C in the presence of sodium formaldehyde sulfoxide or sodium bisulfite (32). The lecithin produced by this method is also reported to be colorless. Sterile lecithin can also be prepared by a filtration technique (33).

## Refined Lecithin

### Oil-free Lecithin

Refined grades of lecithin are produced by extraction methods which remove or concentrate one or more of the crude lecithin components. "Oil-free" grades are obtained by extraction of the crude lecithin batchwise or continuously with acetone (34). Besides the oil, acetone will remove fatty acids and sterols. Products as high as 99% AI are commercially available but 95% AI material is more typical. It is advantageous in most applications to retain the small amounts of oil that remain as the oil greatly enhances the oxidative stability of the phosphatides.

Oil-free lecithins are dry powders or granules. A small percent of an anticaking agent may be added to insure that the product remains free-flowing. Oil-free lecithin is easily added to dry-blended products.

Oil-free phosphatides may be blended with special ingredients to prepare products that resemble the compounded lecithins (4). Blends with cocoa butter for use in chocolate and with organic solvents for use in paints have been prepared, for example. Blends with deodorized oils have been prepared to produce a fluid lecithin that is more bland than the crude product.

### Alcohol Fractionated Phosphatides

Crude lecithin or the oil-free product may be fractionated to concentrate one or more of the phosphatides. The usual solvent employed is ethanol. Phosphatidylcholine (PC) is alcohol soluble and phosphatidylinositol (PI) is alcohol insoluble, while phosphatidylethanolamine (PE) is slightly soluble in alcohol. Fractionation of oil-free lecithin will give an alcohol soluble fraction containing 40–60% PC and an alcohol insoluble portion containing 40–60% PI. Both fractions will contain PE. The PC-rich fraction has enhanced oil-in-water emulsifying properties and the PI-rich fraction has enhanced water-in-oil emulsifying properties.

### Purified Phosphatides

To obtain individual phosphatides of greater than 50–60% purity, some form of selective adsorption process is usually required. At the present time, the most sought after phosphatide is PC. A number of possible isolation methods have been patented. Alumina or silica gel are the more frequently cited adsorbents and are used in either slurry or column chromatographic processes.

A fraction containing 67% PC was obtained by slurrying one part crude phosphatides (containing 19.5% PC) contained in 1.5 parts dichloromethane-ethanol with five parts of alumina (35). The process was repeated with additional alumina. The alumina was eluted with ethanol to give the 67% PC fraction. PC of up to 92% purity reportedly has been obtained using a silica gel column (36, 37). Still another reported variation is adsorption of one part crude lecithin onto 20 parts of silica gel and elution with ethanol-hexane, ethanol and finally ethanol containing three percent ammonia (38). Evaporation of the last solvent gave 0.114 parts of pure PC. The starting fraction contained 0.18 part PC. Variations in which acylated phosphatides are present have also been reported (39, 40).

Pure PC is not easily handled but it may be dissolved in ethyl lactate or similar esters to give sterilizable, fluid solutions containing 20% to 85% PC (41). The calcium chloride salt of PC may also be prepared. This powder is obtained by drum drying ethanol solutions of PC and calcium chloride (42).

## Conclusions

A wide variety of methods are available for modifying the properties of crude commercial lecithin. The major process developments at the present time are in the fractionation of lecithin. Markets appear to be developing in the pharmaceutical and cosmetic industries, particularly for concentrated forms of PC. De-oiled, granular lecithins have firmly established markets and new markets are developing which should ensure the growth of these products. A number of chemically modified and compounded forms of lecithin have established markets. In some cases, such as for hydrogenated or halogenated lecithin, the technology for their preparation is on hand, but large markets for these products have not been developed.

## References


1. Sullivan, D. R., and B. F. Szuhaj, Presentation, AOCS Meeting, Dallas, Texas, April 27–30, 1975.
2. National Soybean Processors' Assoc., Yearbook and Trading Rules, 1981–82.
3. Wittcoff, H., The Phosphatides, Reinhold Publishing Co., New York, 1951.
4. Stanley, J., in Soybeans and Soybean Products, edited by K. S. Markley, Vol. 2, Interscience Publishers Inc., New York, 1950, pp. 593–647.
5. Eichberg, J., in Kirk-Othmer's Encyclopedia of Chemical Technology, 3rd edn., Vol. 14, 1981, pp. 250–269.
6. Wiesehahn, G. A., U.S. Patent 2,194,842 (1940).
7. Acid Value, AOCS Method, Ja 6-55, 1973.
8. Davis, P. F., U.S. Patent 3,357,918 (1967).
9. Fed. Reg. 48 (216), 51149–51150, Nov. 7, 1983.
10. Orthoefer, F. T., U.S. Patent 4200551 (1980).
11. Van Nieuwenhuyzen, JAOCS 58:886 (1981).
12. Hayes, L. P., and H. Wolff, U.S. Patent 2782216 (1957).
13. Eichberg, J., U.S. Patent 3,359,201 (1967).
14. Julian, P. L., U.S. Patent 2,629,662 (1953).
15. Davis, P. F., U.S. Patent 3,576,831 (1971).
16. Unilever N. V., Netherlands Patent Application 68 05527 (1968).
17. Betzing, H., Ger. Offen. 1902607 (1969).
18. Jacini, G., U.S. Patent 2,870,179 (1959).
19. Cole, R. D., U.S. Patent 2,907,777 (1959).
20. M. H. Arveson, U.S. Patent 2,295,192 (1942).

21. Takai, T., et al., Japan Kokai 77 83,911 (1977).
22. R. B. Denham, U.S. Patent 2,166,286 (1939).
23. Makabe, K., Japan 70 15,490 (1970).
24. Daiichi Yakuhin Sangyo KK, British Patent 1081310 (1967).
25. Kembara, R., and W. Rosenthal, British Patent 518,194 (1940).
26. De Groote, M., and B. Keiser, U.S. Patent 2,310,679 (1943).
27. Kronstein, M., and J. Eichberg, U.S. Patent 2,997,398 (1961).
28. Suzuka, A., et al., Jpn. Kokai Tokkyo Koho 79,126206 (1979).
29. Harigaya, E., et al., Jpn. Kokai Tokkyo Koho 79,127408 (1978).
30. Matsueda, Y., et al., Jpn. Kokai Tokkyo Koho 79,107530 (1979).
31. Food Processing, Nov. 1983, p. 68.
32. Celento, R., and C. Riffkin, Ger. Offen 2022898 (1970).
33. Arons, A., et al., Ger. Offen. 2023427 (1970).
34. Flider, F., Food Development, Nov. 1981, pp. 27–31.
35. Betzing, H., and H. Eikermann, S. African Patent 6805407 (1969).
36. Guenther, B. R., European Patent Application 54770 (1982).
37. Guenther, B. R., Ger. Offen 3023814 (1982).
38. Takasaki, A., Japan Kokai 74 93,400 (1974).
39. Guenther, B. R., European Patent Application, 54,768 (1982).
40. Unilever N. V., Netherlands Patent Application 68 17,199 (1969).
41. Materne, C., et al., Ger. Offen 2949849 (1981).
42. Materne, C., et al., British Patent Application 2065659 (1981).

# Chapter Eleven
## Chemical Reactions of Phosphatides

**E. H. Pryde**
Northern Regional Research Center
Agricultural Research Service
US Department of Agriculture
1815 N. University Street
Peoria, IL 61604

### Introduction

Chemical reactions of phosphatides described in the literature provide few surprises and often are predictable from the chemistry of their component parts. One notable exception appears to be the selectivity that occurs during hydrolysis under appropriate conditions, and a major part of this chapter is devoted to hydrolysis and its analogous reaction with alcohols, i.e., alcoholysis. The chemistry of the saturated fatty acid components remains the same with respect to reactions of the carboxyl group and of the unsaturated fatty acids with respect to reactions of the double bond, whether the fatty acid is in triglyceride or phospholipid form. Differences in rates of reaction may occur depending on the state of aggregation of the phospholipid.

Reviews describing phosphatide chemistry include those on newer developments in structure determination of glycerides and phosphoglycerides by Kuksis (1), on chemistry of glycerolipids by Snyder (2), on the chemistry of phospholipids by Strickland (3), and on lecithin by Eichberg (4). Other reviews of interest include those of Hawthorne and Kemp (5) and Van Deenen and DeHaas (6).

### Autoxidation

Autoxidation of polyunsaturated phospholipids takes the same course as that of free fatty acids or esters, with conjugated diene hydroperoxides being primary products (7). For details on the autoxidation of unsaturated fatty acids, consult the reviews by Frankel (8, 9).

A Maillard browning reaction seems to occur when soybean lecithin is heated with water in liquid paraffin at 180 C, but coloration is due more to aldol condensation reactions than to a Maillard reaction (10–12). The main carbonyl products formed as intermediates were 9-oxo-10,12-octadecadienoic and 13-oxo-9,11-octadecadienoic acids (13). The browning reaction is the main process for changes in color and flavor (14). The kinetics of the reaction between phospholipids and acetaldehyde have been described (15). Phosphatidylethanolamine catalyzes the condensation of longchain aldehydes to 2,3-dialkylacroleins; no aldol (hydroxy aldehydes) products were detected (16). However, a series of pyridinium ring compounds has been detected whose formation involves the primary amino group in phosphatidylethanolamine (17). As a model reaction for nonenzymic browning, the reaction of 1,4-benzoquinone with phosphatidylethanolamine has been investigated; again the primary amino group is involved (18).

Catalytic action of transition metals is of importance in the rapid oxidation of phospholipids. When copper is present, autoxidation is markedly reduced by the addition of ethylenediaminetetraacetic acid, well known for its chelating properties (19).

Antioxidant properties of soybean phospholipids, and specifically phosphatidylethanolamine, may be attributable to chelation of copper ions (20).

The kinetics of autoxidation of phosphatidylcholine in chlorobenzene solution have been compared to those of a bilayer dispersion in 0.1 M aqueous NaCl (21, 22). Autoxidations were initiated by thermal decomposition of di-*tert*-butyl hyponitrite. Reduced efficiency of initiation in the bilayer was attributed to high microviscosity in the bilayer. Reduction in oxidizability in the bilayer was attributed to diffusion of the polar, peroxyl radical center to the nonautoxidizable, polar surface region.

The hydroperoxides formed by oxidation of phosphatidylcholine with singlet oxygen have been examined by NMR and GC–MS (23). $\beta$-Carotene or 1,4-diazabicyclo-[2.2.2]-octane serve as singlet oxygen quenchers to prevent liposome damage initiated by photosensitization or radiofrequency discharge (24).

As shown by fluorescence polarization, it is necessary to keep lecithin liposomes under nitrogen to prevent oxidative deterioration (25).

Other reports describe the use of EPR (26), fluorescence (27), chemiluminescence (28) and polarography (29) for the study of autoxidation. Glushchenko et al. (29) state that the rate of autoxidation for phosphatidylethanolamine is much greater than that for lecithin, although the overall mechanism is the same.

Peroxidation of lecithin and phosphatidylethanolamine and decomposition of the hydroperoxides have been followed by chemiluminescence (30,

31). Tocopherol inhibits peroxidation, as shown by decreases in the intensity of chemiluminescence; anthracene and acridine orange stimulate peroxidation and chemiluminescence (32). Removal of the choline group significantly enhances autoxidation, even in the presence of tocopherol (33).

The peroxidic products, including malonaldehyde, formed during autoxidation of phosphatidylcholine in human plasma react with thiobarbituric acid to form complexes having strong absorption at 532 nm (34).

Autoxidation of phosphatidylcholine (from egg lecithin) in benzene, 96% ethanol, and 0.1 M NaCl solutions showed differences due to molecular associations (35). For concentrations below 30 mg/ml, the solutions of lecithin in ethanol are molecular, and in this state the rate of autoxidation is reduced.

As the amount of water is increased in lecithin, a transition occurs from a crystalline state to more rapid packing of the hydrocarbon chains, and a decrease in rate of autoxidation occurs up to a water content of ca. 70% (36).

## Complexes

### Metal Ion Complexes

Phosphatidylcholine binds readily with $Ca^{2+}$, $Mg^{2+}$ and $Ce^{3+}$ ions in a 1:1 molar ratio in methanol, but binding is competitively inhibited by water, and at a 4:1 molar ratio of water to the phospholipid, cation binding is prevented entirely (37). The following association constants were reported: $Ca^{2+}$, 0.0286 $mM^{-1}$; $Mg^{2+}$, 0.0244 $mM^{-1}$; and $Ce^{3+}$, 2.5 $mM^{-1}$. Kinetic analysis with phospholipase $A_2$ indicated that the hydrated form of lecithin was the substrate in methanol/water solutions.

Calcium ion binds to both phospholipids and proteins. Ruthenium red undergoes substantial changes in the resonance Raman spectrum when added to calcium ion binding agents; the spectral changes are reversed when calcium ion is added. The spectra distinguish between different types of binding and permit the assignment of low-affinity calcium binding sites to phospholipids (38).

Calcium ion induces phase separation of phosphatidylcholine-phosphatidylserine bilayer membranes into a solid phase of phosphatidylserine aggregates bridged by calcium and a fluid phase of phosphatidylcholine (39). Spin labeling of the phosphatidylcholine made possible observation of the exchange broadening in the electron spin resonance spectrum as an indication of phase separation.

The binding of $Ca^{2+}$ to lecithin in a monolayer increases the surface potential; the binding varies with the degree of unsaturation in the lecithin, which suggests that the fatty acid characteristics have significant influence on the interaction (40). The influence of $Na^+$, $K^+$, $Li^+$, $Mg^{2+}$, $Ba^{2+}$, $Sr^{2+}$ and $Al^{3+}$ as well as $Ca^{2+}$ on surface area/pressure curves was determined.

Radioactive calcium ($^{45}Ca$) has been used to measure binding to monolayer films of lecithin (41).

The binding of divalent cations is stronger than that of monovalent cations. Anions also have an effect, and they impart decreasing stability to phospholipid emulsions in the following order: $SCN^- > NO_3^- > Cl^- > F^-$ (42).

As determined by $^{31}P$ NMR spectra, cobalt ions have weak bonding with glycerophosphorylcholine, used as a model head group region in membranes. Within the limits of pH 4.5–7.5, pH has no effect on the binding; however, with increasing temperature, the binding becomes stronger (43). The glycerophosphorylcholine apparently forms an inner-sphere complex and displaces water from the first hydration sphere of the cobalt.

Electrophoretic mobility has been used to measure the dissociation constants for several divalent cations adsorbed to phosphatidylcholine bilayer membranes (44) (see Table I).

**TABLE 11-I.**
**Dissociation Constant Measured by Electrophoretic Mobility**

| | Dissociation constant (M) |
|---|---|
| $Mn^{2+}$ | 0.3 |
| $Mg^{2+}$ | 1.0 |
| $Ca^{2+}$ | 1.0 |
| $Co^{2+}$ | 1.2 |
| $Ni^{2+}$ | 1.2 |
| $Sr^{2+}$ | 2.8 |
| $Ba^{2+}$ | 3.6 |

Phosphatidylinositol forms clear, stable dispersions in water on mild ultrasonication (45). Interaction of such dispersions with metallic cations has been studied by hydrogen ion release, turbidity and coagulation methods. In coagulation, cation selectivity was 2.4 to 1 for $Ca^{2+}$ compared to $Mg^{2+}$ and 1.2 for $K^+$ compared to $Na^+$. The reduced viscosity of phosphatidylinositol dispersions increased with dilution. Addition of small amounts of NaCl produced significant reduction in viscosity.

Fluidity of commercial lecithin can be improved by the addition of calcium or magnesium chloride, nitrate or acetate (46).

Precipitation of lecithin with magnesium sulfate is a potential commercial route to products useful in diet therapy in foods or as an emulsifier (47).

Lecithin forms a 3:1 complex with molybdophosphoric acid, and this complex can serve as a basis for clinical analysis (48).

**Iodine Complexes**

Iodine in aqueous solution changes color in the presence of micelles formed from amphipathic compounds, including lecithin; it does not undergo color change in presence of unassociated molecular species. The color change coincides with the critical micelle concentration of the substance, and is due to formation of the triodide ion, $I_3^-$ (49). In nonaqueous media, a similar complex is formed, perhaps as $HI_3$. Iodine forms a molecular charge-transfer complex with the lecithin both in aqueous solution and in the solid state (50).

Egg lecithin bilayers in distilled water exhibit resistance values that are reduced by 3 or 4 orders of magnitude when iodine is added to the water (51). Spectroscopic data confirmed the donor-acceptor complex concept.

Concentration of $I_3^-$ for the same concentration of lecithin in aqueous solutions varies with the amount of unsaturation and with solvent polarity (52). Breaks in the curves relating $I_3^-$ to lecithin concentrations occur at the critical micelle concentration.

Laser Raman studies of molecular interactions with phosphatidylcholine and iodine, iodide and triiodide indicated that the reaction occurred with iodine rather than the ionic forms (53). The reduction in resistance must then be related to iodine-phospholipid interaction and penetration of iodine into the bilayer membrane rather than to an ion transport process.

Interaction of iodine with monomolecular films at an air/water interface has been measured by changes in pressure/area isotherms. The behavior is markedly different with unsaturated lecithin compared to dipalmitoyllecithin, for which no area increase was observed (54). Iodine forms a weak charge transfer complex with the double bonds of the hydrocarbon chain in which the electron is not completely transferred, and vander Waal contribution to complex stability is probably important. Changes in absorption bands for the complex in light vs dark exposure are reversible. Conduction in the iodine-doped bilayers is probably due to an electronic rather than an ionic mechanism.

### Other Complexes

Phosphatidylcholine hydrates more easily than phosphatidylethanolamine or phosphatidic acid in degumming operations, but the last two compounds are more easily hydrated in the presence of phosphatidylcholine as the result of phospholipid-phospholipid interactions (55). Phosphatidylcholine forms mixed micelles with the other compounds, and the interaction was confirmed by $^{31}P$ nuclear magnetic resonance spectroscopy.

Phosphatidylcholine also forms complexes with soy protein (56). After mild sonication, radiolabeled ($^{14}C$) phosphatidylcholine was found to be nonspecifically bound to either 7S or 11S proteins, as shown by a linear sucrose density gradient centrifugation analysis.

## Halogen Addition

A colorless, wax-like substance that melts above 100 C with decomposition and is soluble in chloroform, ethanol and diethyl ether but hardly soluble in acetone or water has been obtained by the reaction of egg lecithin with bromine in chloroform solution at temperatures of 30 C or below (57). Small amounts of a mixture of bromine and iodine add to lecithin to give a product said to be useful for pharmaceutical and industrial emulsifiers (58).

Iodized lecithin has been prepared from lecithin and iodine in carbon tetrachloride solution with iron, tin or antimony periodide as catalyst (59). The product is said to be stable, not hygroscopic and useful for pharmaceutical purposes.

Soybean lecithin iodide may be prepared by reaction of iodine-potassium iodide with lecithin in urea solution at 60 C for 20 hr (60).

The addition of iodine to lecithin in the dark has been followed by nuclear magnetic resonance spectroscopy over a period of 10 days in chloroform or deuteriochloroform solution at 20 C (61). An ionic, *trans* addition-*cis* elimination was postulated to explain formation of intermediate iodinated *trans* double bonds.

## Hydration

Phosphatidylcholine binds one molecule of $H_2O$ per mole very tightly to form a monohydrate. The monohydrate itself is hygroscopic and absorbs more moisture. Anhydrous phosphatidylcholine can be obtained by heating the monohydrate at 90 C for 4 hr under vacuum and also by slow crystallization (of distearoylphosphatidylcholine) from dry ether/chloroform solution (62). Crystallization from chloroform/methanol solution yields two different hydrated forms; a third hydrated form is produced by precipitation with ether from a chloroform solution.

The saturated phosphatidylcholines are polymorphic; thermotropic and lyotropic mesomorphism were examined by several different techniques including light microscopy, differential thermal analysis, and scanning calorimetry; X-ray diffraction, infrared and nuclear magnetic resonance spectroscopy (62).

The extent of hydration, for which the polar head group is responsible, has been estimated by several different procedures. Gravimetric methods for adsorption of water vapor indicated 2.5–2.7 molecules for each phosphatide molecule in a first hydration layer, with a second hydration layer up to a total of 5 moles/mole (63). Further hydration expands the space between biomolecular leaflets and results in a colloidal solution.

Calorimetric techniques showed that ca. 10 molecules of $H_2O$ are bound per mole (62).

Nuclear magnetic resonance studies have indicated a minimum of 11–12 molecules of water per phospholipid molecule (64) and a first hydration shell of 4–5 molecules and a weaker secondary shell of up to a total of 10 molecules (65), the latter study having been carried out on dipalmitoylphosphatidylcholine and its perdeuterated derivative.

Interaction of water with the polar region of phosphatidylcholine micelles in benzene has been investigated also by means of nuclear magnetic resonance (66). The water binding results in changes of the width of the quaternary ammonium resonance. The data indicated only one population of water and one of the quarternary ammonium group, signifying that all the micelles have the same concentration of water and that the water moves rapidly over the various parts of the micelle. A first layer of water is more strongly absorbed than the second layer, resulting in broad proton resonance at low concentrations of $H_2O$ and a sharpening as the second layer forms.

Deuteron magnetic resonance spectroscopic studies on the hydration of egg lecithin resulted in the conclusion that the lecithin dispersions contain water in four kinds of binding: bulk, trapped, loosely bound and tightly bound (67). Bulk water is present when $n>23$ (molar ratio of water to lipid) and exchanges slowly with water associated with lipid. Trapped water, up to 11 molecules, undergoes rapid isotropic motion and exchanges rapidly with bound water. Loosely bound water consists of a main hydration shell of 11 molecules. Tightly bound water consists of the one water molecule in a monohydrate, which is bound symmetrically between neighboring phosphate groups.

A more recent study on egg lecithin in benzene solution has been carried out with 60 MHz and 100 MHz nuclear magnetic resonance spectroscopy, infrared spectrophotometry and phase contrast microscopy (68). Free water was found to be present when more than 9 water molecules per lecithin monohydrate were present. The first added water is bound strongly to the phosphate group. Additional water molecules are less strongly bound but still have restricted motion compared to free water.

Phosphatidylcholine absorbs more water vapor than does phosphatidylethanolamine, and the latter adsorbs 4–5 times more $H_2O$ than tristearin (69).

Results obtained by a resonating quartz crystal microbalance method compare favorably with earlier methods (70). A general equation has been derived, but the assumptions of identical and noninteracting primary sorption sites probably are not valid.

Variations in the thickness of the water layer, produced by different ionic environments, effect remarkable changes in permeability of phospholipid membranes (71).

## Hydrogenation

### Heterogeneous Catalysis

Hydrogenated lecithins are more stable, more easily bleached to a light color, and hence more useful as emulsifiers than the natural, highly unsaturated lecithin from soybean oil. Although the phosphatides are inhibitors for the hydrogenation of soybean oil, they may be hydrogenated at ordinary temperatures and pressures in solvents containing a chlorinated hydrocarbon, e.g., a mixture of ethanol and methylene chloride (72). Catalysts such as 10% palladium on charcoal or platinum oxide can be used. Cyclohexane is not a satisfactory solvent. The products were dark, but bleaching with 1–3% of 50% hydrogen peroxide gave light-colored materials having granular properties, excellent keeping qualities, and undiminished emulsifying power compared to the unhydrogenated lecithin.

Soybean lecithin may be hydrogenated with 5% flaked Ni catalyst and 0.05% Pt catalyst at 100 atm and 100 C for 4 hr (73). A stabilized emulsifier for parenteral feeding mixtures is produced by controlled hydrogenation of soybean lecithin with 1% of $PtO_2$ catalyst at 230 psi for 4 hr at 65 C to an IV of 50 (74).

Prehydrogenation of the catalyst (10% palladium on carbon) is said to prevent formation of a green color characteristic of most hydrogenated lecithin products (75). The hydrogenations were carried out as the acetylated lecithins in benzene solution.

Hydrogenation of O-alk-1-enylglycerols occurs readily over $PtO_2$ catalyst (76, 77).

Hydrogenation of phosphatides from sunflower oil proceeded at a faster rate in ethanol than in hydrocarbon solvents, with decreasing phosphatide content, with increased amount of catalyst, increased pressure and higher temperature. Catalysts used included nickel, 2% palladium on carbon, and reduced Ni-Cu on diatomite (78). Stationary catalysts are more stable than suspended ones (79). Continuous hydrogenation was successful at 80–120 C, at pressures greater than 15 atm, and with a fixed bed Ni-Cr catalyst for 10–15% concentrations in ethanol. In mixed alcohol solvents, the rate of hydrogenation increased with increasing ethanol concentration (80). Further work showed that the hydrogenation rate was highest with methanol and slowest with isopropyl alcohol and that hydrogenation with 10% pure palladium was faster than with 10% mixed platinum-palladium catalyst (81–83).

$PtO_2$ serves to catalyze hydrogenation of lecithin or diglyceride at ambient temperatures, atmospheric pressure, and a time of 90 min (84).

### Homogeneous Catalysis

Aqueous dispersions of unsaturated phosphatidylcholines may be hydrogenated by water-soluble, amphiphilic rhodium catalysts such as chlorotris (sodium diphenylphosphenobenzene-*m*-sulfonate) rhodium (**I**) (85). The hydrogenation proceeds under mild conditions (1.2 atm pressure, 37 C and pH 6.9). The phospholipid dispersion was prepared by sonication. Previously, hydrogenation had been carried out with homogeneous but water-insoluble rhodium catalysts that required the use of tetrahydrofuran or dimethyl sulfoxide to permit penetration of the membrane (86). Hydrogenation of aqueous bilayer dispersions of phosphatidylcholine proceeds at satisfactory rates (2 hr) with $RhCl\ (PPh_3)_3$ catalyst by increasing hydrogen pressure up to 1 MPa (87).

## Hydrolysis and Alcoholysis

### Introduction

Several types of ester functionality, all capable of hydrolysis, are present in the component parts of vegetable phosphatides (see Scheme 1). These may be hydrolyzed totally by chemical methods or selectively by either chemical or enzymatic methods.

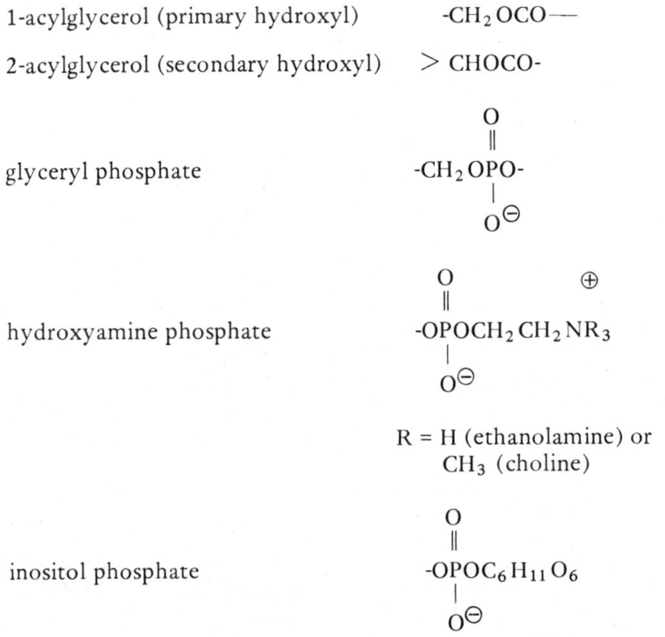


**SCHEME 1**

Total hydrolysis into each of the component parts of the phospholipid can be accomplished by strong acid (HCl, $H_2SO_4$) catalysis in 6 N aqueous or 5–10% methanolic solutions (2, 88). Hydrolysis in strong alkali results in deacylation and cleavage of the hydroxamine phosphate bond, but cyclic phosphate esters formed during the reaction are resistant to hydrolysis:

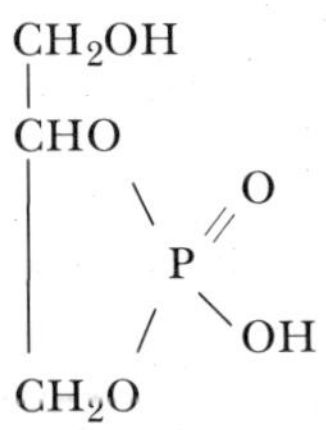

Almost complete saponification within a few minutes is said to take place in a solution of 1 part chloroform and 2 parts saturated methanolic KOH (89).

Kinetics and mechanism for hydrolysis in 2 N HCl at 120 C have been described (90). Deacylation occurs first, followed by formation of a cyclic phosphate triester as an intermediate to cyclic glycerophosphate and choline. The cyclic glycerophosphate is eventually converted to an equilibrium mixture of $\alpha$- and $\beta$-glycerophosphate.

Both chemical and enzymatic hydrolyses are discussed in the following paragraphs. Enzymatic hydrolysis is by far the more selective in lecithin hydrolytic reactions.

Myher (91) and Strickland (3) have reviewed the hydrolysis of phospholipids in general. Dawson et al. (92, 93) give an excellent account of the stepwise hydrolysis of phosphatides for analytical purposes.

### Acylglycerol Bonds

Deacylation takes place under a variety of conditions and reagents to produce *sn*-glycerol-3-phosphorylethanolamine (**I**) or *sn*-glycerol-3-phosphorylcholine (**II**) and fatty acid, salt or ester.

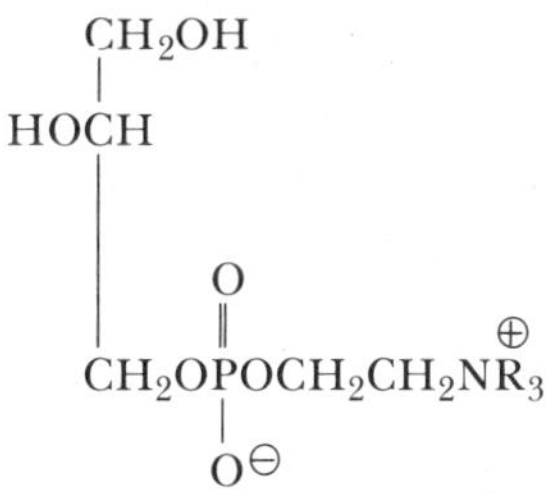

**I**, R = H

**II**, R = $CH_3$

These phosphorylglycerol esters are valuable intermediates for the synthesis of various phospholipids (94).

Deacylation may be carried out either by hydrolysis or by methanolysis (transesterification) in the absence of water.

Kensil and Dennis (95) have studied in some detail the alkaline hydrolysis of phospholipids in model membranes and the dependence on their state of aggregation. Partial hydrolysis with 0.132 M NaOH at 40 C to the 50% point produced 2-acyl and 1-acyl lysophosphatidylcholine as intermediates in the production of glycerolphosphorylcholine. The hydrolysis was pseudo first-order, with hydroxide ion in large excess, and was dependent on the state of aggregation (Table II) as well as the specific polar group (Table III). The fatty acid chain length in the phospholipid, at least in the range of 12:0, 14:0 and 16:0, had only a minor effect in increasing the rate of hydrolysis with decreasing chain length.

**TABLE 11-II**
**Rates and Activation Energies of Alkaline Hydrolysis of Phosphatidylcholine as a Function of Aggregation State (95)**

| | Phosphatidylcholine type | | |
|---|---|---|---|
| Item | Dihexanoyl | Egg | Dipalmitoyl |
| Reaction rates (k × $10^3 s^{-1} M^{-1}$) | | | |
| Monomeric | 370 | — | — |
| Mixed micelles | 77 | 14.7 | 13.8 |
| Vesicles | | | |
| 100% accessibility | — | 1.4 | 0.7 |
| 68% accessibility | — | 2.2 | 1.1 |
| Multibilayers | — | 4.1 | — |
| Activation energies (kcal/mol) | | | |
| Triton X-100 micelles | — | 10.2 | 11.5 |
| Vesicles | — | 19.4 | 17 |
| Monomers | 12.6 | — | — |

**TABLE 11-III.**
**Effect of Polar Group in Phospholipids on Rates of Alkaline Hydrolysis (95)**

| Egg phosphatidyl compound | Rate ($s^{-1}M^{-1}$) | Relative rate |
|---|---|---|
| Choline | 14.0 | 1.0 |
| Ethanolamine | 4.6 | 0.3 |

Hydrolysis is carried out readily with the addition of 1 M methanolic NaOH to a solution of the phospholipid in $CCl_4/CH_3OH/H_2O$ and incubation of the mixture at 37 C for 15 min. After neutralization, washing and extraction of water-insoluble materials, the aqueous solution is taken to dryness under reduced pressure at temperatures below 50 C (96).

Mild alkaline hydrolysis for deacylation of phospholipids usually involves some undesired hydrolysis of the choline-phosphate bond as well. This side reaction can be avoided by the use of a polar solvent. The hydrolysis is carried out with 0.1 M lithium hydroxide for 15 min at room temperature in a chloroform/methanol (2:8, v/v) solution (97).

Calcium, barium and strontium hydroxides, but not magnesium hydroxide, are also effective for hydrolysis. Heating lecithin with 0.005 N barium hydroxide in methanolic solution for 30 min at 37 C produced high yields of L-$\alpha$-glycerylphosphorylcholine (98).

Basic hydrolysis (alcoholysis) with mild alkali treatment for 15 min produces small amounts of a cyclic glycerophosphate. Substantial amounts of the cyclic ester are produced after 20 hr in 0.1 N methanolic potassium hydroxide at 37 C (99) (Scheme 2). The cyclic 1,2-glycerophosphate has been synthesized and characterized as a crystalline barium salt.

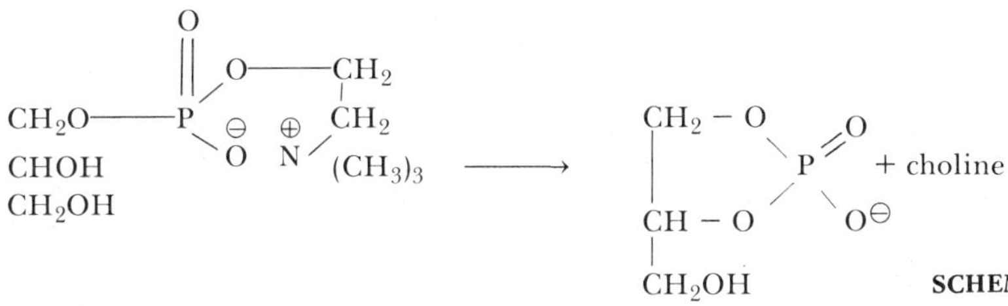


**SCHEME 2**

Hydrolysis may also be carried out with hydroxylamine in aqueous ethanol at room temperature for 75 sec (100). Glycerylphosphorylcholine is isolated as the $CdCl_2$ complex in 65% yield. The presence of excess base (NaOH or $NaOCH_3$) appears to be necessary (101, 102).

Hydrolysis under the influence of mercuric chloride has also been described (103), but this method is cumbersome and unattractive.

Alcoholysis will occur if the reaction takes place in an alcoholic solvent, such as methanol, instead of water to produce the methyl esters of the fatty acids. Unless mild conditions are used, such compounds as cyclic D-1,2-glycerophosphate or methyl glycerolphosphate will be formed. It is essential that an excess of triglyceride or acidic phospholipid be absent to avoid neutralization of catalyst.

Methanolysis may be catalyzed with the quaternary base tetrabutylammonium hydroxide (1.0 M in methanol) to produce an 84% yield of glycerolphosphorylcholine from purified egg yolk when allowed to react with the phosphoglyceride in ether solution for 1 hr at room temperature. The product separates as a glassy precipitate from the solution (104). Tetraethylammonium and tetramethylammonium hydroxides are equally effective (105). The quaternary ammonium hydroxides catalyze methanolysis without racemization, as does lithium methoxide, and in better yields than obtained with the latter.

Methanolysis with methanolic NaOH (0.5 N) is essentially complete

(88%) within 10 min at room temperature (102). The reaction is complete only if the molar ratio of phosphorus to alkali does not exceed one.

Methanolic lithium methoxide (1.0 M) catalyzes room temperature transesterification of phosphatidylethanolamine dissolved in dry ether to yield a sticky precipitate of glycerolphosphorylethanolamine after 2 hr. Work-up of the product resulted in a white, lipid-free, hygroscopic solid (106). Methanolysis of egg-yolk powder by either methanolic lithium or sodium methoxide, sodium hydroxide or tetrabutylammonium hydroxide produces a crude mixture of deacylated phospholipids that can be separated chromatographically (silicic acid) into pure glycerolphosphorylethanolamine and glycerolphosphorylcholine fractions. Transesterification must be carried out in anhydrous solvent; ether is used because the deacylated products are insoluble in it. To protect the amino group in the ethanolamine moiety during subsequent synthetic reactions, the phthalimide formed by reaction with phthalic anhydride may be formed initially and carried through the deacylation reaction (107).

Instead of the work-up procedures described above, the deacylated product can be isolated by precipitation as the $CdCl_2$ complex, which can be used directly for subsequent syntheses (108). The $CdCl_2$ complex can be converted to the pure glycerolphosphorylcholine by passage through an ion change column ($H^+$ form) in 60% yields. Further purification reduces the yield to 40% (103).

Synthesis of saturated phosphatidylcholines can be facilitated by the reaction of the imidazolides of the fatty acids with the $CdCl_2$ complex of glycerolphosphorylcholine (109).

Ethanolaminolysis has been carried out on lecithin and glycerides, and this reaction has been suggested for determining the ethanolamine index number for these compounds by determining the excess of ethanolamine after reaction (110) (Scheme 3).

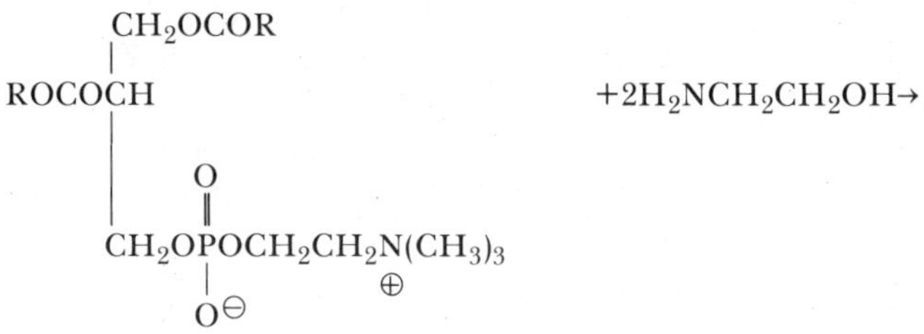


**SCHEME 3**

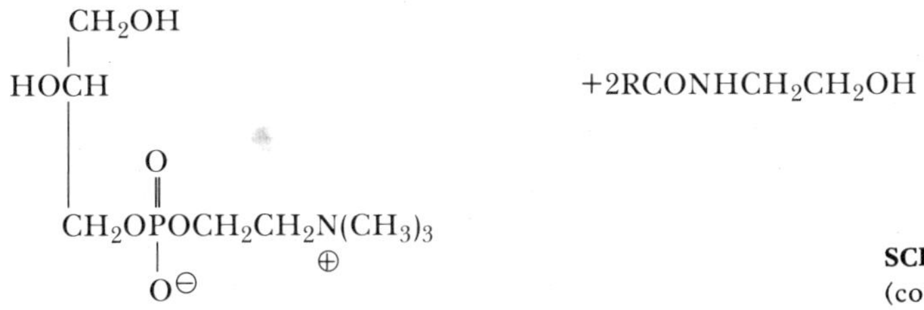

**SCHEME 3**
(continued)

Enzymatic hydrolysis of the acylglycerol bond can be carried out selectively at either the 1-position with phospholipase $A_1$ or the 2-position with phospholipase $A_2$. Sources and preparation of phospholipase $A_1$ have been described in detail, e.g., from rat brain (111) and rat liver liposomes (112).

Sources of phospholipase $A_2$ include the pancreas (113), snake venom (114), sheep red cell membranes (115), bee venom (116) and cobra venom (117).

Enzymatic hydrolysis can be carried out by numerous enzyme preparations other than those already mentioned, including those from rat small intestine mucosa (118), *Escherichia coli* (119), *Crotalus atrox* (120), cobra (*Naja naja naja*) (121), and *Crotalus adamanteus* venom (122); and *Saccharomyces cerevisiae* plasma membranes (123).

Tinker and Wei (120) have discussed the mechanism of heterogeneous catalysis by phospholipase $A_2$, taking surface effects into account; the catalytically active form of the enzyme is a dimer.

The small intestinal mucosa from the rat contains two major phospholipid deacylating enzymes, one of which has been characterized as having phospholipase $A_2$ activity. The latter was active in the presence of deoxycholate and at an optimum pH of 9.6. The other enzyme, as yet uncharacterized as to specificity, had a pH optimum of 5.9 and was active in the absence of deoxycholate (118).

A secretory protein of rabbit parotid gland has been found to have phospholipase $A_2$ activity. The enzyme had an apparent molecular weight of 16,000; it was active on phosphatidylcholine, phosphatidylethanolamine, phosphatidylinositol and phosphatidylserine (124).

In contrast to the phospholipase A activity of several snake venoms, that from *Ophiophagus hannah* does not exhibit preferential rates of hydrolysis for different fatty acids. Furthermore, its action is faster than others and is quite suitable for distinguishing acyl groups in the 2-position from those in the 1-position (125).

Rat hepatic lipase has been isolated as a single enzyme molecule that has triglyceride lipase, monoacylglyceride lipase and phosphatidylcholine lipase activities (126). The phosphodiesters are not hydrolyzed further. However, the intermediary lysophospholipids are hydrolyzed by alternative pathways to glycerol and ethanolamine or choline (127).

The $\alpha$-phospholipase $A_2$ activity of *Crotalus adamanteus* (Eastern diamond back rattlesnake) (114) for monolayers of 1,2-dioctanoyl-*sn*-glycero-3-phosphorylcholine requires the dimeric form of the enzyme (122). The dissociation constant for the dimer was $2 \times 10^{-9}$ M at pH 6.0 and $7 \times 10^{-3}$ M calcium ion. The hydrolytic reaction is first-order with respect to the surface concentration of the phospholipid and to the concentration of the enzyme dimer and requires reversibly bound $Ca^{2+}$.

Hydrolysis of monoacylglycerolphosphorylcholines can be carried out enzymatically by the action of *Penicillium notatum* in yields of 40–65%. Hydrolysis occurs much more rapidly with unsaturated than with saturated fatty acids in the 2-position (128). An enzyme having phospholipase B activity for 2-acyl and then 1-acyl groups in sequence has been isolated from *P. notatum* (129). The enzyme was characterized as having an MW of 116,000, an isoelectric point at pH 4.0, and a carbohydrate content of 30%.

## Glyceryl Phosphate Bonds

Hydrolysis of phosphatidylinositol to diglyceride and inositol phosphate occurs with 2 N HCl or glacial acetic acid at 100 C. Glycerolphosphatidylcholine hydrolyzes under strong acid conditions with intermediate cyclic compounds to glycerol and phosphatidylcholine. Strong acid hydrolysis, 6 N HCl (aqueous) or 2 N HCl (methanolic), results in complete hydrolysis to fatty acids, glycerolphosphate and the amino alcohols or inositol. Glycerol phosphates hydrolyze on prolonged treatment with strong acid at high temperatures (130).

Enzymatic hydrolysis of the glyceryl phosphate bond takes place with phospholipase C to make the diglyceride and the phosphoryl hydroxybase compound (Scheme 4). Sources of phospholipase C include *Bacillus cereus* (131, 132) and *Clostridium perfringens* (133). Phosphatidylinositol-specific phospholipase C can be isolated from *Bacillus cereus* and *Bacillus thuringiensis* (134) and *Staphylococcus aureus* (135).

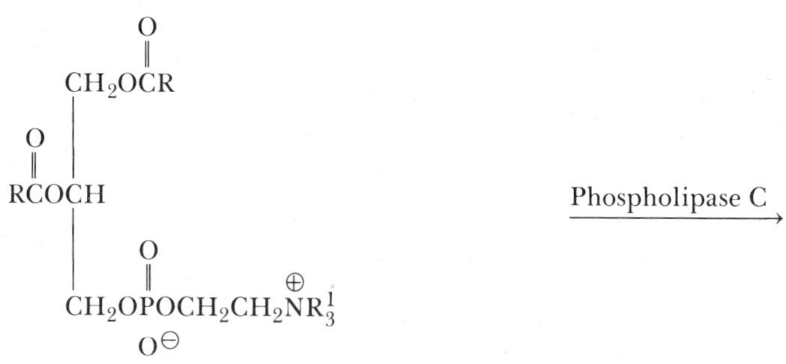


**SCHEME 4**

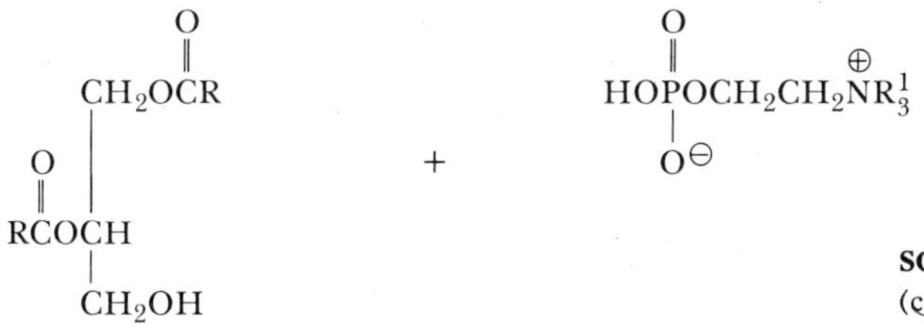

**SCHEME 4**
(continued)

1,2-Diglyceride acetates can be produced by acetolysis directly (88, 136, 137) as well as by hydrolysis (by phospholipase C) followed by acetylation with pyridine and acetic acid. The direct acetolysis route, however, cannot be recommended for subsequent fatty acid positional analysis, since there is considerable rearrangement of the acyl groups (138, 139) (Scheme 5).

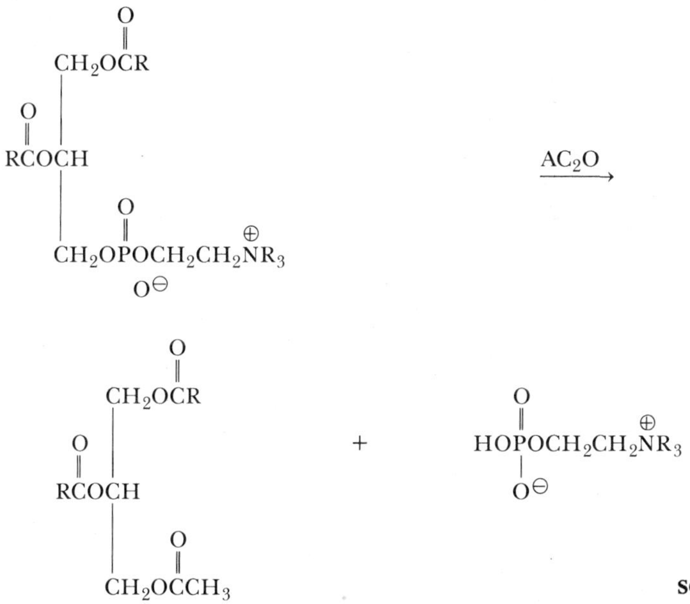

**SCHEME 5**

Small amounts of acetic anhydride, when caused to react with dry lecithin, form a product that emulsifies more readily and is more stable than conventional lecithin (140). The dispersibility of lecithin in water is improved by treatment with anhydrides having less than 10 carbon atoms followed by hydroxylation with hydrogen peroxide (141).

Dephosphorylation of phospholipids from *Escherichia coli* K12 has been carried out by dispersing the phospholipid with sonication in 10 mM Tris-HCl, pH 7.5, then adding phospholipase C (type V from *Bacillus cereus,*

EC. 3.1.4.3) and benzene and agitating for 2 hr at 25 C (142). The benzene solution resulting after centrifugation was evaporated to recover the diacylglycerol, which was converted to the $\rho$-methoxybenzoate for further analysis. The latter derivative had improved storage properties compared to the $\rho$-nitrobenzoates previously used (143).

The phosphoroglycerol bond can be hydrolyzed selectively by phospholipase C from *Clostridium welchii* to produce the diglyceride and the phosphorohydroxyamine fragments. The conditions for this hydrolysis, as described by Holub and Kuksis (144), include the presence of sphingomyelin, trishydroxymethylaminoethane and calcium ion; the digestion is complete at room temperature at pH 7.3 after 3 hr in diethyl ether containing 1% water.

Phospholipase C dephosphorylates glycerolipids containing ether groups, and the acyl group can be removed by saponification to produce alk-1-englycerols or alkylglycerols (145, 146). The dephosphorylation is carried out in 0.05 M phosphate buffer (pH 7.1) at room temperature for 2.5 hr. Alk-1-englycerols are determined by acid hydrolysis to produce aldehydes, which are conveniently converted to acetals (147) (Scheme 6).

$$\begin{array}{l} CH_2OCH=CHR \\ | \\ CHOH \\ | \\ CH_2OH \end{array} \xrightarrow{H^+,\ H_2O} \begin{array}{l} CH_2OH \\ | \\ CHOH \\ | \\ CH_2OH \end{array} + OCHCH_2R$$

$$RCH_2CHO \xrightarrow{H^+,\ CH_3OH} RCH_2CH(OCH_3)_2$$

**SCHEME 6**

Phosphodiesterase is a dominant enzyme in the sheep pancreas (but not the pancreatic juices) and is capable of producing inositol monophosphate and its cyclic derivative from phosphatidylinositol in the presence of $Ca^{2+}$ (148). The pH optimum was ca. 6.0, with greater amounts of the cyclic ester being formed at low pH and lesser amounts at high pH.

The phosphoroglycerol bond can also be hydrolyzed by thermal cleavage of hydrated phosphatidylcholine, -ethanolamine, -serine and -inositol (149). The thermal cleavage can be carried out in diphenyl ether at 250 C for a short period; after 1 min, mostly 1,2-diglycerides are formed but conversion is incomplete; after 5 min, the conversion is complete but the diglycerides have rearranged to mainly 1,3-diglycerides. The diglycerides are then converted to silylated derivatives for GC–MS analysis. Alternatively, the hydrated phospholipid could be heated for several hours at 60 C with *bis*-trimethylsilylacetamide and trimethylchlorosilane to form the derivatized diglycerides for analysis, but the yields are lower than those obtained by the diphenyl ether method. The hydrolytic cleavage was thought to occur through an *ortho* acid structure for the phosphorus (see Scheme 7).

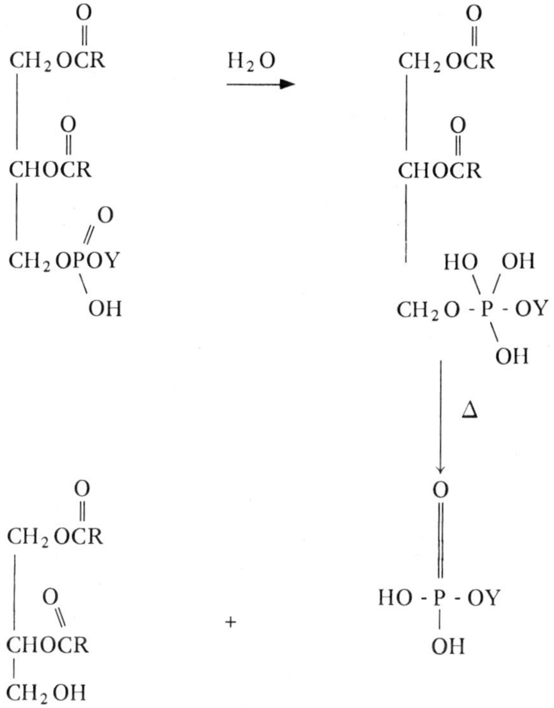

SCHEME 7

**Hydroxamine Phosphate Bonds**

Enzymatic hydrolysis of this bond to form phosphatidic acid and choline or ethanolamine is carried out with phospholipase D (Scheme 8). The latter is available from plastid preparations from carrot root or chloroplast preparations from spinach, cabbage or sugar beet leaves (150), peanut seeds (151), Savoy cabbage (152) and rat brain (153).

Phosphorohydroxyamine bonds are also cleaved in 1 N HCl when heated at 100 C for 18 min, e.g., (96), to give the free hydroxyamine and glycerophosphoric acid.

Diazomethane reacts with several types of phosphatides to produce the dimethyl ester of phosphatidic acid (154). Phosphatidylethanolamine reacted with diazomethane to give N-methylated products to some extent, a tendency that was enhanced by the presence of alcohols.

Apparently, the enzyme from Savoy cabbage catalyzes both hydrolysis and alcoholysis (transphosphatidylase activity) (155). Incubating phosphatidylcholine with the enzyme in the presence of ethanol, ethanolamine and glycerol produces the respective ethyl phosphatidate, phosphatidylethanolamine and phosphatidylglycerol esters plus choline.

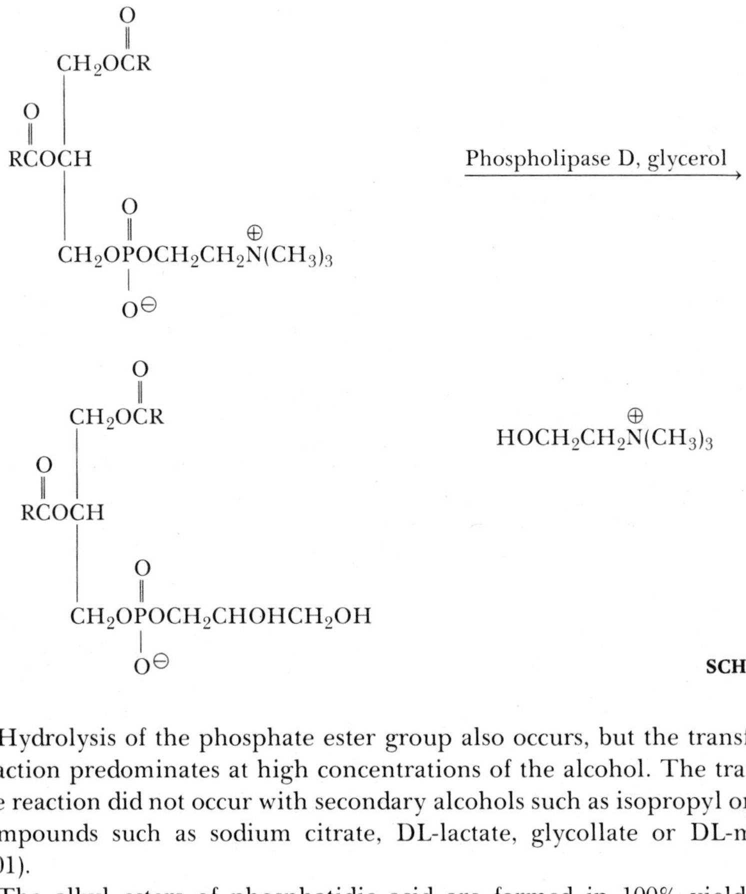


**SCHEME 8**

Hydrolysis of the phosphate ester group also occurs, but the transferase reaction predominates at high concentrations of the alcohol. The transferase reaction did not occur with secondary alcohols such as isopropyl or with compounds such as sodium citrate, DL-lactate, glycollate or DL-malate (101).

The alkyl esters of phosphatidic acid are formed in 100% yield with alkanols having up to 3 carbons, lesser yields with alkanols having 4–6 carbons, and in 0% yield with alkanols having 7–10 carbons (156). With the latter alkanols, the main product was phosphatidic acid. This effect is to be expected, since the higher alcohols would have little solubility in the aqueous phase of the ether/water heterogenous mixture where the enzyme is active. A buffer consisting of sodium acetate and calcium chloride adjusted to pH 5.6, a reaction period of 2 hr, and a temperature of 20 C were used.

Alkanediols (2–4 carbons), unsaturated (allyl and propargyl) alcohols, 2-fluoroethanol, ethanolamine, propanolamine and methoxypropanol also can be used as acceptors, but iodoethanol and mercaptoethanol cannot.

Choline and phosphatidic acid are readily produced by hydrolysis of phosphatidylcholine with phospholipase D from peanut seeds, when the phospholipid is dispersed by sodium dodecylsulfate and when calcium ions are present (157). The active enzyme species has an MW of 200,000. The activity depended on the molar ratio of surfactant to phospholipid and on size of the aggregates.

Thorium and zirconium ions can dephosphorylate several phosphoric acid esters (158). For example, thorium ions hydrolyze phosphorylcholine to the extent of 50% and glycerophosphate to 70% at pH 9.5. Lecithin is dephosphorylated to the extent of 10% at pH 7.

**Inositol Phosphate Bonds**

For a review on early work with phosphatidylinositol, Hawthorne (159) should be consulted.

Complete hydrolysis of phosphatidylinositol occurs with 4 N HCl after refluxing for 40 hr; inositol is stable to this acid concentration for at least 46 hr but is partially destroyed by 6 N HCl. Use of 2 or 3 N HCl is insufficient for consistent conversion (160).

Alkaline hydrolysis (0.5 N NaOH at 100 C for 40 min) of glycerylphosphorylinositol yields a mixture of inositolmonophosphate (30–40%) and glycerophosphate (161).

Most enzymes that are active toward phosphatidylinositol do not affect the inositol phosphate bond. Phospholipase C preparations from *Bacillus cereus* attack the glycerol phosphate bond and at the same time produce an inositol cyclic phosphate by a phosphotransferase mechanism (162). A phosphatidylinositol specific phospholipase C that was inactive toward phosphatidylcholine and phosphatidylethanolamine has been isolated in purified form from rat liver (163). The enzyme had an MW of 68,000–70,000 and was activated by unsaturated fatty acids and $Ca^{2+}$ in synergistic fashion. The water-soluble products of the hydrolysis contained 56% of 1,2-cyclic phosphate and 54% of *myo*-inositol 1-phosphate.

Rat gastric mucosa homogenates contain two enzymatic systems, one of which is capable of hydrolyzing phosphatidylinositol to 1,2-diacylglycerol and inositol phosphate; the other is a deacylating enzyme (164).

## Hydroxylation

Hydroxylation of the double bonds in the unsaturated fatty acids of lecithin improves the stability of the lecithin and the dispersibility in water and aqueous media. Hydroxylation has been carried out with hydrogen peroxide in glacial acetic acid or sulfuric acid at moderate temperatures in accordance with the general equation (165) shown in Scheme 9. Such products are useful in candy manufacture in which sharp moldings can be obtained when the hydroxylated product is used with starch molds (166).

$$-CH=CH- + H_2O_2 \xrightarrow{H^+} -\underset{OH}{\overset{}{CH}}-\underset{OH}{CH}-$$

**SCHEME 9**

Hydroxylation has also been carried out with sodium perborate, sodium carbonate peroxide, sodium phosphate peroxide and potassium permanganate (167). The products were useful as softeners for starch sizing of textile fabrics.

Subsequent treatment of the hydroxylated lecithin with a weak sulfonating agent such as the ethylene chloride- or dioxane-sulfur trioxide complex produces the sulfonated compound. The latter is said to emulsify more readily in water, will not precipitate readily in the presence of acid, and is effective as a detergent and in increasing film strength in lubricating oils (168).

Partial hydroxylation, which can be carried out with perlactic or peracetic acid to the extent of 10–25%, is said to produce emulsifiers of optimum quality (169).

## Oxidation

Periodate oxidation of O-alkylglycerols (146) and colorimetric determination of the product O-alkylglycolic aldehyde with fuchsin reagent (170) provide a convenient method of analysis for these compounds.

The study of the oxidation of biomembrane lipids is of interest because of its importance in physiological processes such as aging. One study has been conducted with the decomposition products of potassium peroxychromate (171). The oxidation may have involved singlet oxygen as one of the decomposition products, but singlet oxygen traps did not prevent oxidation of bilayer dispersions of phosphatidycholine. The oxidation was limited to less than 50% of the unsaturated fatty acids present. The phospholipid was present as multilaminar liposomes. With a surfactant (Tween 80) present, the micellar structures permitted more complete oxidation.

Double bond position in dimethoxyglyceryl ethers of fish liver oils has been determined by oxidation with chromic acid (172) (see Scheme 10).

The dimethoxyglycerol ethers were made from the deacylated ethers by reaction with diazomethane in the presence of $BF_3$ etherate. The glycerol ethers included those having the side chains 16:0, 16:1, 18:0, 18:1, 18:2, 20:1 and 22:2, with the double bond in the 9,10-position for the 16- and 18-carbon chains and in the 13,14-position for the 20- and 22-carbon chains.

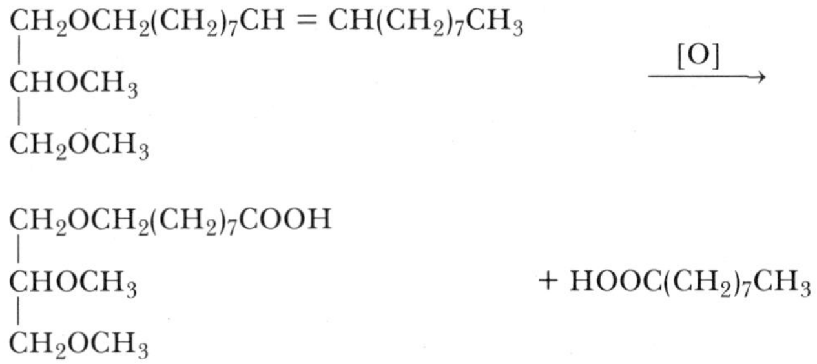

**SCHEME 10**

Structural analysis of lecithins from several sources (egg, soybean, wheat germ and beef spinal cord) has been carried out by reductive ozonolysis of the unsaturated fatty acids present. TLC was applied both to the ozonides for separation into classes, depending on the number of ozonide groups present, and to the aldehyde "cores" resulting from catalytic hydrogenation of the ozonides with Lindlar catalyst (173). The TLC plates were analyzed quantitatively by densitometry.

Reductive ozonolysis was used to determine the double bond location in the acyl, alkyl and alk-1-enyl hydrocarbon chains of the various lipid classes in the preputial glands of mice (76).

Epoxidation of L-α-1-stearoyl-2-oleoyl phosphatidylcholine and the palmitoleoyl isomer has been carried out by *m*-chloroperoxybenzoic acid in chloroform solution, the reaction taking 24 hr at room temperature for completion (174,175). The phospholipid epoxides were a poor substrate for epoxide hydrolase but a good substrate for phospholipase $A_2$. In contrast, *cis*-9,10-epoxy-stearic acid was rapidly hydrated by microsomal and cytosolic epoxidase hydrolase. The phospholipid epoxides are among the products formed during peroxidation of biomembranes (Scheme 11).

$$-CH=CH- \xrightarrow{RCOOOH} -\overset{O}{\widehat{CH-CH}}-$$

**SCHEME 11**

## Radical Reactions

Phosphatidylcholine bilayer vesicles react with hydroxyl radicals generated by pulse radiolysis of $N_2O$-saturated aqueous solutions containing the lecithin (176). Head groups of the model membrane are damaged, permitting increasing water penetration, and the membrane becomes crosslinked, thus increasing the rigidity of the membrane. Reaction of $e^{-}_{aq}$ with the membrane was very slow.

Phenylcarbene and adamantylidene generated photochemically from their diazirines react both with the C–H bond in saturated fatty acids and with the double bonds in unsaturated fatty acids in phospholipid vesicles, thus providing a means of labeling the phospholipid and providing a means for identification of intrinsic hydrophobic sections in membrane proteins (177) (Scheme 12). Phenylnitrene, however, provided only low yields of labeled products (178).

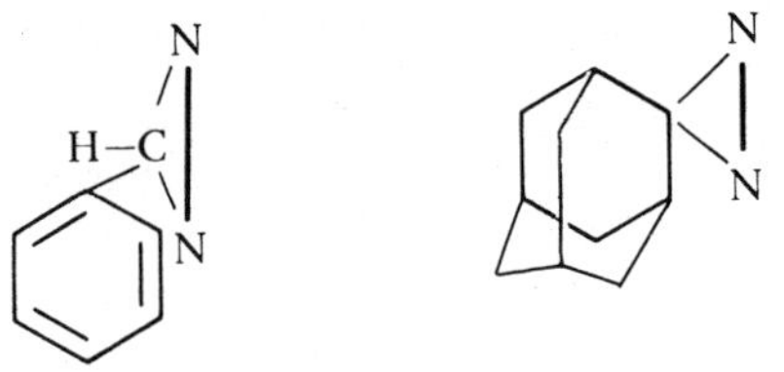


**SCHEME 12**

ApoHDL (high density lipoprotein) has been reconstituted with phosphatidylcholine substituted with azido fatty acids (179). Photoactivation of the azide groups resulted in a crosslinked apolipoprotein in which the phospholipid was crosslinked to both apolipoprotein and to acyl groups of the same lecithin molecule. The physical properties of the crosslinked lipoprotein were not changed by the crosslinks, and hydrophobic interactions of the two materials have been demonstrated by chemical means.

The products from $\alpha$-irradiation (30–1000 rad) of lecithin have been characterized (180). Radiolysis of lecithin emulsions with 30–100 kR from $^{60}Co$ resulted in formation of conjugated peroxides and other products (181).

**Reduction (Hydrogenolysis)**

Hydrogenolysis of lecithin has been carried out with lithium aluminum hydride under carefully controlled conditions in ether solution to produce L-$\alpha$-glycerophosphorylcholine and the fatty alcohols corresponding to the original fatty acid moieties in the 1- and 2-positions (182,183) (Scheme 13). Use of a large molar excess of $LiAlH_4$ resulted in ca. 50% release of choline (184).

**SCHEME 13**

$$-\mathrm{CH_2O\overset{\overset{O}{\|}}{C}R} \xrightarrow{\text{LAH}} -\mathrm{CH_2OH} + \mathrm{HOCH_2R}$$

Under the conditions used by Thompson and Lee (185), both the phosphate diester and the acyl ester bonds were cleaved. Lithium aluminum hydride has no effect on alkyl or alkenyl ether bonds or on acetal groups (e.g., isopropylidene glycerol) in plasmalogen phospholipids (186).

Vitride[R] [$NaH.HAl(OCH_2CH_2OCH_3)_2$] has the advantage over lithium aluminum hydride in permitting higher recovery of reaction products. Both the phosphonyl base and the acyl groups are removed from alkyl or alkenylglycerols. For lipids labeled with $^{14}C$, phospholipase C and saponification gave more quantitative recovery of radioactivity than did the method employing Vitride[R] (77,146).

When sodium borohydride is caused to react with lecithin for 60 min at 37 C in tetrahydrofuran/water mixtures, 94% of the acyl groups were reduced to fatty alcohols. Reduction or isomerization of the double bonds was not detected. Hydrogenolysis does not occur with neutral lipids, and the hydrogenolysis that occurred with lecithin must be attributed to its highly polar nature (187).

## Miscellaneous Reactions

Ethoxylation or propoxylation of lecithin, wherein the oxyalkylating reagent presumably reacts with the hydroxy groups in the inositol or ethanolamine moieties, makes emulsifying products useful for cosmetic and other industrial applications (188) (Scheme 14).

**SCHEME 14**

$$>\mathrm{CHOH} + \mathrm{CH_2\overset{O}{-}CH_2} \longrightarrow >\mathrm{CHOCH_2CH_2OH}$$

Reaction of the hydroxyl groups with isocyanates or diisocyanates produces waxy or granular products that are readily dispersed in water (189) (Scheme 15). The dispersions thus produced are stable and do not undergo fermentation or develop a rancid odor characteristic of the unmodified lecithins.

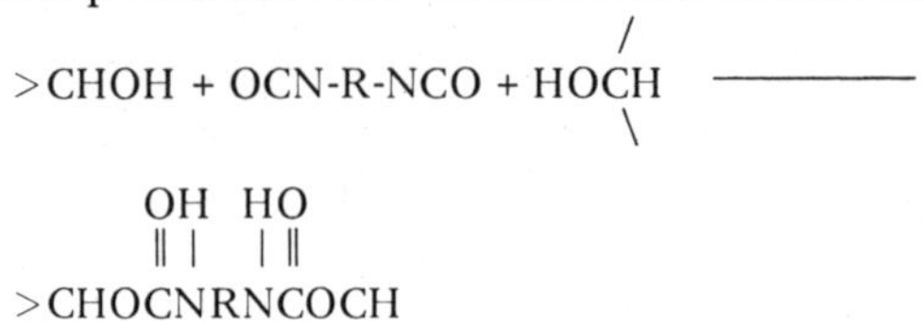

**SCHEME 15**

The reaction product from an N-aliphatic hydrocarbon-substituted trimethylene diamine and lecithin minimizes carburetor icing in automobiles (190). Such compositions are also useful as softening and wetting agents for textiles (191).

Trinitrobenzenesulfonic acid frequently has been used as a means of locating amino groups in the phospholipids of erythrocyte membranes, but an investigation of its reaction with phosphatidylethanolamine has shown the reaction to be only ca. 80% complete (192) (Scheme 16). Furthermore, if cardiolipin or phosphatidylglycerol are present, a further 20% decrease in yield occurs.

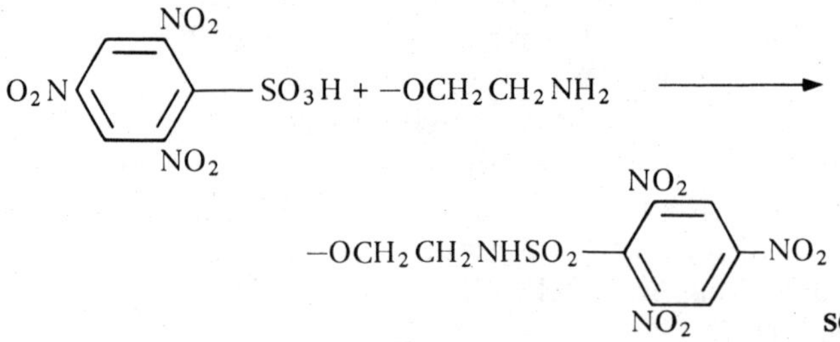

**SCHEME 16**

Reaction of lecithin with cobalt, lead, manganese or zinc hydroxides, acetates or carbonates for 1–4 hr at 70 C produces metal salts that are useful in coatings as paint driers, tackiness removers, enhancers of vehicle stability, pigment wetting agents, gloss control, etc. (193).

The choline moiety of phosphatidylcholine, which is a constituent of animal tissue and is present in many foods as an additive, can react with added nitrites to produce dimethylnitrosamine, a potential carcinogen (194). Boiling lecithin with nitrite in buffer solutions of pH 3.5–7 is sufficient to produce trace amounts of volatile nitrosamines (195).

## References


1. Kuksis, A., Prog. Chem. Fats Other Lipids 12:5 (1972).
2. Snyder, F., in Methods in Cancer Research, Vol. 6, edited by H. Busch, Academic Press, New York, 1971, pp. 410–419.
3. Strickland, K. P., in Form and Function of Phospholipids, edited by G. B. Ansell, J. N. Hawthorne and R. M. C. Dawson, Elsevier, Amsterdam, 1973, chap. 2.
4. Eichberg, J., in Kirk-Othmer Encyclopedia of Chemical Technology, 3rd edn., Vol. 14, John Wiley & Sons, New York, 1981, pp. 255–256.
5. Hawthorne, J. N., and P. Kemp, Adv. Lipid Res. 2:127 (1964).
6. Van Deenen, L. L. M., and G. H. DeHaas, Ibid. 2:167 (1964).
7. Porter, N. A., R. A. Wolf and H. Weenen, Lipids 15:163 (1980).
8. Frankel, E. N., in Fatty Acids, edited by E. H. Pryde, American Oil Chemists' Society, Champaign, IL, 1979, chap. 18.
9. Frankel, E. N., Prog. Lipid Res. 19:1 (1980).
10. Tomioka, F., and T. Kaneda, Yukagaku 23:777 (1974); Chem. Abstr. 82:96613 (1975).
11. Tomioka, F., and T. Kaneda, Kagaku To Seibutsu 14:509 (1976); Chem. Abstr. 86:15426 (1977).
12. Tomioka, F., and T. Kaneda, Yukagaku 25:784 (1976); Chem. Abstr. 86:42075 (1977).
13. Tomioka, F., and T. Kaneda, Ibid. 23:782 (1974); Chem. Abstr. 82:96614 (1975).
14. El-Tarras, M. F., M. Abdel Moety, A. K. S. Ahmad and M. M. Amer, Oleagineux 31:229 (1976); Chem. Abstr. 85:122107 (1976).
15. Suyama, K., and A. Susumu, Agric. Biol. Chem. 43:1981 (1979).
16. Schmid, H. H. O., and T. Takahashi, Hoppe-Seylers' Z. Physiol. Chem. 349:1673 (1968).
17. Nakanishi, T., and K. Suyama, Agric. Biol. Chem. 38:1141 (1974).
18. Pokorný, J., H. Zwain, B. Hegedüsová, N-t. Luân and G. Janiĉek, Z. Lebensm. Unters. Forsch. 159:287 (1975).
19. Morita, M., and M. Fujimaki, Agric. Biol. Chem. 36:1751 (1972).
20. Krog, N., and O. Tolboe, Fette Seifen Anstrichm. 65:732 (1963); Chem. Abstr. 60:4699 (1964).
21. Barclay, L. R. C., and K. U. Ingold, J. Am. Chem. Soc. 102:7792 (1980).
22. Barclay, L. R. C., and K. U. Ingold, Ibid. 103:6478 (1981).
23. Terao, J., Y. Hirota, M. Kawakatsu, and S. Matsushita, Lipids 16:427 (1981).
24. Anderson, S. M., N. I. Krinsky, M. J. Stone and D. C. Clogett, Photochem. Photobiol. 20:65 (1974).

25. Faucon, J. F., and C. Lussan, C. R. Acad. Sci., Ser. C 277:591 (1973); Chem. Abstr. 80:2400 (1974).
26. Odinakova, G. G., O. A. Azizova, A. G. Kotov, A. N. Remizov and D. I. Roshchupkin, Fiz. Khim. Osn. Funkts. Nadmolekulyarnykh Strukt. Kletki, Mater. Vses. Simp., edited by V. E. Kagan, 2:44 (1974); Chem. Abstr. 83:73984 (1975).
27. Orlov, S. N., and V. S. Danilov, Ibid. 2:49 (1974); Chem. Abstr. 83:73985 (1975).
28. Orlov, S. N., V. S. Danilov and Yu. N. Shvetsov, Ibid. 2:51 (1974); Chem. Abstr. 83:73986 (1975).
29. Glushchenko, N. N., A. F. Yakovlev, V. V. Obraztsov, V. E. Kagan, V. S. Danilov and Yu. P. Kozlov, Tr. Mosk. Ova. Ispyt. Prir. 52:206 (1975); Chem. Abstr. 83:143349v (1975).
30. Orlov, S. N., V. G. Rebrov, N. N. Glushchenko, V. E. Kagan, T. N. Myagkova, R. P. Evstigneeva and Yu. P. Kozlov, Ibid. 52:223 (1975); Chem. Abstr. 83:159511 (1975).
31. Ivanov, I. I., S. I. Galkina, M. N. Merzlyak and Yu. N. Kaurov, Ibid. 52:241 (1975); Chem. Abstr. 83:159512 (1975).
32. Ivanov, I. I., S. K. Buzas and N. I. Goldshtein, Ibid. 39:52 (1972); Chem. Abstr. 78:155484 (1973).
33. Merzlyak, M. N., A. P. Bezrukov, I. I. Ivanov and V. P. Shvets, Ibid. 52:217 (1975); Chem. Abstr. 83:143350 (1975).
34. Gutteridge, J. M. C., and T. R. Tickner, Anal. Biochem. 91:250 (1978).
35. Schreiber, J., Pharmazie 34:36 (1979); Chem. Abstr. 90:181927 (1979).
36. Kaurov, Yu. N., A. S. Vzdykhańko, I. L. Ivanov, and V. I. Shvets, Fiz. Khim. Osn. Funkts. Nadmolekulyarnykh Strukt. Kletki, Mater. Vses. Simp. edited by V. E. Kagan, 1:99 (1974); Chem. Abstr. 83:73989 (1975).
37. Misiorowski, R. L., and M. A. Wells, Biochemistry 12:967 (1973).
38. Friedman, J. M., D. L. Rousseau, G. Navon, S. Rosenfeld, P. Glynn and K. B. Lyons, Arch. Biochem. Biophys. 193:14 (1979).
39. Ohnishi, S., and T. Ito, Biochemistry 13:881 (1974).
40. Shah, D. O., and J. H. Schulman, J. Lipid Res. 6:341 (1965).
41. Kimizuka, H., and K. Koketsu, Nature 196:995 (1962).
42. White, M. S., and N. Lakshminarayanaiah, Curr. Mod. Biol. 3:39 (1969).
43. McLaughlin, A. C., C. Grathwohl and R. E. Richards, J. Magn. Reson. 31:283 (1978).
44. McLaughlin, A. C., C. Grathwohl and S. McLaughlin, Biochim. Biophys. Acta 513:338 (1978).
45. Abramson, M. B., G. Colacicco, R. Curci and M. M. Rapport, Bio-

chemistry 7:1692 (1968).
46. Davis, P. F., U.S. Patent 3,357,918 (1967).
47. Debay, A., and O. Sackur, U.S. Patent 2,899,449 (1959).
48. Worth, H. G. J., and D. J. Wright, Clin. Chem. (Winston-Salem, N.C.) 23:1995 (1977).
49. Ross, S., and V. H. Baldwin, Jr., J. Colloid Interface Sci. 21:284 (1966).
50. Bhowmik, B., G. L. Jendrasiak, and B. Rosenberg, Nature (London) 215:842 (1967).
51. Rosenberg, B., and G. L. Jendrasiak, Chem. Phys. Lipids 2:47 (1968).
52. Jendrasiak, G. L., Ibid. 4:85 (1970).
53. Loshchilova, E., and B. Karvaly, Ibid. 19:159 (1977).
54. Szundi, I., Ibid. 22:153 (1978).
55. Kanamoto, R., Y. Wada, G. Miyajima, and M. Kito, JAOCS 58:1050 (1981).
56. Kanamoto, R., M. Ohtsuru and M. Kito, Agric. Biol. Chem. 41:2021 (1977).
57. Bergell, P., U.S. Patent 803,541 (1905).
58. Wörner, E., U.S. Patent 864,219 (1907).
59. Hayashi, S., U.S. Patent 3,072,689 (1963).
60. Fujimoto, M., Jpn. Patent 73.38,686 (1973); Chem. Abstr. 80:133833 (1974).
61. Zanger, M., and J. L. Rabinowitz, J. Org. Chem. 40:248 (1975).
62. Chapman, D., R. M. Williams and B. D. Ladbrooke, Chem. Phys. Lipids 1:445 (1967).
63. Elworthy, P. H., J. Chem. Soc. 1961:5385.
64. Henrickson, K. P., Biochim. Biophys. Acta 203:228 (1970).
65. Veksli, Z., N. J. Salsbury and D. Chapman, Ibid. 183:434 (1969).
66. Walter, W. V., and R. G. Hayes, Ibid. 249:528 (1971).
67. Finer, E. G., J. Chem. Soc. Faraday Trans. II 69:1590 (1973).
68. Davenport, J. B., and L. R. Fisher, Chem. Phys. Lipids 14:275 (1975).
69. Elworthy, P. H., J. Chem. Soc. 1962:4897.
70. Gascoyne, P. R. C., and R. Pethig, J. Chem. Soc. Faraday Trans. I 73:171 (1977).
71. Cerbon, J., Biochim. Biophys. Acta 144:1 (1967).
72. Cole, R. D., U.S. Patent 2,907,777 (1959).
73. Jacini, G., U.S. Patent 2,870,179 (1959).
74. Dailey, J. P., R. L. Colescott and Z. F. Chmielewicz Br. Patent 825,936 (1959).
75. Davis, P. F., U.S. Patent 3,026,341 (1962).
76. Snyder, F., and M. L. Blank, Arch. Biochem. Biophys. 130:101 (1969).
77. Snyder, F., M. L. Blank, and R. L. Wykle, J. Biol. Chem. 246:3639 (1971).

78. Melamud, N. L., G. V. Chebotareva, T. A. Danilova, L. T. Smirnova and D. M. Maiorov, Tr. Vses. Nauchno. Issled. Inst. Zhirov 26:295 (1967); Chem. Abstr. 69:37336 (1968).
79. Melamud, N. L., G. V. Chebotareva, L. T. Smirnova, T. A. Danilova and D. M. Maiorov, Ibid. 26:304 (1967); Chem. Abstr. 69:37337 (1968).
80. Sokol'skii, D. V., and T. Omaraliev, Izv. Akad. Nauk Kaz. SSR, Ser. Khim. 16:64 (1966); Chem. Abstr. 66:39050 (1967).
81. Sokol'skii, D. V., and T. Omaraliev, Katal. Reakts. Zhidk, Faze, Tr. Vses. Konf., 2nd, Alma-Ata, Kas. SSR 1966:63; Chem. Abstr. 68:96992 (1968).
82. Sokol'skii, D. V., and T. Omaraliev, Sb. Statei Aspir. Soiskatelei Minist. Vyssh. Sredn. Spets. Obraz. Kaz. SSR Khim. Khim. Tekhnol. 6:14 (1967); Chem. Abstr. 71:103429 (1969).
83. Sokol'skii, D. V., and T. Omaraliev, Izv. Akad. Nauk Kaz. SSR, Ser. Khim. 17:44 (1967); Chem. Abstr. 67:109896 (1967).
84. Long, C., and M. F. Maguire, Biochem. J. 57:223 (1954).
85. Farin, F., H. L. M. VanGaal, S. L. Bonting and F. J. M. Daemen, Biochim. Biophys. Acta 711:336 (1982).
86. Chapman, D., and P. J. Quinn, Proc. Natl. Acad. Sci. U.S.A. 73:3971 (1976).
87. Quinn, P. J., D. Chapman, C. Vigo and B. R. Boar, Biochem. Soc. Trans. 5:1132 (1977); Chem. Abstr. 88:33485 (1978).
88. Hanahan, D. J., J. Ekholm and C. M. Jackson, Biochemistry 2:630 (1963).
89. Husek, P., M. Kralikova and V. Felt, Z. Klin. Chem. Klin. Biochem. 11:509 (1973); Chem. Abstr. 80:107956 (1974).
90. DeKoning, A. J., and K. B. McMullan, Biochim. Biophys. Acta 106:519 (1965).
91. Myher, J. J., in Handbook of Lipid Research. I. Fatty Acids and Glycerides, edited by A. Kuksis, Plenum Press, New York and London, 1978, chap. 3.
92. Dawson, R. M. C., N. Hemington and J. B. Davenport, Biochem. J. 84:497 (1962).
93. Dawson, R. M. C., in Lipid Chromatographic Analysis, 2nd edn. revised and expanded, Vol. 1, edited by G. V. Marinetti, Marcel Dekker, Inc., New York, 1976, chap. 4.
94. Jensen, R. G., and D. T. Gordon, Lipids 7:611 (1972).
95. Kensil, C. R., and E. A. Dennis, Biochem. 20:6079 (1981).
96. Dawson, R. M. C., Biochim. Biophys. Acta 14:374 (1954).
97. Brockerhoff, H., J. Lipid Res. 4:96 (1963).

98. Okui, S., Y. Suzuki, H. Koyama and O. Umezawa, Yakugaku Zasshi 84:1206 (1964); Chem. Abstr. 62:7630 (1965).
99. Maruo, B., and A. A. Benson, J. Biol. Chem. 234:254 (1959).
100. Hanahan, D. J., Biochem. Prep. 9:55 (1962).
101. Dawson, R. M. C., Biochem. J. 102:205 (1967).
102. Hübscher, G., J. N. Hawthorne and P. Kemp, J. Lipid Res. 1:433 (1960).
103. Tattrie, N. H., and C. S. McArthur, Biochem. Prep. 6:16 (1958).
104. Brockerhoff, H., and M. Yurkowski, Can. J. Biochem. 43:1777 (1965).
105. Chadha, J. S., Chem. Phys. Lipids 4:104 (1970).
106. Robles, E. C., and G. F. M. Roels, Ibid. 6:31 (1971).
107. Aneja, R., J. S. Chadha, E. Cubero Robles and R. Van Daal, Biochim. Biphys. Acta 187:439 (1969).
108. Brandt, A. E., and W. E. M. Lands, Ibid. 144:605 (1967).
109. Hermetter, A., and F. Paltauf, Chem. Phys. Lipids 28:111 (1981).
110. Wicek, H. B., Oleagineux 25:473 (1970).
111. Gatt, S., and Y. Barenholz, Methods Enzymol. 14:167 (1969).
112. Waite, M., R. Hanumantha Rao, H. Griffin, R. Franson, C. Miller, P. Sisson and J. Frye, Ibid. 71:674 (1981).
113. Magee, W. L., and J. F. Uthe, Ibid. 14:170 (1969).
114. Wells, M. A., and D. J. Hanahan, Ibid. 14:178 (1969).
115. Zahler, P., and R. Kramer, Ibid. 71:690 (1981).
116. Cottrell, R. C., Ibid. 71:698 (1981).
117. Deems, R. A., and E. A. Dennis, Ibid. 71:703 (1981).
118. Takagi, R., and T. Sasaki, J. Biochem. 85:29 (1979).
119. Proulx, P., G. Nantel and G. Baraff, Can. J. Biochem. 56:319 (1978).
120. Tinker, D. O., and J. Wei, Ibid. 57:97 (1979).
121. Roberts, M. F., R. A. Deems and E. A. Dennis, J. Biol. Chem. 252:6011 (1977).
122. Shen, B. W., F. H. C. Tsao, J. H. Law and F. J. Kézdy, J. Am. Chem. Soc. 97:1205 (1975).
123. Witt, W., H.-J. Brüller, G. Falker and G. F. Fuhrmann, Biochim. Biophys. Acta 711:403 (1982).
124. Castle, A. M., and J. D. Castle, Ibid. 666:259 (1981).
125. Nutter, L. J., and O. S. Privett, Lipids 1:258 (1966).
126. Jensen, G. L., B. Daggy and A. Bensadoun, Biochim. Biophys. Acta 710:464 (1982).
127. Kunze, H., B. Hesse and E. Bohn, Ibid. 711:10 (1982).
128. Uziel, M., and D. J. Hanahan, J. Biol. Chem. 220:1 (1956).
129. Kawasaki, N., J. Sugatani, and K. Saito, J. Biochem. (Tokyo) 77:1233 (1975).

130. Ansell, G. B., J. N. Hawthorne and R. M. C. Dawson, Editors, Form and Function of Phospholipids, 2nd edn., (B.B.A. Library, Vol. 3), Elsevier Scientific Publishing Co., Amsterdam, 1973.
131. Ottolenghi, A. C., Methods Enzymol. 14:188 (1969).
132. Little, C., Ibid. 71:725 (1981).
133. Takahashi, T., T. Sugahara, and A. Ohsaka, Ibid. 71:710 (1981).
134. Ikezawa, H., and R. Taguchi, Ibid. 71:731 (1981).
135. Low, M. G., Ibid. 71:741 (1981).
136. Bevan, T. H., D. A. Brown, G. I. Gregory and T. Malkin, J. Chem. Soc. 1953:127.
137. Renkonnen, O., JAOCS 42:298 (1965).
138. Renkonnen, O., Lipids 1:160 (1966).
139. Nutter, L. J., and O. S. Privett, Ibid. 1:234 (1966).
140. Eichberg, J., U.S. Patent 3,359,201 (1967).
141. Szuhaj, B. F., and J. R. Yaste, U.S. Patent 3,962,292 (1976).
142. Batley, M., N. H. Packer and J. W. Redmond, Biochim. Biphys. Acta 710:400 (1982).
143. Batley, M., N. H. Packer and J. W. Redmond, J. Chromatogr. 198:520 (1980).
144. Holub, B. J., and A. Kuksis, Lipids 4:466 (1969).
145. Mavis, R. D., R. M. Bell and P. R. Vagelos, J. Biol. Chem. 247:2835 (1972).
146. Blank, M. L., E. A. Cress, C. Piantadosi and F. Snyder, Biochim. Biophys. Acta 380:208 (1975).
147. Davenport, J. B., and R. M. C. Dawson, Biochem. J. 84:490 (1962).
148. Dawson, R. M. C., R. F. Irvine, K. Hirasawa and N. L. Hemington, Biochim. Biophys. Acta 710:212 (1982).
149. Horning, M. G., G. Casparrini and E. C. Horning, J. Chromatogr. Sci. 7:267 (1969).
150. Kates, M., and P. S. Sastry, Methods Enzymol. 14:197 (1969).
151. Heller, M., N. Mozes and E. Maes, Ibid. 35:226 (1975).
152. Yang, S. F., Ibid. 14:208 (1969).
153. Taki, T., and J. N. Kanfer, Ibid. 71:746 (1981).
154. Crone, H. D., Biochim. Biophys. Acta 84:665 (1964).
155. Yang, S. F., S. Freer and A. A. Benson, J. Biol. Chem. 242:477 (1967).
156. Kovatchev, S., and H. Eibl, Adv. Exp. Med. Biol. 101:221–226. (1978).
157. Heller, M., P. Greenzaid and D. Lichtenberg, Ibid. 101:213 (1978).
158. Sir, S., and A. Rezek, Enzymologia 31:335 (1966); Chem. Abstr. 66:43826 (1967).
159. Hawthorne, J. N., J. Lipid Res. 1:255 (1960).
160. Taylor, W. E., and J. M. McKibbin, J. Biol. Chem. 201:609 (1953).

161. Hawthorne, J. N., Biochem. J. 75:495 (1960).
162. Michell, R. H., and D. Allan, FEBS Lett. 53:302 (1975).
163. Takenawa, T., and Y. Nagai, J. Biol. Chem. 256:6769 (1981).
164. Wassef, M. K., and M. I. Horowitz, Biochim. Biophys. Acta 665:234 (1981).
165. Wittcoff, H., U.S. Patent 2,445,948 (1948).
166. Glabe, E. F., U.S. Patent 2,513,638 (1950).
167. Gaver, K. M., E. P. Lasure and L. M. Thomas, U.S. Patent 2,621,133 (1952).
168. Hennessy, D. J., and R. J. Moshy, U.S. Patent 2,952,694 (1960).
169. Julian, P. L., H. T. Iveson and M. L. McClelland, U.S. Patent 2,629,662 (1953).
170. Gray, G. M., Methods Enzymol. 14:678 (1969).
171. Edwards, J. C., and P. J. Quinn, Chem. Phys. Lipids 28:89 (1981).
172. Hallgren, B., and S. Larsson, J. Lipid Res. 3:31 (1962).
173. Privett, O. S., and M. L. Blank, JAOCS 40:70 (1963).
174. Sevanian, A., R. A. Stein and J. F. Mead, Biochim. Biophys. Acta 614:489 (1980).
175. Sevanian, A., R. A. Stein and J. F. Mead, Lipids 16:781 (1981).
176. Barber, D. J. W., and J. K. Thomas, Radiat. Res. 74:51 (1978).
177. Bayley, H., and J. R. Knowles, Biochemistry 17:2420 (1978).
178. Bayley, H., and J. R. Knowles, Ibid. 17:2414 (1978).
179. Stoffel, W., and P. Metz, Hoppe-Seyler's Z. Physiol. Chem. 360:197 (1979).
180. Goida, E. A., Zh. Fiz. Khim. 46:225 (1972); Chem. Abstr. 77:95314 (1972).
181. Mochalina, A. S., and T. P. Klimova, Radiobiologiya 17:711 (1977); Chem. Abstr. 88:47038 (1978).
182. Urakami, C., and H. Okura, Bull. Chem. Soc. Jpn. 31:779 (1958).
183. Urakami, C., and K. Yamaguchi, Ibid. 36:1049 (1963).
184. Kameyama, H., and C. Urakami, Ibid. 38:1081 (1965).
185. Thompson, G. A., Jr., and P. Lee, Biochim. Biophys. Acta 98:151 (1963).
186. Wood, R., and F. Snyder, Lipids 3:129 (1968).
187. Nichols, B. W., and R. Safford, Chem. Phys. Lipids 11:222 (1973).
188. DeGroote, M., and B. Keiser, U.S. Patent 2,310,679 (1943).
189. Jones, E. P., and J. C. Cowan, U.S. Patent 2,636,889 (1953).
190. Standard Oil Co. (Ohio), Br. Patent 846,693 (1960).
191. Sincroft, D. E., and E. F. Sipos, U.S. Patent 2,987,527 (1961).
192. Bishop, D. G., E. M. Bevers, G. van Meer, J. A. F. op den Kamp and L. L. M. Van Deenen, Biochim. Biophys. Acta 551:122 (1979).

193. Kronstein, M., and J. Eichberg, U.S. Patent 2,997,398 (1961).
194. Pensabene, J. W., W. Fiddler, R. C. Doerr, L. Lakritz and A. E. Wasserman, J. Agric. Food Chem. 23:979 (1975).
195. Moehler, K., and E. Hallermayer, Z. Lebensm. Unters. Forsch. 151:52 (1973); Chem. Abstr. 78:122824 (1973).

# Chapter Twelve

## Traditional and Novel Approaches to the Analysis of Plant Phospholipids

**William N. Marmer**
Agricultural Research Service,
US Department of Agriculture
Eastern Regional Research Center
600 East Mermaid La.
Philadelphia, PA 19118

### Introduction

Due to the importance of phosphorus analysis in the quantitation of phosphatides, this chapter begins with a detailed survey of wet chemical and instrumental methods of phosphorus analysis. Many of these analytical procedures have not yet been evaluated for routine analyses. This chapter includes a discussion of enzymatic analysis of phosphatides. The value of such procedures will depend on the availability of commercial enzyme kits such as one now available from Japan, and also on the broadening of the scope of such analysis to include not only the choline-containing species but other phosphatides as well. The chapter's final section covers the analysis of the phosphatides' subclasses, including their isolation from tissue, their chromatographic separation, identification and quantitation, and the determination of their individual molecular species.

### Analysis of Phosphorus Content

Traditionally, phosphatides have been quantitated by analysis of their phosphorus content. As most of the material removed from crude oil by degumming is phosphatide, the residual phosphorus content of refined oil is a good indicator of the extent of degumming. Degumming of soy oil, for example, can reduce its phosphorus content from as high as 1000 ppm to 3–9 ppm (1). Maximum allowable phosphorus content in degummed soy oil is 250 ppm, according to the trading rules of the National Soybean Processors' Association (2).

For nearly 60 years, phosphorus analysis has been accomplished by one variation or other of the wet chemical analysis of Fiske and SubbaRow (3), in which molybdenum blue from molybdate, orthophosphate, and a reducing agent is measured colorimetrically. In 1947, Raymond Reiser wrote (4), "The results of phospholipid determinations in plant materials are as numerous as the authors, all of whom use a different procedure." The 1970s saw the development of instrumental methods, some of which expanded upon the wet chemical ones by automation, others of which followed completely different approaches. Some of these new methods discussed below already are in use, whereas others are presented to show the line of thought of researchers who still strive for innovation in phosphorus analysis. When the analyst considers a method, he must know its sensitivity, accuracy, and repeatability. He also must balance its convenience with its expense.

**Molybdenum blue and molybdovanadophosphate yellow.**

The various methods for complexing inorganic phosphate with molybdate all involve three routines. Initially, organic material must be digested and phosphorus converted to orthophosphate. Then color is developed, usually under controlled pH. Finally, a colorimetric determination is run, either on the same aqueous solution, or on an organic extract containing the colored species.

*Digestion to Inorganic Phosphate.* The digestion process, which eliminates all organic material, recently has been the object of two critical studies. Yee (5,6), who worked with amniotic and serum lipids, compared five acid digestions, all run at 225 C for 30 min: (a) perchloric acid; (b) perchloric acid with molybdate; (c) sulfuric acid/hydrogen peroxide/urea; (d) sulfuric acid/perchloric acid; and (e) sulfuric acid/perchloric acid/vanadium pentoxide. He concluded that one was as good as another, although digestion "e" worked best on sphingolipids. The analyst who wishes to mineralize a large oil matrix, however, must be warned that perchloric acid alone is unsuitable; most frequently, the perchloric acid/oil mixture will explode on heating. Acid digestion was not highly regarded, however, in the other comparative study. Daun and coworkers (7) applied four highly different digestion techniques to canola oil: (a) sulfuric acid/perchloric acid—the method of Rouser and coworkers (8), which gave poor reproducibility; (b) saponification—the technique of Hartman and coworkers (9), in which saponification by ethanolic sodium hydroxide takes place in an oven in a Teflon crucible and is followed by acidification and partition between hexane and water. This also gave poorly reproducible results and was labor-intensive; (c) the AOCS official Method Ca 12-55 (10)—a method that involves charring and ashing in the presence of zinc oxide, then acidification to dissolve the ash, neutralization, and reacidification. This was "lengthy and cumbersome," taking two days to complete the ashing; and (d) the preferred method, the oxygen bomb technique (11)—ashing in a bomb with zinc oxide, water and oxygen, followed by dissolution in dilute nitric acid. This offered the "best combination of speed, accuracy and precision." In the four methods tried by Daun's group, all but the preferred one ("d") included subsequent development of molybdenum blue for colorimetry, whereas "d" made use of yellow molybdovanadophosphate instead. Daun's preliminary presentation indicated that the great advantage of molybdovanadophosphate over molybdenum blue was the former's lack of sensitivity to the acidity of the medium (12).

*Quantitation of Inorganic Phosphate.* Once the organic material has been digested to orthophosphate, phosphate may be determined as one of two complexes with molybdenum—molybdenum blue or yellow molybdovanadophosphate. To develop the blue color, phosphate is treated, within a specified pH range, with a molybdate solution (usually the ammonium salt) and a reducing agent. Many of the modifications to the Fiske-SubbaRow procedure (3) seek to facilitate the procedure by minimizing the number of solutions required and maximizing the shelf life of the reagents. A good example is the technique of Vaskovsky and coworkers (13), whose combined molybdate-hydrazine reagent is stable for months under refrigeration. All analyses require parallel runs with inorganic phosphate standard solution (usually $KH_2PO_4$). One group (14) has used a scintillation counter to quantitate the blue color. The method relies on the quenching of radiation from an external $^{14}C$ source. Response is linear over a wide range of phosphorus concentration, from 0 through 0.4 mg P. Table I outlines several reports on phosphorus analysis by molybdenum blue determination. Some serve best for analysis of small amounts of isolated phosphatide, while others may be used for determining trace phosphorus in oil matrices. Some procedures have been adapted for automated analysis. Some involve transfer of the colored complex to an organic phase prior to colorimetry. The table attempts to summarize any information given on sensitivity, accuracy and precision.

The development of molybdovanadophosphate yellow instead of molybdenum blue offers the advantage of room temperature color development, together with less sensitivity to pH and to interfering ions. Recent studies with the yellow complex are outlined in Table II. Note that the automated application was for high concentrations of inorganic phosphate.

A programmable desktop calculator is convenient for repetitive calculations of phosphorus content from colorimetric analysis. A program for a Texas Instruments TI-59 calculator/PC-100C printer is available from the author. This program incorporates linear regression analysis of phosphorus standards.

**TABLE 12-I.**
**Phosphorus Analysis by Molybdenum Blue Determination**

| Digestion | Chromogenesis | Sensitivity and repeatability | Reference |
|---|---|---|---|
| HCl/$SnCl_2$, 105°/16 hr (A).[a] | Molybdate/ascorbate/ K(SbO) tartrate (A). | CV < 2.5%, 0.5–20 ppm P. | 15 |
| 3N $HNO_3$/0.1 M $K_2B_4O_7$, then ash (A). | Molybdate/ascorbate (A). | CV = 3% at 0.3 $\mu$g P. Not designed for oil. | 16, 17 |
| — | Molybdate/ascorbate (A) | CV < 1%. Designed for inorganic P. | 18 |
| NaOH/ethanol in Teflon® crucible. | p-Methylaminophenol sulfate/$NaHSO_3$/ $Na_2SO_3$. | CV = 1% at 60 ppm P. In later study (ref. 7): >> ± 5 ppm error. | 9 |
| ZnO, then char, pyrolyze; dissolve in HCl. | Molybdate/hydrazine. | In later study (ref. 7): ± 2 ppm error, but lengthy and cumbersome. | 10 |
| $H_2SO_4$/$HCl0_4$ (4/1). | Molybdate/ANSA/ $NaHSO_3$/ $Na_2SO_3$. All dry-blended in mortar. | — | 19 |
| $H_2O_2$/10 N $H_2SO_4$ | As above, but no blending. Highly cited method. | — | 20 |
| 70% $HClO_4$, heated. | As above. | — | 21 |
| 70% $HClO_4$ at room T. | As above. The original Mo blue method. | — | 3 |
| Several equally applicable to plant lipids. | Molybdate/$SnCl_2$/ hydrazine (A). | CV = 7% at 1.7 $\mu$g P. | 5, 6, 22 |
| 70% $HClO_4$ | Molybdate/hydrazine. | CV = 2% at 0.1 $\mu$g P. | 13 |
| $H_2O_2$/$H_2SO_4$ | Molybdate/ascorbate. Highly cited method. | In later study (ref. 24): CV = 1%–2% at 0.9 $\mu$g P. | 23 |
| *Variations:* | | | |
| No digestion. | Molybdate/HCl/ $H_2SO_4$/Hg in chloroform–methanol. Blue color extracted by nonane. Read at 710 nm. | CV = 5% at 10 $\mu$g P. No interference by inorganic P. | 25 |
| No digestion. | As above, but extracted by chloroform. | Emulsion problems; demonstrated on peanut oil. | 26 |

**TABLE 12-I (Continued).**
**Phosphorus Analysis by Molybdenum Blue Determination**

| Digestion | Chromogenesis | Sensitivity and repeatability | Reference |
|---|---|---|---|
| No digestion. Phosphatides extracted by glacial acetic acid. | Molybdate/$SnCl_2$. Blue color extracted by chloroform. Read at 735 nm. | CV = 0.7% at 1 ppm P. | 47 |
| $HClO_4/HNO_3$ | Molybdate/HCl. Color extracted by n-butyl acetate. Read at 310 nm. | CV = 3%–4% at 0.7–10 $\mu$g P. | 27 |
| — | Similar to above, but tungstate instead of molybdate. Color extracted by n-amyl alcohol. Read at 267 nm. | "Same precision, accuracy, sensitivity as molybdate methods." | 28 |
| — | Mo blue read by scintillation technique. | Very large range: 0–0.4 mg P. | 14 |
| $Mg(NO_3)_2$ pyrolysis. | Physphomolybdate/ malachite green/ Triton X-100® complex. Read at 650 nm after 5 min at room T. | CV = 2%, 0.1–0.5 $\mu$g P. | 29 |
| $H_2O_2$/10 N $H_2SO_4$ | Similar complex to above. Read at 623 nm. | CV = 1%–2% at 30 ng P. | 24 |
| — | Seven dyes studied, incl. malachite green. | Rhodamine B most sensitive and suitable. | 30 |
| — | Phosphomolybdate/ methylene blue complex in 0.9 M $HNO_3$. Color extracted by methyl isobutyl ketone. Read at 655 nm. | CV < 4%, 0.8–8.0 $\mu$g P. | 31 |
| — | Phosphomolybdate/ ethyl violet complex. Color extracted by methyl isobutyl ketone. Read at 602 nm. | CV 1.3%, 0–0.6 $\mu$g P | 32 |

[a](A) = automated procedure.

**TABLE 12-II**
**Phosphorus Analysis as Molybdovanadophosphate**

| Digestion | Chromogenesis | Sensitivity and repeatability | Reference |
|---|---|---|---|
| $ZnO/H_2O/O_2$, bomb (up to 0.8 g oil). | Molybdate/ metavanadate. Read at 400 nm. | Later study (ref. 7): ± 4 ppm P error; lower limit 1–2 ppm P; best of the tested methods. | 11 |
| 4 N $HClO_4$, 95 C (A). | As above. Read at 420 nm, (A) Designed for $P_2O_5$ of fertilizers. | CV < 1% for 7%–43% P (collaborative study; official method). | 33, 34 |
| Cellulose/MgO, ash in muffle furnace. | As above. Read at 400 nm. | CV = 4%–10% at 630 ppm P. | 35 |
| MgO, ash in muffle furnace. | As above. Read at 460 nm. | (Official method) | 36 |
| *Variation:* | | | |
| — | As above; Differential-pulse voltammetric determination at glassy carbon electrode. Flow injection. Measured at +0.32V (A). | CV < 1% at 10 ppm P. 1 ppm measurements are practical. | 37 |

[a](A) = automated procedure.

Calculator input commands are signalled by appropriate alphanumerics and output is designed in that case for "percent phospholipid." A modification to allow output as "ppm phosphorus" is available from the author on request.

### Gravimetric

Some official methods are based on precipitation of phosphorus as a complex. An AOAC micromethod (38) involves the gravimetric determination of an ammonium phosphomolybdate, whereas the same precipitate is titrated in the AOCS method for total phosphorus of vegetable lecithins (39). Other AOAC methods measure quinolinium phosphomolybdate $[(C_9H_7N)_3 \cdot H_3PO_4 \cdot 12MoO_3]$ gravimetrically (40) or by titration (41).

### Specific Ion Electrode

Although there is no specific ion electrode for phosphate ion, an indirect method using a fluoride ion electrode may be used to determine phosphate over the range 0 to $2 \times 10^{-4}$ M (0–20 ppm $PO_4$ ± 0.5 ppm; 0–7 ppm P). Lanthanum nitrate is added to the digested sample to form very insoluble lanthanum phosphate. Excess lanthanum, still in solution, then is titrated with sodium fluoride. If sulfate is in the sample (as it would be were sulfuric acid used in the digestion step), it is precipitated out with barium ion; lanthanum sulfate is soluble (42). Titration of excess $Ce^{4+}$ ion after precipitation of $Ce_4(PO_4)_3$ offers a related approach (43).

### Ion Chromatography

Ion chromatographs are high performance liquid chromatography (HPLC) instruments designed to separate and quantitate ionic species at the ppm and even ppb levels. Predigested samples may be passed through a series of ion exchange columns. Eluates then are analyzed by conductivity detection (44). Typical runs require less than 10 min. The technique may be applied to the analysis of trace phosphatides by detection of phosphate ion after digestion. It is also conceivable that phosphatide analysis may be accomplished by ion chromatographic analysis of phosphatide base residues (choline, ethanolamine, serine, etc.) after sample digestion with phospholipase D (or phosphobase residues after phospholipase C).

### Analysis of Phosphatides without Prior Digestion to Inorganic Phosphorus

Most of the innovative approaches to phosphatide analysis attempt to avoid the initial digestion steps of the traditional methods. Yoshida et al. have experimented with complexation of phosphatides with transition metals. Complexation with tetrathiocyanatocobaltate resulted in analyses heavily skewed toward phosphatidylcholine (PC) and sensitive only to unsaturated molecular species (45). Complexation with dithiocyanatoiron reagent and subsequent extraction of the complex into 1,2-dichloroethane for spectrophotometric analysis at 470 nm showed linear response to concentrations of all PC in the range 20–240 nmol; the method was sensitive to other phosphatide subclasses, although the response of the acidic phosphatides was often considerably diminished (46). This method would not be suitable for the analysis of soy phosphatides, for example, because a major component—phosphatidylinositol (PI)—gave a response only 23% of PC's. Totani et al. have developed a molybdate-based system that is sensitive to undigested phosphatides. The phosphatides are removed from soy oil by extraction with glacial acetic acid. An aliquot of the acetic acid solution then is treated with a molybdate-stannic chloride preparation and the resulting blue color removed by chloroform extraction for colorimetric determination. Sensitivity and repeatability are excellent (CV = 0.7% at 1 ppm P) (47). Note, however, that chloroform is a potential carcinogen and therefore should be used with caution. Other examples of direct analysis of phosphatides are given in subsequent sections of this chapter.

*Atomic Absorption Spectrometry.* The one instrumental technique that has gained wide acceptance as an alternative to the wet chemical methods is atomic absorption (AA) spectrometry. Direct measurement of traces of phosphorus had been unfeasible in the flame mode due to low sensitivity, although excellent precision and accuracy have been obtained for materials such as fertilizers that contain high amounts of phosphorus (48). In the mid-1970s, however, trace amounts of phosphorus were analyzed successfully using a flameless instrument equipped with a graphite furnace and an electrodeless discharge lamp (EDL). Sensitivity of the signal (213.5/213.6 nm doublet) was enhanced sixfold by incorporation of lanthanum ion, so that phosphorus could be detected down to 10 ng (49). For application to the analysis of trace phsophorus in oils, one must look to the two papers by Prevot and Gente-Jauniaux. In the first paper (50), these workers showed that phosphorus could be analyzed in amounts less than 1 ppm in oils by use of flameless atomic absorption, combined with an EDL. Oil samples in methyl isobutyl ketone were injected directly (no need for predigestion). The second paper (51) gives an improved technique, in which lanthanum is included as its oil-soluble cyclohexylbutyrate salt. Sensitivity to 10 ng phosphorus was established with a coefficient of variation (CV) of 10% at 5 ppm phosphorus. This technique has been applied in other studies to gasoline additives (110 ppm phosphorus) (52), and most interestingly, to whole tissue samples (53). In the latter citation, leaves, bovine liver and oyster tissue were analyzed in amounts under 1 mg. Phosphorus was detected in amounts under 1 microgram with CV's 6% to 10%.

Daun et al., in their study of several methods of phosphorus analysis (7), concluded that oil analysis by atomic absorption spectroscopy could be relied on to give repeatability to ±3 ppm (CV = 5%). The authors were hesitant to recommend the technique due to the expense of the instrumentation but praised the technique for its accuracy, precision, sensitivity and rapidity.

*Atomic Emission Spectrometry.* With the exception of one emission technique, namely molecular emission cavity analysis (MECA), emission spectrometry has seen little application for analysis of phosphorus in oils.

MECA instrumentation (54) was designed as an inexpensive supplement to atomic absorption, where those species least sensitive to atomic absorption would be candidates for MECA analysis. Early targets were sulfur and phosphorus, though the developers of the process have concentrated heavily on sulfur. The MECA instrument's small sample container, the cavity, is heated in a hydrogen-rich flame to about 500 C. Phosphorus determinations are done by determining the resulting emission of HPO species at 528 nm. Knowles et al. (55) claimed a CV of 10% for 0 through 50 ppm phosphorus in determining $PO_4^{3-}$ and $P_2O_7^{3-}$ ions, AMP, adenosine-5-triphosphate (ATP) and phytic acid. However, they did not achieve direct fractionation of the different phosphorus species concurrent with the MECA analysis; such fractionation is achievable for different sulfur species (sulfate, sulfite, thiosulfate, etc.). Daun et al. (7) evaluated MECA for the analysis of phosphorus in oils by following the procedure of Rogers and Downey (56) (who reported CV = 10% at 3 ppm) and reported a detection limit of 30 ppm for inorganic phosphorus and two ppm for organic phosphorus. MECA, like atomic absorption, was praised for its rapidity and minimal sample size requirement. Hesitancy was expressed over its cost, however, and MECA was found to have less sensitivity to phosphorus than does atomic absorption. Recently, MECA has been coupled to HPLC for quantitation of organophosphates (57). The authors claimed a sensitivity of 5 ng phosphorus, with a CV of 3% over the range of linear response (10–100 ng). A wheel of 40 cavities collected the column eluate, one drop per cavity. Problems with incomplete burnoff must yet be overcome.

In one study designed to analyze trace phosphorus in water, samples were vaporized to 2800 C in a graphite furnace (as opposed to 500 C in MECA instrumentation), then carried in a nitrogen stream to a flame photometer for measurement of HPO emission. Sensitivity was reported to three ppm, CV 2% to 3% (58). Recently, phosphorus emission was used for detection in an HPLC system. Detection limits were below one ppm, especially for nonionic materials. The system used a dual flame setup—one to reduce or eliminate solvent interferences and the other to bring about HPO emission of the resulting nebulized sample. Flame photometry also has been coupled with gas chromatography (GC). A recent report, for example, describes the analysis of glycerophosphate isomers (silylated) in commercial phospholipids (59).

Plasma emission spectrometers, meanwhile, base their operation on excitation in an electrical arc. Emission analysis of phosphorus is recognized as an AOAC official method (60). A review outlining the differences between two types of plasma emission spectrometers—inductively-coupled plasma (ICP) vs. direct-current plasma—recently was published (61). Other similar reviews cover ICP vs. atomic absorption (62,63). The last reference cites the following detection limits for phosphorus in ppb: a) with atomic absorption—flame, 30,000; graphite furnace, 20. b) with ICP—sequential, 20; simultaneous with other elements, 60. One group claims that phosphorus may be detected at less than one ppb by ICP as reduced molybdoantimonyl phosphoric acid, by extracting the derivative from water by diisobutylketone (64). An ICP device has been coupled to an HPLC system for analysis of trace amounts of nucleotides. Sensitivity was 750 ng, with a CV of 4.5% at 10 micrograms; phosphorus emission was measured at 213.6 nm (65). Direct analysis of phosphorus in powdered rock samples has been achieved with a sensitivity of 100 ppm (CV = 5%–15%) (66). Kantor et al. have analyzed vegetable oils for phosphorus by (a) ashing a 1-g sample with 0.2 g $MgCO_3$ and bringing to 200 mg with graphite powder, then (b) by packing 30 mg of this mixture into a hollow graphite electrode and extruding it into an electrical arc. Phosphorus was determined at 255.3 nm relative to magnesium at 273.3 nm. The detection limit was six ppm phosphorus, with a CV of 4.8% at 30–50 ppm (67). More recently, experiments have been reported on the direct aspiration of vegetable oil (in kerosene solution) into an argon plasma arc. Analysis of phosphorus at 213.6 nm was accomplished in less than a minute down to 0.5 ppm (CV = 20%; CV = 5% at four ppm). The same procedure was used to quantitate other trace elements, such as copper, often at the ppb level. Adjustments had to be made continually for drifting of the output signal (68). It seems that interest in the plasma technique will increase as new instruments appear on the market competitive in pricing to atomic absorption spectrometers.

Another technique, that of X-ray Emission Spectrometry (XES), has been used for direct phosphorus analysis of thin-layer chromatography (TLC) spots and column chromatographic eluates. XES/TLC requires calibration curves for each subclass of phospholipids. Lipids were shown to be detectable with linear response from 0.4 to 8 $\mu$g P (CV = 5%–10%) (69).

*X-ray Fluorescence Spectrometry.* This technique has been used for rapid multi-element analysis of freeze-dried food samples. CV values of 5% were obtained for phosphorus in orchard leaf samples (1,930 ppm) and in bovine liver (10,700 ppm) (70). The technique requires that samples be mixed with cellulose powder and then pressed into a pellet. Because self-supporting pellets are not feasible for samples whose fat content surpasses 60%, this would preclude analysis of oils for trace phosphorus.

*Neutron Activation Analysis.* Neutron activation analysis certainly cannot yet be considered routine. The technique, which measures gamma radiation of decaying species after neutron bombardment, requires a beta emitter. Commercial services are available that accept samples for such analysis. One report showed that 10 to 30 ng of phosphorus could be determined in this way with a CV of 2% to 3% (71).

*Infrared Analysis.* Recent advances in infrared spectrometry allow the examination of carbonyl absorbances of various lipid species in a mixture and subsequent quantitation of the species based on subtle differences in absorbance maxima from species to species. Examination of the absorbance at four different frequencies and consolidation of the data into a matrix of four linear equations allowed the determination of PC in a mixture that also contained triglyceride and sterol ester. Errors ranged from 4% to 30%, but the instrument performed all calculations on-line, thus allowing a rapid analysis (72).

*Chronoamperometric Analysis.* One group has experimented with a dropping mercury electrode for the determination of PC in aqueous methanol or ethanol. The method depends on the adsorption of surfactant-type molecules onto the mercury and the resulting inhibition of the redox process. PC was detected at the 1–8 ppm level of phosphorus, with a CV = 1% (73).

*$^{31}P$ Nuclear Magnetic Resonance (NMR) Spectroscopy.* Biochemists and biophysicists have given considerable attention in the last 10 years to $^{31}P$ NMR analysis of biological tissue and synthetic membranes. Techniques require the use of instrumentation with Fourier-transform capability, and spectra are accumulated with $^{1}H$-$^{31}P$ decoupling. The considerable work of Glonek and his coworkers in analyzing intact animal tissues has appeared in review form (74). Quantitative work using $^{31}P$ signals requires optimizing the spectra by assuring uniform spin-lattice relaxation ($T_1$) and by controlling the nuclear Overhauser effect (NOE). London and Feingenson (75) demonstrated that $^{31}P$ NMR could be used for quantitation of the phospholipid subclasses of 50 mg of soybean lecithin sonicated in $H_2O/D_2O$. Detrimental peak broadening was controlled by addition of ethylenediaminetetraacetic acid (EDTA) and a detergent (such as potassium cholate). Use of aqueous dispersions allowed pH control, useful for shifting the signal of one subclass away from that of another subclass. Relative ratios for the phosphatides PC:PI: phosphatidylethanolamine (PE): phosphatidic acid (PA) in a half hour experiment were 1.00:0.60:1.06:0.31, respectively, vs. 1.00:0.48:0.95:0.26 by two-dimensional TLC and wet chemical analysis of the subclass spots. The nature of the acyl groups on each subclass did not effect the signal shift. The authors claimed a detection limit of three ppm phosphorus. Absolute quantitation was not reported, although it should have been achievable using another phosphorus species as internal or external standard. A representative spectrum is shown in Figure 1. Gurley and Ritchey (76) were able to analyze phosphorus-containing species (organic phosphates, phosphonates and thiophosphates) in chloroform using $^{31}P$ NMR. They, too, optimized conditions for NOE and $T_1$ and used as a relaxant iron ethylene glycol and shift reagent $Gd(fod)_3$. Detection limits (signal-to-noise ratio = 2) for trimethyl phosphate and triethyl phosphate were six ppm and 11 ppm, respectively, in 25-min experiments. Tetraethylammonium phosphate in a precision

capillary tube was used as an external standard. Response factors varied from one phosphate ester to another due to changes in steric bulk and the resulting hindrance to approach by the relaxagent Gd(III).

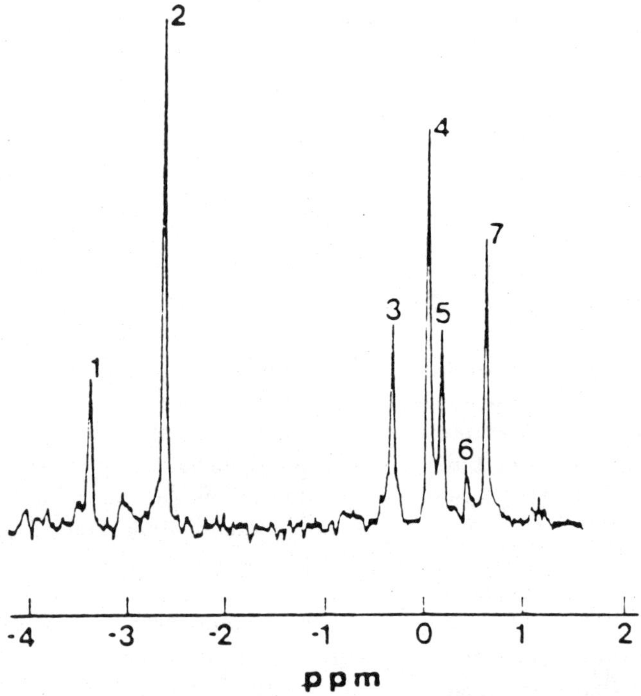


Fig. 12-1. [31]PNMR spectrum of a phospholipid mixture consisting of 5% w/v cholate, 50 mM EDTA, and the following: peak 1, 10 mg of PA; peak 2, 6.6 mg of $K_2HPO_4$; peak 3, 8 mg of DPG; peak 4, 13 mg of PE; peak 5, 12.5 mg of PS; peak 6, PI present as an impurity in PS; peak 7, 12.5 mg of dipalmitoyl PC. Total volume 1 mL, pH ~8. 100 transients were collected with an acquisition time of 2 sec per transient, no delay between transients and a filtering time constant of 1 sec (75).

Kasler and Tierney (77) experimented with phosphorus detection in aqueous sulfuric acid, with special attention given to accuracy and repeatability. Triphenylphosphite in chloroform with the relaxagent chromic acetylacetonate ("Cr(acac)$_3$"), all sealed into a capillary, served as external standard. Quantitation of $KH_2PO_4$ in the 500 ppm range (w/v) with added $CrCl_3$ gave a CV value under 2%. Organic compounds first were digested in the NMR tubes with sulfuric and perchloric acids, then quantitated for phosphorus. CV values were similar to the CV for $KH_2PO_4$.

Although $^{31}P$ NMR has become a respected research tool, its use for routine analysis of trace amounts of phosphorus is not yet established. The more complex spectral patterns of lipid mixtures by $^{13}C$ or $^{1}H$ NMR preclude routine analysis of trace phosphatide content, though specific lipid subclasses may be analyzed by their characteristic signals. The quaternary ammonium group of PC gives rise to distinct $^{13}C$ and $^{1}H$ methyl resonances (78). The $^{1}H$ methyl resonance recently was used to determine the PC content of soy lecithin to the same precision as obtained by HPLC (CV under 7%) (79).

### The Phosphatide/Phosphorus Ratio

All the methods discussed above lead to a quantitation of phosphorus. This is in essence a molar quantitation of phosphatide, because the main phosphatides-PE, PI, PC—bear one phosphorus atom per molecule. Lipid biochemists often use an arbitrary factor of 25 to estimate the weight of phosphatide from the weight of phosphorus. A mixed acyl PC that bears one palmitoyl group and one linoleoyl group would have a molecular weight of 776/(776 g PC/mol PC) (1 mol PC/g-atom P) (1 g-atom P/31.0 g P) = (25.0 g PC/g P), hence, the factor 25. Obviously the problem becomes complex when PC bears a variety of fatty acyl groups, more complex when the phosphatide mixture is more than just PC, and in the case of "vegetable lecithin," even more complex when the product contains other materials mixed with phosphatides (materials such as neutral lipids, glycolipids, saccharides and amino acids). This requires empirically determining the ratio. The results of several empirical determinations appear in Table III.

## Enzymatic Analysis

Quantitation of total phosphatide content by enzymatically assisted reactions is yet a goal for future research, although one group (85) has reported such a feat. Phospholipase C (EC 3.1.4.3) is used to convert phosphatides to 1,2-diglycerides. Lipoprotein lipase (EC 3.1.1.3) then liberates glycerol, which may be detected colorimetrically in the presence of $NAD^+$, glycerol dehydrogenase (EC 1.1.1.6), and phenazine methosulfate nitro blue tetrazolium dye. Corrections must be made for interference by triglycerides in the lipid to be analyzed. This interference has caused another research group to label this procedure as unsatisfactory (86).

Most enzymatically-assisted analyses of phosphatides have been specific for choline-containing lipids PC, lysophosphatidylcholine (LPC) and sphin-

**TABLE 12-III**
**Conversion Factors: The Phosphatide/Phosphorus Ratio**

| | | Acetone insolubles from | |
|---|---|---|---|
| Oil source | Freshly obtained lecithin | Raw oil | Deoiled lecithin |
| Soy | 32.07 ± 1.26[a] | 31.69 ± 1.38[b]<br>30[c]<br>32.7[d]<br>32.1 ± 1.4[e] | 30.7[a] |
| Rape | 31.75 ± 1.14[a] | | |
| Peanut | 24.4[f] | | |

[a]Ref. 80
[b]Ref. 81
[c]Ref. 10
[d]Ref. 82
[e]Ref. 83
[f]Ref. 84, determined from total lipid extract of seeds.

gomyelin (SPH). Choline typically is liberated by phospholipase D (EC 3.1.4.4) (86–94), although this also may be achieved with phospholipase C (EC 3.1.4.3) together with phosphatase (EC 3.1.3.1) (95). Choline then may be used to generate hydrogen peroxide in the presence of choline oxidase (EC 1.1.3.17). Quantitation of hydrogen peroxide may be done by dye coupling in the presence of peroxidase (EC 1.11.1.7) (86,88,90–93). A commercial kit for choline analysis by dye coupling (phospholipase D/choline oxidase/peroxidase/4-aminoantipyrene) is available from Nippon Shoji Company, Osaka, Japan (PL kit K and autoanalyzers PL kit K "f") and is based on the method of Takayama et al. (88). A reaction scheme for this process is shown in Figure 2. Choline quantitation also may be achieved by voltammetric means (89), by a choline dehydrogenase (EC 1.1.99.1) system (87,95) or by a choline kinase (EC 2.7.1.32) system (94,96). A commercial kit based on a choline kinase system (96) is available from Boehringer Mannheim Biochemicals, Indianapolis, IN 46250, and Mannheim, Federal Republic of Germany (Lecithin Kit 529 362). Details of all of these enzymatic methods are outlined in Table IV.

Another method requires phosphorus in inorganic form at the outset. A purine nucleoside phosphorylase (EC 2.4.2.1)/xanthine oxidase (EC 1.2.3.2)/uricase (EC 1.7.3.3) system is used to generate hydrogen peroxide, which is measured as quinoneimine dye after a coupling reaction catalyzed by peroxidase. The four-enzyme system requires 15 min at 37 C and may be used over a range of 0 through 1.6 ppm P (97). The initial reaction—the conversion of orthophosphate and isosine to ribose-1-phosphate and hypoxanthine—is used in a related method that utilizes HPLC for quantitation;

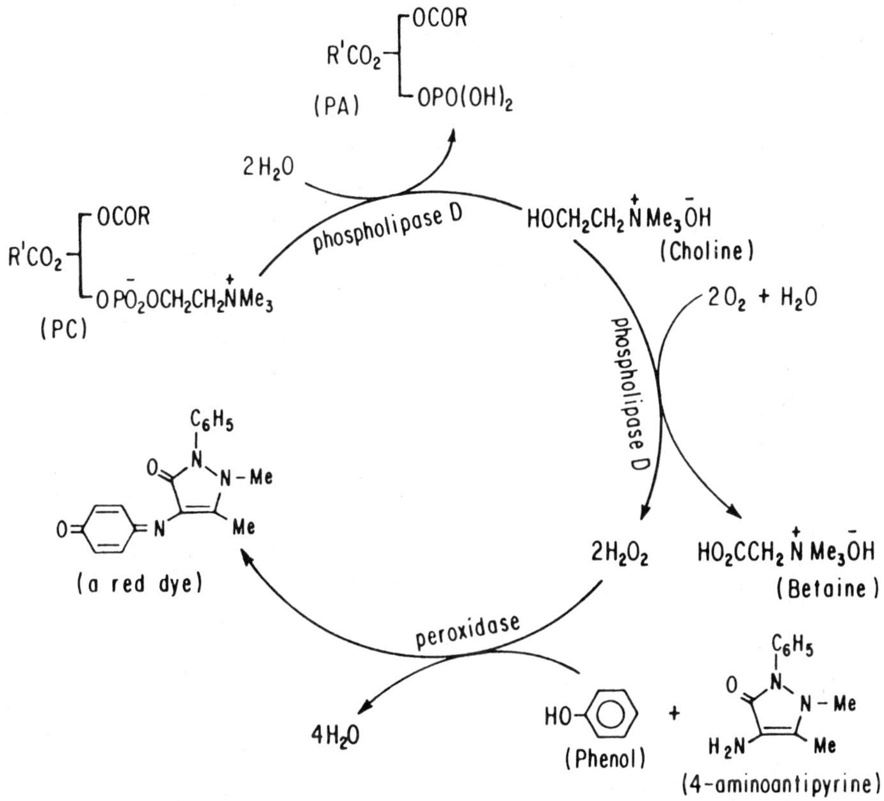


Fig. 12-2. Reaction scheme for enzymatic determination of PC (88).

hypoxanthine is separated from inosine and related to orthophosphate concentration. Linearity is claimed through 30 ppm phosphorus (CV = 10%) and the detection limit is 0.75 ppm (97a).

The future may see applications of enzyme technology to specific non-choline-containing phospholipids. Researchers already have identified PI-specific phospholipases C that show no detectable attack on any other phospholipids (98).

## Subclass Analysis

Because many fine review papers and monographs have been published on the isolation, separation and identification of lipid classes, this review will

**TABLE 12-IV**
**Enzymatic Analysis of Phosphatides (choline-species only, unless specified)**

| Substrate | Choline liberation | Choline reaction | Quantitation | Remarks | Reference |
|---|---|---|---|---|---|
| Serum (Commercial enzyme kit available; cf text). | Phospholipase D. | Choline oxidase liberates 2 $H_2O_2$. | Peroxidase, dye coupling, UV/VIS. | CV .48% at 76 ppm P. CV .45% at 116 ppm P. Range 0–400 ppm P. 97% "recovery" at 14 ppm P. | 88 |
| All phosphatides of serum. | Phospholipase C liberates DG; Lipoprotein lipase then liberates glycerol. | Glycerol dehydrogenase + $NAD^+$ + dye. | — | Comparable to choline oxidase in sensitivity, but cf. ref. 86 below. | 85 |
| Bile (procedure of ref. 88). | Phospholipase D. | Choline oxidase. | Peroxidase, dye coupling. | CV 3%–6% at 15–79 ppm P. Also, found ref. 85 (above) unsatisfactory, due to LPC, DG, TG, in sample. | 86 |
| Bile. | Phospholipase D. | Choline oxidase. | Peroxidase, dye coupling. | Linearity through 300 ppm P. | 93 |
| Bile (procedure of ref. 83). | Phospholipase D. | Choline oxidase. | Peroxidase, dye coupling. | linearity 60–300 ppm P. CV .67% at 60 ppm P. | 92 |
| PC of amniotic fluid. | Phospholipase D from cabbage. | Choline oxidase. | Peroxidase, dye coupling. | CV .87% at 1 ppm $(16{:}0)_2$PC | 90 |
| PC of amniotic fluid. | Phospholipase D from Streptomyces chromofuscus. | Choline oxidase. | Peroxidase, dye coupling. | CV .58% at 4 ppm SPH | 91 |

**TABLE 12-IV (Continued)**
**Enzymatic Analysis of Phosphatides (choline-species only, unless specified)**

| Substrate | Choline liberation | Choline reaction | Quantitation | Remarks | Reference |
|---|---|---|---|---|---|
| "Lecithins" of serum. | Phospholipase D. | Choline dehydrogenase gives betaine aldehyde, and simultaneously: | Reduced dye (OR $NAD^+$ + betaine aldehyde dehydrogenase (ED 1.2.1.8) gives NADH (UV)). | — | 87 |
| PC, SPH of serum | Phospholipase C liberates phosphocholine. Then phosphatase liberates choline. | Choline dehydrogenase gives betaine aldehyde, and simultaneously: | Reduced dye (OR $NAD^+$ + betaine aldehyde dehydrogenase (ED 1.2.1.8) gives NADH (UV)). | — | 95 |
| a. PC in foods<br>b. PC of amniotic fluid | As above | Denature, then choline kinase + ATP → ADP. | ADP quantitation by pyruvate kinase and lactate dehydrogenase. | CV < 1% | 96 |
| PC of serum | Immobilized phospholipase D. | Immobilized choline oxidase. | $H_2O_2$ measured voltammetrically. | CV 5% at 120 ppm P. range 60–140 ppm P. | 89 |
| — | Phospholipase D. | Choline kinase + ATP → ADP. ADP + phosphoenolpyruvate → pyruvate. | Pyruvate + NADH (340 nm)→$NAD^+$ (no 340 nm). | Presented as method of phospholipase D quantification; not yet used for PC/LPC. | 94 |

will focus only on novel developments most relevant to plant phosphatides.

**Extraction**

Extraction of phosphatides from tissue most often has been done with chloroform-methanol systems. Two of the most widely cited articles are the extraction methods of Folch et al. (99) and Bligh and Dyer (100). These methods are relied on for efficient extraction of total lipid and minimal contamination from nonlipid materials. Many define "lipid" as tissue material that is isolated by a Folch or Bligh-Dyer method. Numerous variations have appeared subsequently, including one that shows that chloroform may be replaced by dichloromethane (101). Extraction by a dry column method, using dichloromethane and methanol, eliminates much of the tedium associated with the methods cited above and allows separation of lipids into neutral and polar fractions simultaneously with extraction (102). Although the method was developed for meat products, it also has been used on peanuts (103). Another group has shown that lipids may be extracted efficiently using a hexane-isopropanol system (104).

Two studies have dealt with the problem of isolation of lipid that is tightly associated with protein. One group utilizes proteolytic enzymes to liberate this lipid and to increase the lipid yield in a Folch extraction (105). The other group, mainly concerned with the elimination of non-lipid impurities in a lipid extract, reverses the usual extraction-then-wash sequence by pretreating the tissue (immature soybeans) with hot dilute acetic acid prior to chloroform-methanol extraction. This reversal removes the normal nonlipid contamination of the lipid extract and denatures hydrolytic enzymes (106). Users may encounter difficult emulsion problems using this technique. It should be stressed that the enzyme phospholipase D is abundant in premature seeds and, if not denatured prior to extraction, leads to rapid transphosphatidylation upon contact with methanol (107). It is very probable that many an unknown phosphatide, identified in charts and figures as the ubiquitous PX, is actually the artifact phosphatidylmethanol. A recent study has compared the application of several extraction procedures to soy products (108). The study showed that a Folch-type method was as efficient as two day-long Soxhlet extractions (one with choroform-methanol, the other with benzene-ethanol) and more efficient than a short extraction with hexane-ethanol. A similar study has been reported on such legume seeds as kidney beans, field beans, lentils, chick peas and peas; chloroform-methanol was more efficient than diethyl ether, although acid hydrolysis prior to extraction gave the highest yields of fatty acids (109). Yields of PI suffer from readsorption onto protein during the standard aqueous wash that follows chloroform-methanol extractions. Readsorption is suppressed with higher cation concentrations (110). A recent review (111) criticized the application of most standard extraction techniques for the isolation of vegetable lipids. Lipid losses during aqueous washing could be minimized by clean-up of a crude extract on acid-washed alumina.

The researcher who seeks the extraction of total lipids from an oilseed must follow procedures directly at odds with the procedures applied for commercial isolation of crude oils. The isolation of total lipids implies the maximization of phosphatide yield, whereas the commercial procedures—hexane extraction, the use of supercritical fluids, etc.—minimize phosphatide yield. Reliable lipid data depend on proper technique, a point well made in an article by Kinsella et al. (112).

## Chromatography

Examining the phosphatide fraction of a lipid isolate that consists mainly of neutral lipid requires a way to concentrate the desired fraction. Although the dry column method of extraction does yield a phosphatide concentrate directly (102), those who use a traditional extraction method usually resort to some type of chromatography. (A tedious bulk separation of polar lipids from nonpolar lipids may be done using a series of separatory funnels and a counter-current technique with petroleum ether and aqueous ethanol (113).)

For a general review and critique of phosphatide fractionation, see the work of Nelson (114). Recent developments, especially as they pertain to the separation of vegetable phosphatides, are discussed below; the most significant, perhaps, have been in the use of HPLC to separate the phosphatides by subclass (PC, PE, etc.) and by molecular species (dipalmitoyl-PC, 1-palmitoyl-2-oleoyl-PC, etc.).

*Column Chromatography (Non-HPLC).* An unusual phenomenon has been reported (115) to occur on silica gel. Commercial lecithin or oil was dissolved in petroleum ether and passed through a deactivated silica gel column with petroleum ether. By virtue of micelle formation in this solvent, (a) the phosphatides (and hydrocarbons) passed through the column unadsorbed, while (b) triglyceride and sterol esters, (c) sterols and free fatty acids, and (d) nonlipids were adsorbed. Fraction b was then eluted with petroleum ether/ether, then c with ether, leaving behind d. Experimental results were reported in detail for soy lecithin.

The advantages of using gel chromatography were demonstrated by Biacs et al. for sunflower and rapeseed lipids (116) and for soy (117). Use of Sephadex LH-20 gel allowed good recovery not only of phosphatide fractions but also of glycolipids (including cerebrosides, sterol glycosides).

Richter et al. (118) claimed that six g of egg yolk phospholipid could be fractionated on silica gel to give 4.2 g pure PC, 0.2 g pure LPC, and 1.2 g PE (which contained a trace of cerebroside). However, 4.2 out of six g is equivalent to 70% PC, an unusually high percentage of total phospholipid. Ten gram amounts of egg yolk lecithin were fractionated by a two-column procedure (alumina, then silica gel) by Radin (119), using slight helium pressure for the first column and then 60 psi (by chromatographic pump) for the second. A fraction close to two g was almost all PC, though it included small amounts of sphingomyelin and LPC; lighter column loading allowed the recovery of cleaner PC. Radin avoided the use of toxic solvents, as he also had done in his lipid extraction procedure (104). Instead, he eluted the alumina column with ethanol and the silica gel column with hexane-isopropanol-water mixtures. To dry a lipid sample azeotropically, cyclohexane served well as a substitute for the traditional benzene.

Another group has been able to use a dry column chromatographic method to isolate an intact phosphatide fraction. Up to 800 mg of total lipid is applied to five g silica gel. Elution of the column with isopropanol draws off neutral lipid, sterols and free fatty acid. Then ammoniated methanol elutes the phosphatides (120).

A large-scale isolation of PC from crude egg or soy phosphatide has been achieved on a column of hydroxylapatite ($Ca_{10}(PO_4)_6(OH)_2$). After a column of 300 g of adsorbent has been charged with 10 g crude phosphatide, elution is begun (30 mL/min) with a binary solvent system of acetone/methanol (decreasing in acetone content as elution progresses). The first three liters draws off neutral lipid. PC is eluted with the next ten liters and sphingomyelin, if present, with the next seven. All other phosphatides do not elute (121).

A novel approach to isolate subfractions that are usually the most difficult ones to obtain is the use of immobilized neomycin. Schacht (122), noting neomycin's affinity for molecules with multiple phospho groups, developed a method to isolate polyphosphoinositides free of other phospholipids (including PI). Palmer (123) then expanded the method by taking advantage of neomycin's role as a weak anion exchanger. He demonstrated that the method could be used to isolate preparative amounts of the following lipids: phosphatidylserine (PS), then PA, then PI (which included some sulfatide), then diphosphatidylglycerol (DPG), then PI phosphate, and finally, PI bisphosphate. All neutral lipids and uncharged phospholipids passed through the column without adsorption.

*High-Performance Liquid Chromatography (HPLC).* Progress in HPLC of lipids was reviewed in 1982 (123a). There are several recent reports that demonstrate improved efficiency of separation of phospholipid components, other reports that attack the problem of detection, and finally a new development in HPLC for which there are as yet no reported applications in the lipid area—the use of microbore columns (124). Regarding separation efficiency, most reported techniques are extremely difficult to reproduce; there are too many uncontrollable variables, such as column temperature and column efficiency, to allow successful separations on first attempt. Therefore, the researcher must view any recommendation as a guideline and adjust the HPLC system by experimentation to optimize the separation for his or her particular case. The process of finding a proper solvent system need not be a hit-or-miss one. The use of a simplex optimization procedure has been shown to be a good approach for optimizing the separation of phosphatides by HPLC (125). Such a systematic approach often surprises its users by its efficiency.

Regarding detection, researchers seem to have settled for UV detection in the 205 nm wavelength region as the closest to what may be labelled as standard for lipids. Others have experimented with the attachment of chromophoric groups to various lipid subclasses. Infrared detection (126) has not become widespread, probably because of its lack of sensitivity and the need for aqueous components for HPLC separation of phosphatides. Flame ionization detection (FID) suffered a poor reputation when moving wire detectors were available commercially; Privett and coworkers' results with a perforated band (cf. below) have not yet been adopted commercially. Tracor Instruments has recently introduced its model 945 LC/FID detector, which uses a woven fibrous quartz belt. Successful utilization in phosphatide analysis has not yet been reported.

Two publications demonstrate how HPLC may be used for preparative scale isolation of phospholipid subclasses. In one study, ten grams of egg yolk phospholipid were separated on a 10-m silicic acid column, using a chloroform-methanol gradient elution. Baseline separations were achieved for PE, PI, lysophosphatidylethanolamine (LPE) and PC, in that order, with no accounting for sphingomyelin or PS. Some evidence for molecular species separation also was seen (127). In the other study, radially compressed silica gel columns were used with chloroform-methanol-water systems to obtain baseline separation of five g egg yolk phospholipid into neutral lipid, PE, PC, and sphingomyelin fractions. A 26-g sample of the same mixed lipid was separated in a two-step procedure, where each of two fractions isolated from the first run was reinjected separately for further purification (128).

For analytical purposes, good separations of phosphatides into their subclasses have been reported on silica and ion exchange columns. Rat brain and liver lipids and numerous model lipids were used in a study that optimized separations on silica by use of a gradient elution with hexane-isopropanol-water-sulfuric acid. Detection was at 205 nm. Such an acid system is not suitable for plasmalogen-containing animal lipids (129). Another good separation on silica was achieved using acetonitrile-methanol-85% phosphoric acid, again using rat tissues and standards, with detection at 203 nm (130). Yet another group used acetonitrile-methanol-water and detection at 210 nm to analyze small amounts of PC in chocolate (131); the same group later (79) compared this method to an NMR method (choline methyl absorption) to achieve the same analysis. Excellent results recently were reported by Patton et al., who separated rat liver lipids using an isocratic system of hexane-isopropanol-25 mM phosphate buffer-ethanol-acetic acid (132). Detection was at 205 nm. Soy phosphatides were separated on a silica column using hexane/isopropanol/acetate buffer (or acetic acid) systems and detection at 206 nm (123). The same column (Lichrosorb Si 60) was used by another group to achieve a more impressive separation; the use of toluene or benzene in the latter separation precluded UV monitoring, however, and relied instead on a flow-through radioactivity detector (with radiolabelled lipids) (134).

Others have resorted to separations on ion exchange type columns. Columns of amino functionality bonded to silica ($\mu$Bondapak-$NH_2$, Waters Associates, Milford, MA) were used with gradient elution. One group used a chloroform-methanol-water system on a synthetic mixture of phospholipids with flame ionization detection (135), while another used a hexane-isopropanol-methanol-water system on egg yolk phospholipids with detection at 206 nm (136). A cation exchange column of silica covalently bonded to benzene sulfonate functionality (PXS 10/25 SCX, Whatman, Clifton, NJ) served to separate rabbit myocardial lipids; isocratic and gradient runs with acetonitrile-methanol-water allowed good separations, except for PE/PS. Detection was at 203 nm and quantitation was done by radio-labelling experiments and by phosphorus analysis (137).

Despite these reported successes in class separation by HPLC, it must be stressed that this methodology is by no means routine. The chemist who attempts to reproduce these results will likely encounter considerable difficulty.

There is a trend now to accomplish the separation of intact, as opposed to derivatized, molecular species of phospholipid subclasses. Compton and Purdy (138) focused on sharpening the peak of dioleoyl-PC (reversed phase column, aqueous methanol). Elevated column temperatures and addition of either acid (sulfuric, perchloric, but especially phosphoric or diethylphosphoric) or ion-pairing agents (triethylammonium sulfate) greatly improved efficiency, presumably by suppressing adsorption and thus emphasizing partition on the reversed phase column. Earlier, another group successfully separated a synthetic mixture of saturated PC's using methanol-water-chloroform (no adsorption suppressant) on reversed phase columns; results with egg yolk PC were not so effective. Refractive index detection was used (139). Better results were seen in a later study using egg yolk PC and a methanol-1 mM phosphate buffer system and UV detection at 205 nm (140). To date, the best reported results are those of Patton et al. (132), who obtained sharply defined separations (Fig. 3) of the molecular species of PC, of PE, and of PI (rat liver) using a methanol-acetonitrile-water system that contained choline chloride (20 mM). Similarly good separations were obtained by this group for PS, using methanol-25 mM $KH_2PO_4$-acetonitrile-acetic acid that also contained choline chloride (30 mM). Patton's group used UV detection at 205 nm.

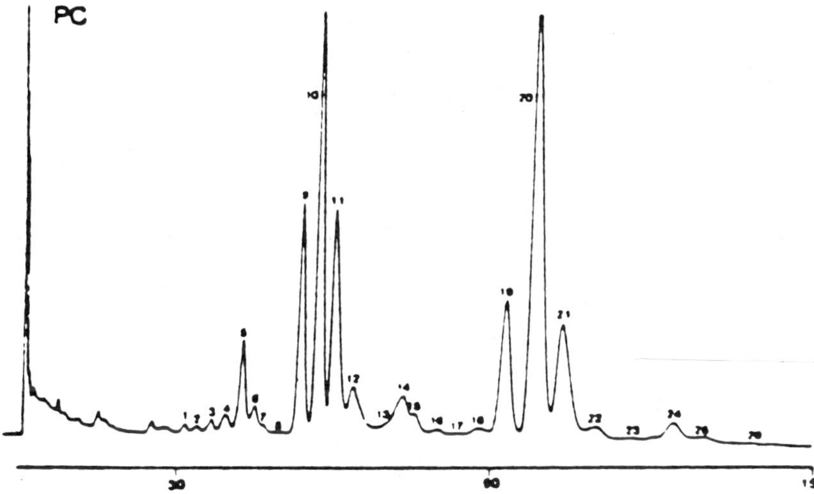


Fig. 12-3. HPLC separation of molecular species of PC (from rat liver) (132). Conditions: C18 reverse phase column (4.6 × 250 mm Ultrasphere ODS, Altex Scientific, Inc., Berkley, CA). Load, 220 $\mu$g. Eluting solvent, 20 mM choline chloride in methanol/water/acteonitrile (90.5:7:2.5). Flow rate, 2.0 mL/min. Detection, 205 nm. Each numbered peak is from one molecular species (two for peaks, 2, 4, 5, 14, 16, 22) (e.g., peak 9 = 16:0–22:6 PC, peak 10 = 16:0-20:4 PC, peak 11 = 16:0-18:2 PC, peak 12 = 18:1-18:2 PC).

Myher and Kuksis, meanwhile, have been able to remove the polar headgroup of phospholipids by hydrolysis in the presence of phospholipase C, then separate the resulting diglycerides as their silyl ether derivatives on a short capillary column (SP-2330) by GC (141,142).

Another derivatization approach was used by Hsieh et al. (143) with the derivatized lipid (egg yolk PC) separated into molecular species by reversed phase HPLC. This group converted PC to phosphatidic acid dimethyl ester ("PAME") by direct reaction with diazomethane, according to the method of Renkonen (144). A methanol-water system was used with refractive index detection, and an acetonitrile-water system was used with flame ionization detection.

The approach of attaching chromophoric groups to lipids to enhance their absorptivity never has produced a universal chromophore. Success in detection of lipid subclasses has been achieved by esterification of hydroxy groups or amidation of amino groups with chromophoric acylation agents, depending on the availability of hydroxy or amino groups on the lipid molecule. Perbenzoylation of hydroxy groups of glycolipids, including cerebrosides, allows detection at 280 nm in ethyl acetate-containing solvents and 230 nm in dioxane systems (145). The amino groups of ethanolamine and serine-containing phospholipids have been reacted either with 4-biphenylcarbonyl chloride to yield derivatives that may be detected at 280 nm (146), or with dansyl chloride (1-dimethylamino-naphthalene-5-sulfonyl chloride) to add a fluorescent chromophore (147).

Already mentioned was the conversion of phosphatides to the dimethyl esters of phosphatidic acid (143) and to silyl ethers of diglycerides (141). One group (148) has even used the enzymatic methodology of Takayama et al. (88) to quantitate choline-containing phosphatides as the lipids elute from an HPLC column. UV detection has been studied in detail at 206 nm with the separation of phospholipids in hexane-isopropanol-water systems (149). Recently a call has been made for monitoring at 215 nm as the best compromise between saturated and unsaturated lipids; using a variable wavelength detector, one may monitor at 195 nm and at 215 nm to determine whether two peaks have the same degree of unsaturation (150).

Privett and coworkers have reported their work with a perforated stainless steel band as a transport system between an HPLC column and both a flame ionization detector and a chemical ionization mass spectrometer. In the FID mode, the column eluate is carried on the band first through an evaporator to remove solvent, then into a reactor where cracking and hydrogenolysis reduce the lipids to hydrocarbons. The hydrocarbons are swept into the FID for quantitation (151,152), or into a mass spectrometer for identification by single ion monitoring in the chemical ionization mode using methane as the reagent gas (153). Using a computer to assemble and interpret the mass spectral data allows a reconstruction of the fatty acid profile of each lipid subclass (154,155).

*Thin-Layer Chromatography (TLC).* Egon Stahl, in his review of "A Quarter Century of Thin-Layer Chromatography—an Interim Report" (156), cites six reasons why TLC is likely to remain the most used separation procedure:

(a) the greatest variety of stationary and mobile phases; (b) a large variety of development techniques; (c) the greatest variety of detection means; (d) the capacity of a single plate for multiple analyses; (e) the ability to couple with extraction and identification procedures; and (f) the simplest and cheapest chromatographic procedure for fast qualitative and semiquantitative separations.

Although some may be attracted to automated variations of TLC, such as TLC with programmed multidevelopment (157) or "thinchrography" (cf. below), the achievements exemplified below on conventional TLC plates show the versatility of conventional TLC. In fact, following simple optimization rules (158) greatly facilitates TLC and makes meaningless any general comparisons between TLC and HPTLC (high performance TLC).

Two citations are given for two-dimensional separation of phosphatides on silica gel because both are specific for soybean phosphatides. In each one, distinct separations of PC, PE, PI, PS and PA are achieved. In one method (Fig. 4), separations are achieved on 20 × 20 cm plates of Silica gel G by development first in chloroform/methanol/ammonia (65:25:2,v/v) and then (after reactivation under vacuum for 30 min) in chloroform/methanol/acetic acid/water (85:15:10:3) (108). In the other method, separations are achieved on 20 × 20 cm plates of Silica gel 60, using development solvents similar to the first example: chloroform/methanol/ammonia (65:30:5), followed—after air drying for 30 min—by chloroform/methanol/acetic acid/water (75:12:10:3) (159). Another group used two-dimensional TLC on Silica gel 60 plates as small as 5 × 5 cm. Good separations are reported with as little as five min developing time for each solvent (tetrahydrofuran/acetone/methanol/water (50:20:40:8), then chloroform/acetone/methanol/acetic acid/water (50:20:10:15:5) (160). Pork lung phosphatides were separated unidimensionally with a discontinuous gradient of moisture on Silica gel H; high humidity at the lower end of the plate favors separation of most phosphatides by partition rather than by adsorption, while lower humidity at the top favors separation of diphosphatidylglycerol from neutral lipids (161). The importance of humidity control was stressed in an earlier report (162), in which erythrocyte phosphatides were separated unidimensionally on silica by chloroform/methanol/petroleum ether/water (8:8:6:1). Disaturated PC was separated from mono- and diunsaturated PC on Silica gel HR with sodium carbonate by developing at −70 C ("cryochromatography") (163). Another study took advantage of the large void areas in two-dimensional TLC. Samples were applied in an oblique line and the separated subclasses isolated later as oblique bands themselves. Subclasses may be separated on a preparative scale on plates of conventional thickness (164).

A method that separates not the phosphatides but their polar degradation products or precursors has been run on silica gel HPTLC plates using two-dimensional TLC with methanol/water/ammonia/chloroform/acetone (5:2:1:2:2), followed by methanol/water/ammonia (5:3:1). This technique separated inorganic phosphate, choline, choline phosphate, glycerophosphorylcholine (GPC), glycerophosphate and betaine (165).

Rapid and efficient separation of phosphatides was achieved unidimensionally on (diethylamino) ethyl—modified silica gel using chloroform/methanol/water/pyridine/ammonia (130:55:8:4:4). This method was an extension of applications of DEAE-silica gel to column chromatography (166). Excellent separation of eight phosphatides was achieved unidimensionally on boric acid-impregnated silica plates using chloroform/methanol/water/ammonia (125:75:6:2) for development (167). Argentation TLC has been applied to the resolutions of the molecular species of PE and of PS as their trifluoroacetamide derivatives (168). Separations were achieved on silica gel/20% silver nitrate using a chloroform/methanol/water (65:25:4) system. Finally, an older reference teaches the method of separating the phosphatides on basic alumina (with 10% $CaSO_4$). The chief advantage of using alumina over silica gel is the solubility of the adsorbent in strong acid. The TLC spots may be scraped and digested for phosphorus analysis without any need for centrifugation prior to colorimetric analysis (169).

Although molybdate reagents commonly used for phosphorus analysis may be used as TLC visualization reagents (13), some require heating for color development while some also are unstable on the TLC plate. The use of ammonium pentachlorooxomolybdate, prepared from ammonium paramolybdate and HI/HCl, overcomes the problems of molybdate reagents and is sensitive to 40 ng phosphorus. Phospholipids are seen as blue spots on a yellow background; inorganic phosphorus is inert (170). On the other hand, phospholipid may be converted on the plate to inorganic phosphorus by spraying with perchloric acid and then heating the plate. The plate then may be treated with a molybdate reagent and Malachite green to reveal bright green spots on a light orange background (171). Care must be taken to guard against hazardous buildup of perchloric acid. Another group praises the use of an acid-fuchsin-uranyl nitrate reagent for visualization (172). Two nondestructive reagents are recommended for their ease of removal from the lipid spots. One is N-phenyl-1-naphthylamine, which reveals lipid as blue spots under short wavelength UV light (173). The other is 1,6-diphenylhexatriene, which reveals lipid as violet spots under long wavelength UV light (174).

The weak element of TLC always has been quantitation. Phosphatides contain a marker by virtue of their phosphorus atom and therefore may be

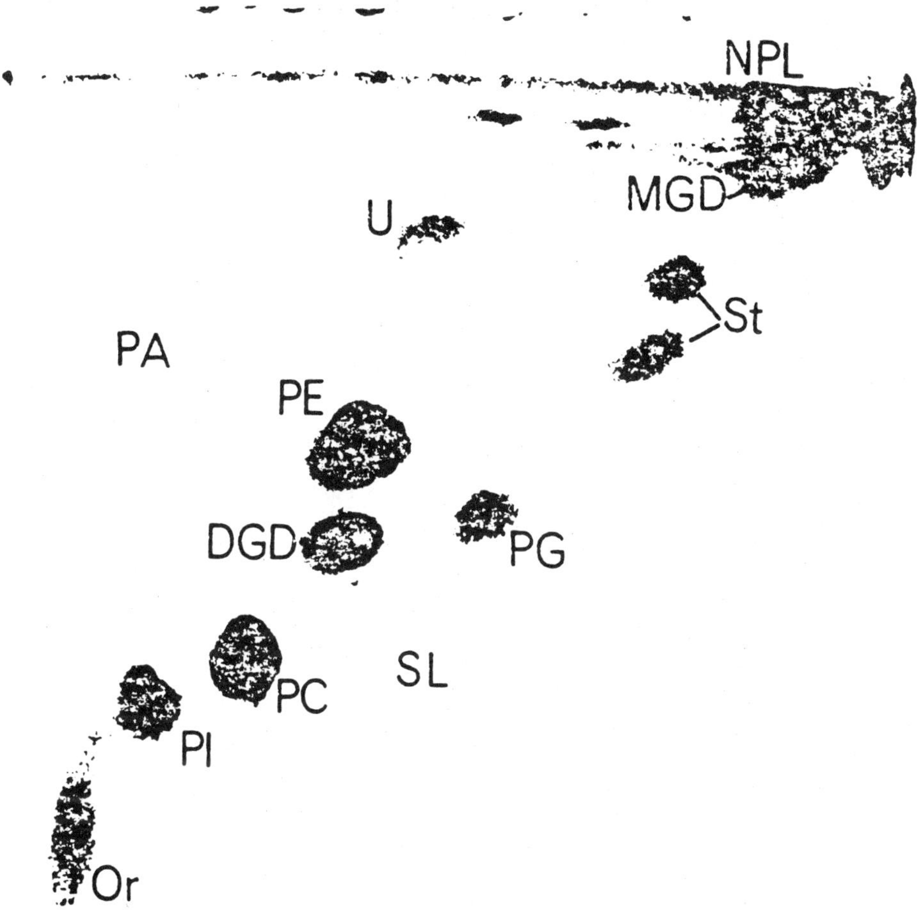


Fig. 12-4. Two-dimensional TLC separation of polar lipids from 100 mg fresh weight immature soybean cotyledons (steam-killed prior to extraction) (107). Conditions as per text. Or = origin; SL = sulfolipid; MGD, DGD = mono-, digalactosyl diacylglycerol; U = unknown; St = sterols; NPL = nonpolar lipids. Other abbreviations as per text.

analyzed by scraping the TLC spots and proceeding with the wet chemical or X-ray emission methodology discussed at the beginning of this chapter. A similar approach uses the fatty acids as a marker. Transmethylation is followed by GC analysis of the resulting methyl esters in the presence of an internal standard. Peak areas are translated into weights and then related to phospholipid weight by the factor of 1.371 (regardless of subclass) (175,176).

Alternatively, one may visualize the TLC spots and quantitate by some type of scanning. Densitometric scanning, however, has its problems, especially with regard to variations in charring or staining caused by spot size, shape and intensity. Furthermore, variations are introduced by the degree of unsaturation and by differing responses from lipid subclass to subclass.

A recent analysis of densitometric scanning of charred TLC spots showed that improvements in repeatability by use of the newer "flying spot" technique over the older "split scanning" were minimal [CV = 10.3%, 11.2%, respectively, for 7 $\mu$g $(18{:}1)_2$PC spots], although the flying spot technique was best for irregularly shaped spots. Reflectance measurements were preferred over transmittance measurements because the latter produced higher background noise. Maximum absorbance occurred at 350 nm wavelength and linearity was experienced over the range 0.1–10 $\mu$g lipid (177). Charring intensity may be doubled by using copper sulfate, instead of the commonly used acetate, in phosphoric acid (178). An earlier study, which used the flying spot technique in the transmittance mode on red blood cell phosphatides, required hydrogenation of the lipids prior to TLC to cancel the effects of varying unsaturation on charring intensity (179). Phosphatides separated by HPTLC were quantitated in the slit scanning mode to achieve CV values generally under 10% for nanogram amounts of phosphatides (180).

A novel approach to quantitation of charred spots involves not densitometric scanning of spots on plates, but instead the quenching produced when the scraped charred adsorbent is suspended in a scintillation counting gel that contains a radioactive source (radium-226 external standard pellet) (181). Lipids could be determined rapidly in a range of 10–1000 $\mu$g with a CV under 5%. The authors claimed that variation in response due to degree of unsaturation was minimal. Response was nonlinear with respect to concentration, and therefore required the preparation of standard curves. Later work with such a procedure showed linear response for 2–20 $\mu$g of ceramide (182).

Others have favored fluorometric densitometry over the charring techniques. Fluorescence has been induced by treatment of the spots either with fluorescent dyes or by heat treatment in the presence of a reagent. In one example of dye treatment, 2′,7′-dichlorofluorescein (DCF) allowed one group to detect 10–50 $\mu$g of lipid with a CV = 5%–15% by scanning emitted light at 550 nm during excitation at 365 nm. Lipid subclass responses were linear with concentration, but each subclass had its own slope (183). Later, the same DCF dye was used in acetone solution into which TLC plates were dipped. This dipping technique led to the staining of the background of the plate.

Lipid concentration then was related to the diminution of fluorescence. Dipalmitoyl-PC then could be quantitated in the range of 15 ng-1 μg with a CV under 6% by relating the response to a standard curve. Response was measured at 504 nm with excitation at 312 nm wavelength (184).

Another team recommended ammonium 8-anilino-1-naphthalenesulfonate (ANS) because, unlike DCF or rhodamines, it produced fluorescent spots on a nearly dark nonfluorescent background. PC could be detected to 200 ng (185). This dye, however, had been criticized (173) for its difficulty of removal should the researcher wish to recover the intact lipid from the adsorbent. An alternative is 6-p-toluidino-2-naphthalenesulfonic acid, which is more sensitive than ANS, and—for cholesterol at least (no experiments were performed with phosphatides)—is removable from the lipid and allows linear quantitation over the range of 60–1000 ng (186).

As has been mentioned already, with detection in HPLC, chromophoric groups may be attached to amino groups of such lipids as PE and PS. In one example (187), lipids were reacted with fluorodinitrobenzene and then subjected to TLC. The resulting derivatives could be seen as yellow spots against a white background. However, densitometric response was linear in the range 0.02–0.4 μmol of derivative, considerably less sensitive than the fluorescence methods above.

Good sensitivity was achieved by inducing the lipids to fluoresce by heating the developed silica gel-alumina (2:1, no binder) plate to 110–150 C for 2–12 hr in a sealed tank with solid ammonium bicarbonate. Apparently a Schiff base forms with malonaldehyde released during oxidation of polyunsaturation. PC response (455 nm emission) as the square root of peak area was linear in the range of 2–100 μg upon excitation at 380 nm. Application of the reagent in the vapor state allowed uniform treatment of the plate (188). This technique was used in a later study on Silica gel 60 HPTLC plates. Excellent repeatability (CV = 2%) was seen down to 1 nmol of lipid though each phosphatide subclass required its own calibration curve (189).

The best sensitivity was seen by a group that heated its developed HPTLC plates at 110 C for 30 min after dipping them in a dilute perchloric acid solution (acetone/conc. perchloric acid/conc. hydrochloric acid, 250:10:6). Fluorescence of lipid spots was detected above 500 nm upon excitation at 360 nm. Quantitation could be achieved with as little as 20 pmol of PI with a CV under 5%. Fluorescence required polyunsaturation in the lipid spot both for this method and the ammonium bicarbonate one (190), these authors concluded.

Thinchrography (Rod TLC), meanwhile, is a method that aims to combine the principles of TLC with quantitation by flame ionization detection (FID). Development takes place not on TLC plates but instead on thin quartz rods coated with sintered silica or alumina. For detection, the developed rods are scanned in an instrument (Iatroscan TH-10 Analyzer, Iatron Laboratories Inc., Tokyo) that moves them through the flame of the FID, at which time the separated components are vaporized, burned and detected. The instrument's concept was discussed by Padley over a decade ago (191). The literature on the method and a discussion on the optimization of the instrument's performance are subjects of a recent extensive review (192). Rods have even been impregnated with boric acid (193) and silver nitrate (194) for specialized separations. A problem encountered during the analysis of oilseed lipids is the buildup of noncombustible material derived from salts of phytic acid (195).

The utility of thinchrography as a qualitative and "semiquantitative" method has been confirmed, but its value for quantitative determinations is controversial. Operational factors that can lead to improved results include proper humidity control of unspotted rods, very careful spotting technique, the use of an internal standard, proper adjustment of hydrogen flow in the FID, adequate equilibration with solvent vapor, good evaporation of solvent after development, proper scanning speed and adequate rod cleaning. Nevertheless, inadequacies still persist. There is variability from rod to rod and in the orientation of the rod within the flame. The range of sensitivity is approximately 1–30 $\mu$g lipid, thus precluding the detection of components under 3% of a mixture (unless a large spot is burned off selectively and more sample spotted and developed). Incomplete burnoff of large spots can occur.

Some representative analyses of error may be cited. One group experienced a CV = 5% for 5 $\mu$g spots of olive oil that had been predeveloped to concentrate the spot and to assure its uniform radial distribution, although a CV of 10% had been experienced in routine operations (196). Another group also claimed a CV of 5% but noted a great difference in response factors for different classes of lipids (197). The advantage of using internal standards has been cited. 1-Octadecanol has been used as an internal standard for neutral lipid separations, in which a CV of 10% was reported (198). In a study of polar lipid separation, the introduction of LPC as an internal standard improved precision from 12%–20% (20–5 $\mu$g) to 7% (199). Other groups, reporting quantitative errors in excess of 10%, have hesitated to recommend the use of thinchrography in quantitative applications (200–202).

The last reference points out significant problems with response factors. Aside from nonlinearity problems and highly divergent values from one report to another, this group also noted that response factors varied with the position of the spot on the rod. This was a problem because $R_f$ values changed with repeated use of the same rod.

There is a tendency to contrast the effort involved in using thinchrography to the facility of carrying out routine TLC. Proponents of the Iatroscan point out that comparisons to TLC are unfair, that analogies to TLC are only superficial. When accuracy and precision are important, researchers will probably find their time better spent using TLC techniques until such time that HPLC is suitable for routine analysis of phosphatides.

*Mass Spectrometry.* Interesting developments in mass spectrometry (MS) have led to methods for examining intact and derivatized phospholipids to determine the distribution of molecular species. MS methodology also has been used for rapid determination of fatty acid profiles. While MS has proven to be a remarkable tool for structural elucidation, a major shortcoming is its inability to provide good quantitation of components of a mixture.

MS in the field desorption mode (FD–MS) has enabled the examination of underivatized relatively nonvolatile molecules, including several phosphatides. Wood and coworkers have studied the FD–MS fragmentation pattern of 50 model phospholipids and optimized the source voltage and temperature to achieve the highest intensity of mass ions $(M + 1)^+$ and minimal fragmentation. This group with some success experimented with natural mixtures including egg yolk and amniotic fluid PC's. Difficulties included problems with $(M + 1)^+$ intensity with phosphatidylglycerol (PG), failure in the analysis of PS and difficulties with contamination by metal ions (203–205). Recent work with glycosphingolipids has shown how FD–MS can be used to determine the carbohydrate sequence of these complex lipids (206). A review of FD–MS and other "soft-ion" MS methods [chemical ionization (CI–MS) and field ionization] for lipid analysis demonstrates the advantages of these methods over the more conventional electron-impact MS (EI–MS) (207). Another review focuses on the electronics and mechanics of these soft-ion systems (208).

The work of Privett's group, in which a CI–MS instrument is coupled to another instrument (HPLC, GC), has been discussed in the section on HPLC detection. The group's identification of fatty acids by specific ion monitoring (154) offers the most rapid means of establishing a fatty acid profile, although the system is incapable of handling unidentified constituents.

Molecular species of phospholipid subclasses have been identified by the MS of derivatized lipids. For example, one group studied PC after the lipid underwent conversion to the acetylated diglyceride. GC/CI–MS with $NH_3$ allowed examination of bacterial PE, PG, and DPG (209). Earlier work with the acetylated products of egg yolk PC (210) and of total phospholipids from bacteria (211) allowed molecular species identification by "conventional" EI–MS.

Other groups have examined phospholipids as their diglyceride silyl ethers. Saito's group has examined molecular species of fetal lung PC as the trimethylsilyl ethers, and noted that no isomerization from the 1,2-diglyceride form to the 1,3-diglyceride form occurred (212). Slight isomerization was noted by those who examined the acetylated derivatives discussed above. Later work by Saito's group showed the use of *tert*-butyldimethylsilyl ether derivatives in molecular species analyses by GC–MS (213).

A technique that has not yet been applied to phosphatides is the combination of TLC with MS. A recent report (214) described the direct introduction of phenols, steroids, nucleosides and derivatized amines and amino acids together with their TLC adsorbent (polyamide) into the ion source of the spectrometer. The method should be applicable to the analysis of phosphatide bases, although general use with lipids is restricted by the nature of the adsorbent. Earlier attempts with silica gel and alumina failed, the report noted, because vaporization in the source always led to decomposition.

*Gas Chromatography (GC).* Methodology using GC has been mentioned throughout this review, in the notes on molecular species separation (in the discussion of HPLC) (141), in notes on TLC quantitation (175,176), in conjunction with mass spectral analysis (154,209,212,213) and with flame photometric detection of phosphorus (59).

Advances in GC and data handling technologies have made fatty acid profiling considerably more sophisticated than was the case a decade ago. With the increasing detail that is seen by use of capillary columns comes the concomitant task of handling more and more data. Esterification processes have not changed greatly since they were reviewed critically in 1975 (215). My coworkers and I have found one particular transmethylation reagent—KOH/methanol—to be highly satisfactory when free fatty acid content is negligible. The method (216) is protective of enol ether groups of the plasmalogens commonly found in the polar lipid fractions of animal, but not plant, tissues. However, it is incapable of transmethylating the fatty acyl moieties of ceramide lipids. When free fatty acids are present, they form soaps which then must be esterified with an acidic system such as one that uses $BF_3$/methanol (217). A recently reported alternative process uses tetramethylammonium hydroxide and methanol in ether. As transesterification proceeds, the glycerol byproduct forms a lower layer and retains any free fatty acids of the original lipid as their tetramethylammonium soaps. Because these soaps pyrolyze in a GC injector to give methyl esters, injection of the glycerol layer leads to a profile of the free fatty acid fraction. On the other hand, injection of the ether layer leads to a profile of the transesterified glycerides. Alternatively, the two layers may be made miscible by additional methanol, after which a total lipid profile may be obtained (218). Analogous to the reaction with tetramethylammonium hydroxide is transmethylation with trimethylsulfonium hydroxide; reaction in methanol is instantaneous, and the by-product dimethyl sulfide does not interfere with the subsequent chromatography (218a).

We also have found that considerably greater detail in fatty acid profiling may be achieved by examining the profile of the polar lipid fraction separately from the neutral lipid fraction (219). To accomplish this easily, we used the dry column extraction technique (102), which produced a neutral and a polar fraction concomitantly with lipid isolation.

Two warnings should be mentioned with regard to transesterification. First, chloroform should be avoided as a solvent during this process, because the ethanol present in chloroform as a stabilizer will lead to the formation of ethyl ester artifacts (220). Second, a solution of methyl esters should not be subjected to a stream of gas for concentration. Lower homologs will evaporate preferentially to leave behind a solution altered in its relative concentration of one ester over another (221).

Because most of us researchers are tied to sample introduction by syringe, we should be aware of the phenomenon of "discrimination," i.e., that high boiling components leave the syringe needle in lower proportions than the volatile ones. A recent study (222) implies that discrimination may be minimized by using larger sample volumes (e.g., 2 $\mu$L instead of 1 $\mu$L) and by preheating the empty syringe needle in the injector for three–five sec prior to depressing the plunger. Finally, in assembling the voluminous data that usually arises during multiple fatty acid profiling, one should beware that automated peak identification by retention time windows is often faulty, especially when several peaks elute with similar retention times. This problem usually arises because retention times just are not so reproducible. This makes it mandatory that raw profiles be assigned their peak identities manually (219,223) or semiautomatically (i.e., with manual editing) (224). My coworkers and I have found that in repetitive profiling, manual assignments are facilitated by generation of a modified report that substitutes the more reproducible *relative* retention times (RRT) for the initially reported actual retention times. A commonly occurring peak, conveniently chosen within a bunch of peaks that have similar retention times, is assigned RRT = 1. In our work, the palmitoleate peak served well (219).

## Acknowledgments

I thank the editor of the Journal of Lipid Research for permission to reproduce Figures 1 and 2. Similar permission was obtained from the respective authors of all the figures. Special thanks are due to Dr. George M.

Patton and Dr. P. Grattan Roughan for providing copies of the original artwork of Figures 3 and 4, respectively.

## Appendix

### Sources of Published Information:

1. Primary literature:
   a. Selected reading of original publications.
   b. Current Contents (Institute for Scientific Information, Philadelphia) (search by keyword and by scanning of contents page facsimiles of selected publications).
2. Secondary literature:
   a. Current Awareness Literature Service (National Agricultural Library, USDA) (Chemical Abstracts, Biological Abstracts, Food Science and Technology Abstracts, etc., continual searching by keyword combination).
3. Personal file:
   a. Input from above sources, collected on cards indexed by keywords, and searchable by keyword combination using the in-house computer facilities at ERRC.

## References

1. Norris, F. L., in Bailey's Industrial Oil and Fat Products, Vol. 2, 4th edn., edited by D. Swern, John Wiley and Sons, New York, NY, 1982, p. 253.
2. National Soybean Processors' Association, Yearbook and Trading Rules, 1978–79, Washington, D.C., 1978.
3. Fiske, C. H., and Y. SubbaRow, J. Biol. Chem. 66:375 (1925).
4. Reiser, R., JAOCS 24:199 (1947).
5. Yee, H. Y., Microchem. J. 23:426 (1978).
6. Yee, H. Y., Microchem. J. 23:497 (1978).
7. Daun, J. K., L. K. Davidson, J. A. Blake and Wo Yuen, JAOCS 58:914 (1981).
8. Rouser, G. A., N. Siakotos and S. Fleischer, Lipids 1:85 (1966).
9. Hartman, L., M. Cardoso-Elias and W. Esteves, Analyst (London) 105:173 (1980).
10. Official and Tentative Methods of the American Oil Chemists' Society, AOCS, Champaign, IL, 1979, Method CA 12–55.
11. Yuen, W., and P. C. Kelly, JAOCS 57:359 (1980).
12. Daun, J. K., L. D. Davidson, J. A. Blake and Wo Yuen, JAOCS 57:127A,

Abstract 104 (1980).
13. Vaskovsky, V. E., E. Y. Kostetsky and I. M. Vasendin, J. Chromatogr. 114:129 (1975).
14. Noble, R. C., J. H. Shand and I. G. West, Lab. Practice 28:393 (1979).
15. Flores, J., P. Nieto, G. C. Clemente and M. C. Miralles, Rev. Agroquimica y Tecnologia de Alimentos 21:121 (1981).
16. Kaitaranta, J. K., and S. P. Bessman, Anal. Chem. 53:1232 (1981).
17. Bessman, S. P., Anal. Biochem. 59:524 (1974).
18. Malmstadt, H. V., K. M. Walczak and M. A. Koupparis, American Laboratory, pp. 27–40, Sept. 1980.
19. Portoukalian, J., R. Meister and G. Zwingelstein, J. Chromatogr. 152:569 (1978).
20. Bartlett, G. R., J. Biol. Chem. 234:466 (1959).
21. Eng, L. F., and E. P. Noble, Lipids 3:157 (1968).
22. Yee, H. Y., Microchem. J. 23:460 (1978).
23. Morrison, W. R., Anal. Biochem. 7:218 (1964).
24. Petitou, M., F. Tuy and C. Rosefeld, Anal. Biochem. 91:350 (1978).
25. Sandhu, R. S., Clin. Chem. 22:1973 (1976).
26. Raheja, R. A., C. Kaur, A. Singh and I. S. Bhatia, J. Lipid Res. 14:695 (1973).
27. Lowry, R. R., and I. J. Tinsley, Lipids 9:491 (1975).
28. Subramanian, K. S., and A. Corsini, Can. J. Chem. 56:1729 (1978).
29. Duck-Chong, C. G., Lipids 14:492 (1979).
30. Kratochvila, J., and L. Sommer, SCR Fac. Sci. Nat. Univ. Purkynianae Brun 10:53 (1980); cf. also Debruyne, I., Anal. Biochem. 130:454 (1983).
31. Matsuo, T., J. Shida and W. Kurihara, Anal. Chim. Acta 91:385 (1977).
32. Motomizu, S., T. Wakimoto and K. Toei, Anal. Chim. Acta 138:329 (1982).
33. Johnson, F. J., J. Assoc. Off. Anal. Chem. 61:533 (1978).
34. AOAC Official Methods of Analysis, 13th edn., Washington, DC, 1980, Methods 2.032–2.039.
35. Brammel, W. S., J. Assoc. Off. Anal. Chem. 64:808 (1981).
36. IUPAC, Standard Methods for the Analysis of Oils, Fats, and Derivatives, 5th edn., edited by C. Paquot, Pergamon Press, Oxford, 1979, Method II.D.16.2.
37. Fogg, A. G., and N. K. Bsebsu, Analyst (London) 106:1288–95 (1981).
38. AOAC Official Methods of Analysis, 13th edn., Washington, DC, 1980, Method 47.033.
39. Ibid. Method Ja 5–55.
40. Ibid. Methods 2.026–2.028.

41. Ibid. Methods 2.029–2.031.
42. Orion Research, Inc., Newsletter/Specific Ion Electrode Technology III(1, 2):6 (1971).
43. Hu, Y., Fen Hsi Hua Hsueh 7:161 (1979).
44. Hoover, T. B., Separ. Sci. Technol. 17:295 (1982).
45. Yoshida, Y., E. Furuya and K. Tagawa, J. Biochem. (Tokyo) 86:825 (1979).
46. Yoshida, Y., E. Furuya and K. Tagawa, J. Biochem. (Tokyo) 88:463 (1980); cf. also Stewart, J. C. M., Anal. Biochem. 104:10 (1980).
47. Totani, Y., H. E. Pretorius and L. M. duPlessis, JAOCS 59:162 (1982).
48. Hoft, D., J. Oxman and R. C. Gurira, Agr. Food Chem. 27:145 (1979).
49. Ediger, R. D., At. Absorpt. Newsl. 15:145 (1976).
50. Prevot, A., and M. Gente-Jauniaux, Rev. Fr. Corps Gras 24:493 (1977).
51. Gente-Jauniaux, M., and A. Prevot, Rev. Fr. Corps Gras 26:325 (1979).
52. Vigler, M. S., A. S. Strecker and A. Varnes, Appl. Spectrosc. 32:60 (1978).
53. Langmyhr, F. J., and I. M. Dahl, Anal. Chim. Acta 131:303 (1981).
54. Hieftje, G. M., T. R. Copeland and D. R. de Olivares, Anal. Chem. 48:156R (1976).
55. Knowles, D. J., P. Marriott and S. J. E. Slater, Proc. Analyt. Div. Chem. Soc. 15:62 (1978).
56. Rogers, L. J., JAOCS 52:441A (Abstract 28) (1975).
57. Cope, M. J., and A. Townshend, Anal. Chim. Acta 134:93 (1982).
58. Campbell, D. R., U.S. N.T.I.S., PB Rep., PB-252952, 1976, 46 pp. (Chem. Abstr. 86:78391y).
59. Handa, Y., M. Higashimoto, K. Matano, A. Tokumura and H. Tsukatani, Yakugaku Zasshi 101:749 (1981).
60. AOAC Official Methods of Analysis, 13th edn., Washington, D.C., 1980, Methods 49.001–49.005.
61. Zander, A. T., Industrial Res. Dev., Feb. 1982, pp. 146–50.
62. Hanson, T. J., and R. D. Ediger, American Laboratory, March 1980, pp. 116–23.
63. Kahn, H. L., Industrial Res. Dev., Feb. 1982, pp. 156–60.
64. Miyazaki, A., A. Kimura and Y. Umezaki, Anal. Chim. Acta 138:121 (1982).
65. Heine, D. R., M. B. Denton and T. D. Schlabach, Anal. Chem. 54:81 (1982).
66. Savinova, E. N., A. V. Karyakin and T. P. Andreeva, Geokhim. Anal. Metody Izuch. Veshchestv. Sostava Osad. Porod Rud 2:26 (1974); see also Ivanova, M. S., and N. G. Fridlender, Anal. Metody Izuch. Veshchestv. Sostava Osad. Porod Rud. 2:24 (1974).

67. Kantor, T., B. Tarnay, T. Cserfalvi and A. Borodi, Period. Polytech., Chem. Eng. 21:237 (1977).
68. Dijkstra, A. J., and D. Meert, JAOCS 59:199 (1982).
69. Chapman, L. R., J. Chromatogr. 66:303 (1972).
70. King, R. T., J. Sci. Food Agr. 28:631 (1977).
71. Guinn, V. P., JAOCS 45:767 (1968).
72. Kisner, H. J., C. W. Brown and G. J. Kavarnos, Anal. Chem. 54:1479 (1982).
73. Sanchez-Perez, A., J. Hernandez-Mendez, and J. Armesto-Gomez, Anal. Lett. 14:629–48 (1981).
74. Glonek, T., in Phosphorus Chemistry Directed Towards Biology, edited by W. J. Stec, Pergamon Press, Oxford, 1980, pp. 157–74.
75. London, E., and G. W. Feigenson, J. Lipid Res. 20:408 (1979).
76. Gurley, T. W., and W. M. Ritchey, Anal. Chem. 48:1137 (1976).
77. Kasler, F. K., and M. Tierney, Mikrochim. Acta II:411 (1978).
78. Birdsall, N. J. M., J. Feeney, A. G. Lee, Y. K. Levine and J. C. Metcalfe, J. Chem. Soc., Perkin Trans. 2:1441 (1972).
79. Press, K., R. M. Sheeley, W. J. Hurst and R. A. Martin, Jr., J. Agric. Food Chem. 29:2096 (1981).
80. Pardun, H., Fette, Seifen, Anstrichm. 83:240 (1981); cf. also Arens, M., and H. Pardun, Ibid. 84:28 (1982).
81. List, G. R., A. J. Heakin, C. D. Evans and T. L. Mounts, JAOCS 55:521 (1978).
82. Scholfield, C. R., and H. J. Dutton, J. Biol. Chem. 214:633 (1955).
83. Pardun, H., Fette, Seifen, Anstrichm. 66:467 (1964).
84. Singleton, J. A., and H. E. Pattee, JAOCS 58:873 (1981).
85. Sugiura, M., T. Oikawa and K. Hirano, Clin. Chim. Acta 89:447 (1978).
86. Gurantz, D., M. F. Laker and A. F. Hofmann, J. Lipid Res. 22:373 (1981).
87. Hayashi, H., K. Watanabe and T. Tatano, Japan. Kokai 77,125,395 (1977) (Chem. Abstr. 88:47242n).
88. Takayama, M., S. Itoh, T. Nagasaki and I. Tanimizu, Clin. Chim. Acta 79:93 (1977).
89. Karube, I., K. Hara, I. Satoh and S. Suzuki, Anal. Chim. Acta 106:243 (1979).
90. Artiss, J. D., T. F. Draisey, R. J. Thibert and K. E. Taylor, Microchem. J. 24:239 (1979).
91. Artiss, J. D., T. F. Draisey, R. J. Thibert, B. Zak and K. E. Taylor, Microchem. J. 25:153 (1980).
92. Annoni, G., M. L. Dioguardi, A. Tripodi and M. Zuin, Clin. Chem. Acta 26:669 (1980).

93. Qureshi, M. Y., G. M. Murphy and R. H. Dowling, Clin. Chim. Acta 105:407 (1980).
94. Carman, G. M., A. S. Fischl, M. Dougherty and G. Maerker, Anal. Biochem. 110:73 (1981).
95. Hayashi, H., K. Watanabe and T. Tatano, Japan Kokai 78,07,396 (1978) (Chem. Abstr. 88:185817e).
96. Anon., Kakao und Zucker 33:284 (1981).
97. Machida, Y., and T. Nakanishi, Agric. Biol. Chem. 46:807 (1982).
97a. Morgan, D. K., and N. D. Danielson, J. Chromatogr. 262:265 (1983).
98. Shukla, S. D., Life Sci. 30:1323 (1982), and refs. 13, 14, 16–19 therein.
99. Folch, J., M. Lees and G. H. Sloane Stanley, J. Biol. Chem. 226:497 (1957).
100. Bligh, E. G., and W. J. Dyer, Can. J. Biochem. Physiol. 37:911 (1959).
101. Chen, I. S., C.-S. J. Chen and A. J. Sheppard, JAOCS 58:599 (1981).
102. Marmer, W. N., and R. J. Maxwell, Lipids 16:365 (1981).
103. Adnan, M., C. J. Argoudelis, J. Tobias, W. N. Marmer and R. J. Maxwell, JAOCS 58:550 (1981).
104. Hara, A., and N. S. Radin, Anal. Biochem. 90:420 (1978).
105. Olifson, L. E., and N. D. Osadchaya, Lab. Delo. 155–157 (1981).
106. Phillips, F. C., and O. S. Privett, Lipids 14:949 (1979).
107. Roughan, P. G., C. R. Slack and R. Holland, Lipids 13:497 (1978).
108. Melton, S. L., R. E. Moyers and C. G. Playford, JAOCS 56:489 (1979).
109. Wan Pee, W., L. Boni, P. Peeters, M. Amine and H. Hendrikx, Riv. Ital. Sost. Grasse 56:293 (1979).
110. Palmer, F. B. St. C., Biochim. Biophys. Acta 231:134 (1971).
111. Zhukov. A. V., and A. G. Vereshchagin, Sov. Plant Physiol. (Engl. Transl. Fiziol. Rast. (Moscow) 27 (1, part 2):140 (1980).
112. Kinsella, J. E., L. Posati, J. Weihrauch and B. Anderson, Crit. Rev. Food Tech. 5:299 (1975).
113. Galanos, D. S., and V. M. Kapoulas, J. Lipid Res. 3:134 (1962).
114. Nelson, G. J., in Analysis of Lipids and Lipoproteins, edited by E. G. Perkins, AOCS, Champaign, IL, 1975, p. 70ff.
115. Sen Gupta, A. K., Fette, Seifen, Anstrichm. 78:111 (1976).
116. Biacs, P., A. Erdélyi, É. Kurucz-Lusztig and J. Holló, Acta Alimentaria 7:181 (1978).
117. Erdélyi, A., É. Kurucz-Lusztig, P. Biacs and J. Holló, Acta Alimentaria 6:21 (1977).
118. Richter, H., Ch. Srey, K. Winter and W. Fürst, Pharmazie 32:H.3 (1977).
119. Radin, N. S., J. Lipid Res. 19:922 (1978).
120. Gentner, P. R., I. Dietrich and M. Bauer, Fette, Seifen, Anstrichm.

84:15 (1982).
121. Primes, K. J., R. A. Sanchez, E. K. Metzner and K. M. Patel, J. Chromatogr. 236:519 (1982).
122. Schacht, J., J. Lipid Res. 19:1063 (1978).
123. Palmer, F. B. St.C., J. Lipid Res. 22:1296 (1981).
123a. Aitzetmüller, K., Prog. Lipid Res. 21:171 (1982).
124. Kucera, P., ed., Microcolumn High-Performance Liquid Chromatography, Elsevier-North Holland Publishing Co., Amsterdam, 1984.
125. Rainey, M. L., and W. C. Purdy, Anal. Chim. Acta 93:211 (1977).
126. Parris, N. A., J. Chromatogr. Sci. 17:541 (1979).
127. Fager, R. S., S. Shapiro and B. J. Litman, J. Lipid Res. 18:704 (1977).
128. Patel, K. M., and J. T. Sparrow, J. Chromatogr. 150:542 (1978).
129. Yandrasitz, J. R., G. Berry and S. Segal, J. Chromatogr. 225:319 (1981).
130. Chen, S. S.-H., and A. Y. Kou, J. Chromatogr. 227:25 (1982).
131. Hurst, W. J., and R. A. Martin, Jr., JAOCS 57:307 (1980).
132. Patton, G. M., J. M. Fasulo and S. J. Robins, J. Lipid Res. 23:190 (1982).
133. Nasner, A., and Lj. Kraus, J. Chromatogr. 216:389 (1981).
134. Alam, I., J. B. Smith and M. J. Silver, J. Chromatogr. 234:218 (1982).
135. Kiuchi K., T. Ohta and H. Ebine, J. Chromatogr. 133:226 (1977).
136. Hanson, V. L., J. Y. Park, T. W. Osborn and R. M. Kiral, J. Chromatogr. 205:393 (1981).
137. Gross, R. W., and B. E. Sobel, J. Chromatogr. 197:79 (1980).
138. Compton, B. J., and W. C. Purdy, J. Liq. Chromatogr. 3:1183 (1980).
139. Porter, N. A., R. A. Wolf and J. R. Nixon, Lipids 14:20 (1979).
140. Smith, M., and F. B. Jungalwala, J. Lipid Res. 22:697 (1981).
141. Myher, J. J., and A. Kuksis, JAOCS 59:308A (Abstract 294) (1982); Can. J. Biochem. 60:638 (1982).
142. Kuksis, A., J. J. Myher, K. Geher, N. A. Shaikh, W. C. Breckenridge, G. J. L. Jones and J. A. Little, J. Chromatogr. 182:1 (1980).
143. Hsieh, J. Y.-K., D. K. Welch and J. G. Turcotte, Lipids 16:761 (1981).
144. Renkonen, O., Biochim. Biophys. Acta 152:114 (1968).
145. Ullman, M. D., and R. H. McLuer, J. Lipid Res. 18:371 (1977).
146. Jungalwala, F. B., R. J. Turel, J. E. Evans and R. H. McLuer, Biochem. J. 145:517 (1975).
147. Chen, S. S.-H., A. Y. Kou and H.-H. Y. Chen, J. Chromatogr. 208:339 (1981).
148. Okazaki, M., N. Hagiwara and I. Hara, J. Biochem. (Tokyo) 91:1381 (1982).
149. Geurts van Kessel, W. S. M., W. M. A. Hax, R. A. Demel and J. de Gier, Biochim. Biophys. Acta 486:524 (1977).

150. Herslöf, B. G., and T. J. Pelura, JAOCS 59:308A (Abstract 295) (1982).
151. Privett, O. S., and W. L. Erdahl, Anal. Biochem. 84:449 (1978).
152. Phillips, F. C., and O. S. Privett, JAOCS 58:590 (1981).
153. Erdahl, W. L., and O. S. Privett, Lipids 12:797 (1977).
154. Erdahl, W. L., W. Beck, C. Jones, D. E. Jarvis and O. S. Privett, Lipids 16:614 (1981).
155. Erdahl, W. L., F. C. Phillips, W. R. Anderson and O. S. Privett, JAOCS 59:308A (Abstract 296) (1982).
156. Stahl, E., Angew. Chem. Int. Ed. Eng. 22:507 (1983).
157. Jupille, T. H., and J. A. Perry, Science 194:288 (1976).
158. Siouffi, A. M., F. Bressole and G. Guiochon, J. Chromatogr. 209:129 (1981).
159. Biverstedt, A., and L. Rydag, Proceedings of the 9th Scandinavian Symposium on Lipids 268–74 (1977).
160. Portoukalian, J., R. Meister and G. Zwingelstein, J. Chromatogr. 152:569 (1978).
161. Nouvelot, A., G. Sezille, P. Dewailly and J. C. Fruchart, Rev. Franc. Corps Gras 24:365 (1977).
162. Peter, H. W., and H. U. Wolf, J. Chromatogr. 82:15 (1973); see also Watkins, T. R., J. High Resol. Chromatogr. Chromatogr. Commun. 5:104 (1982).
163. Henderson, R. F., and M. H. Clayton, Anal. Biochem. 70:440 (1976).
164. Rodríguez de Turco, E. B., and N. G. Bazán, J. Chromatogr. 137:194 (1977).
165. Latyshev, N. A., and V. E. Vaskovsky, J. High Resolut. Chromatogr. Chromatogr. Commun. 3:478 (1980).
166. Do, U. H., P. T. S. Pei and S. Ramachandran, 31st Pittsburgh Conference on Analytical Chemistry and Applied Spectroscopy, Atlantic City, NJ, Paper No. 763, March 10–14, 1980.
167. Fine, J. B., and H. Sprecher, J. Lipid Res. 23:660 (1982).
168. Young, S. K. F., A. Kuksis, L. Marai and J. J. Myher, Lipids 12:529 (1977).
169. Rao, P. V., S. Ramachandran and D. G. Cornwell, Lipids 3:187 (1968).
170. Kundu, S. K., S. Chakravarty, N. Bhaduri and H. K. Saha, J. Lipid Res. 18:128 (1977).
171. Vaskovsky, V. E., and N. A. Latyshev, J. Chromatogr. 115:246 (1975).
172. Michalec, C., and J. Reinisová, J. Chromatogr. 162:229 (1979).
173. Bernard, D. M., and R. E. Vercauteren, J. Chromatogr. 120:211 (1976).
174. Hyslop, P. A., and D. A. York, Anal. Biochem. 101:75 (1980).
175. Radwan, S. S., J. Chromatogr. Sci. 16:538 (1978).
176. Christie, W. W., R. C. Noble and J. H. Moore, Analyst 95:940 (1970).

177. Bitman, J., and D. L. Wood, J. Liq. Chromatogr. 4:1023 (1981).
178. Bitman, J., and D. L. Wood, J. Liq. Chromatogr. 5:1155 (1982).
179. Chapelle, S. M., Clin. Chim. Acta 92:11 (1979).
180. Hedegaard, E., and B. Jensen, J. Chromatogr. 225:450 (1981).
181. Shand, J. H., and R. C. Noble, Anal. Biochem. 101:427 (1980).
182. Selvam, R., and N. S. Radin, Anal. Biochem. 112:338 (1981).
183. Ishida-Ichimasa, M., Y. Ichimasa and K. Uranaka, Agric. Biol. Chem. 40:1253 (1976).
184. Freer, D. E., B. E. Statland and G. Sher, Clin. Chem. Acta 25:960 (1979).
185. Vinson, J. A., and J. E. Hooyman, J. Chromatogr. 135:226 (1977).
186. Jones, M., R. W. Keenan and P. Horowitz, J. Chromatogr. 237:522 (1982).
187. Heidemann, G., and D. Lekim, J. Chromatogr. 128:235 (1976).
188. Segura, R., and A. M. Gotto, Jr., J. Chromatogr. 99:643 (1974).
189. Kupke, I. R., and S. Zeugner, J. Chromatogr. 146:261 (1978).
190. Harrington, C. A., D. C. Fenimore and J. Eichberg, Anal. Biochem. 106:307 (1980); cf. also Zhou, L., H. Shanfield and A. Zlatkis, J. Chromatogr. 239:259 (1982).
191. Padley, F. B., J. Chromatogr. 39:37 (1969).
192. Ackman, R. G., Methods in Enzymology 72:205 (1981).
193. Tanaka, M., T. Itoh and H. Kaneko, Lipids 15:872 (1980).
194. Sebedio, J.-L., and R. G. Ackman, J. Chromatogr. Sci. 19:552 (1981).
195. Ackman, R. G., and A. D. Woyewoda, J. Chromatogr. Sci. 17:514 (1979).
196. Harvey, H. R., and J. S. Patton, Anal. Biochem. 116:312 (1981).
197. Bradley, D. M., C. R. Rickards and N. S. T. Thomas, Clin. Chim. Acta 92:293 (1979).
198. Van Tornout, P., R. Vercaemst, H. Caster, M. J. Lievens, W. de Keersgieter, F. Soetewey and M. Rosseneu, J. Chromatogr. 164:222 (1979).
199. Martin-Ponthieu, A., N. Porchet, J.-C. Fruchart, G. Sezille, P. Dewailly, X. Codaccioni and M. Delacour, Clin. Chem. Acta 25:31 (1979).
200. Hammond, E. W., J. Chromatogr. 203:397 (1981).
201. Christie, W. W., in Fats and Oils: Chemistry and Technology, edited by R. J. Hamilton and A. Bhati, Applied Science Publishers, Ltd., London, 1980, pp. 1–23.
202. Crane, R. T., S. C. Goheen, E. C. Larkin and G. A. Rao, Lipids 18:74 (1983).
203. (a) Wood, G. W., P.-Y. Lau, G. Morrow, G. N. S. Rao, D. E. Schmidt, Jr. and J. Tuebner, Chem. Phys. Lipids 18:316 (1977); (b) Wood, G. W., and

S. E. Perkins, Anal. Biochem. 122:368 (1982); cf. also Sugatani, J., M. Kino, K. Saito, T. Matsuo, H. Matsuda and K. Katakuse, Biomed. Mass Spectrometry 9:293 (1982).
204. Wood, G. W., P.-Y. Lau and G. N. Subba Rao, Biomed. Mass. Spectrom. 3:172 (1976).
205. Wood, G. W., and P.-Y. Lau, Biomed. Mass Spectrom. 1:154 (1974).
206. Kushi, Y., and S. Handa, J. Biochem. (Tokyo) 91:923 (1982).
207. Games, D. E., Chem. Phys. Lipids 21:389 (1978).
208. Ligon, W. V., Jr., Science 205:151 (1979).
209. Kagawa, Y., and T. Ariga, J. Biochem. (Tokyo) 81:1161 (1977).
210. Hasegawa, K., and T. Suzuki, Lipids 10:667 (1975).
211. Ohno, Y., I. Yano and M. Masui, Advances Mass Spectrom. Biochem. Med. 2:559 (1976).
212. Ogino, H., T. Matsumura, K. Satouchi and K. Saito, Biomed. Mass Spectrom. 4:326 (1977).
213. Satouchi, K., and K. Saito, Biomed. Mass Spectrom. 6:396 (1979).
214. Kraft, R., A. Otto, A. Makower and G. Etzold, Anal. Biochem. 113:193 (1981).
215. Sheppard, A. J., and J. L. Iverson, J. Chromatogr. Sci. 13:448 (1975).
216. Christopherson, S. W., and R. L. Glass, J. Dairy Sci. 52:1289 (1969).
217. Slover, H. T., and E. Lanza, JAOCS 56:933 (1979).
218. Metcalfe, L. D., and C. N. Wang, J. Chromatogr. Sci. 19:530 (1981).
218a. Butte, W., J. Chromatogr. 261:142 (1983).
219. (a) Maxwell, R. J., and W. N. Marmer, Lipids 18:453 (1983); (b) Marmer, W. N., R. J. Maxwell and J. C. Phillips, Lipids 18:460 (1983).
220. Johnson, A. R., A. C. Fogerty, R. L. Hood, S. Kozuharov and G. L. Ford, J. Lipid Res. 17:431 (1976).
221. Marmer, W. N., Lipids 13:835 (1978).
222. Grob, K., Jr., and S. Rennhard, J. High Resol. Chromatogr. Chromatogr. Commun. 3:627 (1980).
223. Bacon, G. D., J. Chromatogr. 172:57 (1979).
224. Lanza, E., B. M. Golden, J. Zyren and H. T. Slover, J. Chromatogr. Sci. 18:126 (1980).

# Chapter Thirteen

## Lecithin In Baking

**Y. Pomeranz**
U.S. Grain Marketing Research Laboratory
Agricultural Research Service
US Department of Agriculture
Manhattan, KS 66502

### Introduction

The term "lecithin" is the commercial or popular name for a naturally occurring mixture of similar compounds more accurately identified as phosphatides or phospholipids. The principal components of the natural mixture are phosphatidylcholine, phosphatidylethanolamine, inositol phosphatides and related phosphorus-containing lipids (1).

The chemical structures of the three major phosphatides are:

**Phosphatidylinositol**

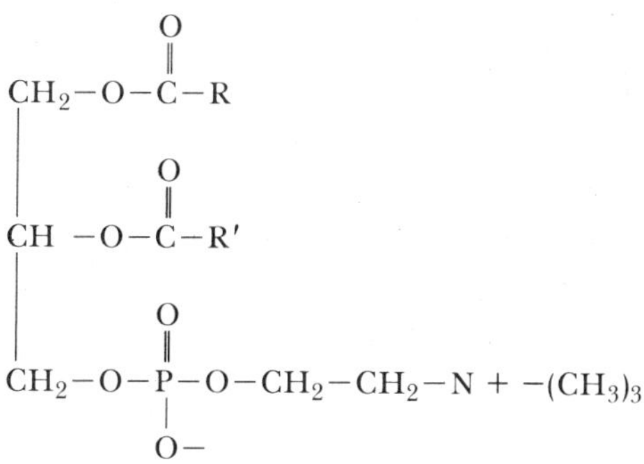

**Phosphatidylethanolamine (Cephalin)**

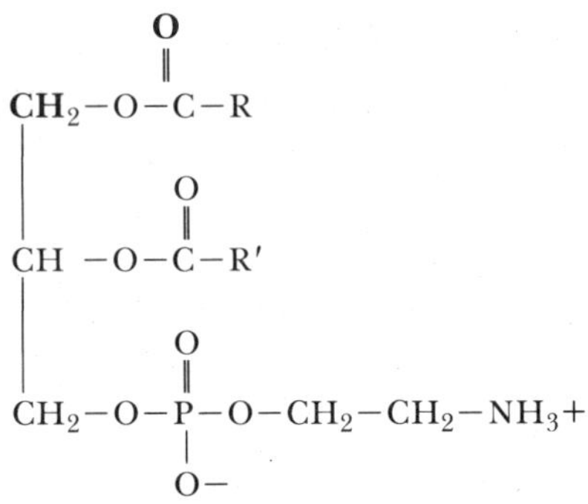

**Phosphatidylinositol**

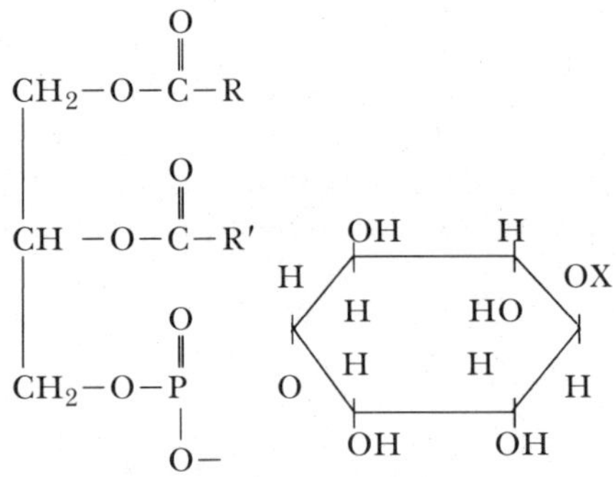

X—One or more phosphate groups linked to one or more sugar molecules.
R,R′—Represent fatty acid constituents.

Lecithin is nature's principal surface active agent. It is found in all living cells, whether of animal or vegetable origin. The highest concentrations generally occur in animal products. For example, fresh egg yolk contains 8 to 10% phosphatides which play a rôle in the remarkable emulsifying properties of the product. The major commercial source of lecithin is the soybean, which contains 0.3 to 0.6% phosphatides. In humans and in animals the phosphatides are concentrated in the vital organs, such as the brain, liver, and kidney; and in vegetables it is highest in the seeds, nuts, and grains (1).

Lecithin is recognized as an edible product which has nutritional value as well as industrial utility.

Lecithin is one of those substances that strikes not only the fancy but also the intellect of man. Numerous reports have attributed to it almost magic intellectual and health powers (2). It is used in a variety of foods, including baked products, because of its emulsifying, wetting, colloidal, antioxidant, and physiological properties (3), (4). According to Eichberg (5), in yeast-leavened doughs 0.1 to 0.3% commercial lecithin improves water absorption, ease of handling, fermentation tolerance, shortening value of fat, volume and uniformity, and shelf life. In biscuits and crackers, pies, and cakes 1 to 3% lecithin—on shortening basis—promotes fat distribution and shortening action, facilitates mixing and acts as a release agent.

This chapter covers the distribution of phosphatides in wheat, flour, and gluten; phosphatides' functional properties; their use in production of various baked goods (bread, cakes, cookies, extrusion products, etc.); and their effects on quality of such goods. The terms phosphatides and phospholipids denote fat-like substances which contain nitrogen and phosphorus. Unless stated otherwise, the terms lecithin and commercial lecithins are used to denote a mixture of phospholipids—generally from soybeans. Whereas the terms lecithin and phosphatidylcholine are often used interchangeably in the literature, the latter is used in this review only to denote 1,2-diacyl glycerylphosphorylcholine.

## Phosphatides in Wheat, Flour, and Gluten

Phosphatides in wheat, milled wheat products, and wheat flour components have been the subject of numerous investigations (6–13). Wheat germ phosphatides were determined by Moruzzi (14) and Pomeranz (15,16). Those and other studies were summarized and reviewed by Pomeranz (17); Mecham (18); Morrison (19–22); and Acker and coworkers (23–26). Lipids of wheat, expressed as % total lipids, are given in Figure 1 (29). Lipid distribution in wheat tissues expressed as % total lipids, are given in Figure 2 (29). Data in parentheses are expressed as % non-starch lipids.

Lipids in the starchy endosperm of wheat can be divided into those inside and those outside the starch granules. Nonstarch lipids include free lipids and lipids bound in residues of endoplasmic reticulum and other membranes and organelles. The diacylphospholipids and N-acyl phosphatidylethanolamines are regarded as membrane lipids and are characterized by high proportions of N-acyl phosphatidylethanolamine, N-acyl lysophosphatidylethanolamine, and phosphatidylcholine (22). Lysophosphatidylcholine can be formed by degradation of phosphatidylcholine but is more likely to be "leaky" starch lipid. The starch lipids are almost exclusively lysophospholipids, and they comprise about 80% of the phospholipids in the whole kernel. There is no N-acyl lysophosphatidylethanolamine in starch, although it is a major lysophospholipid in the non-starch lipids. Flour non-starch lipids consist of 59 to 63% nonpolar lipids, 22 to 26%

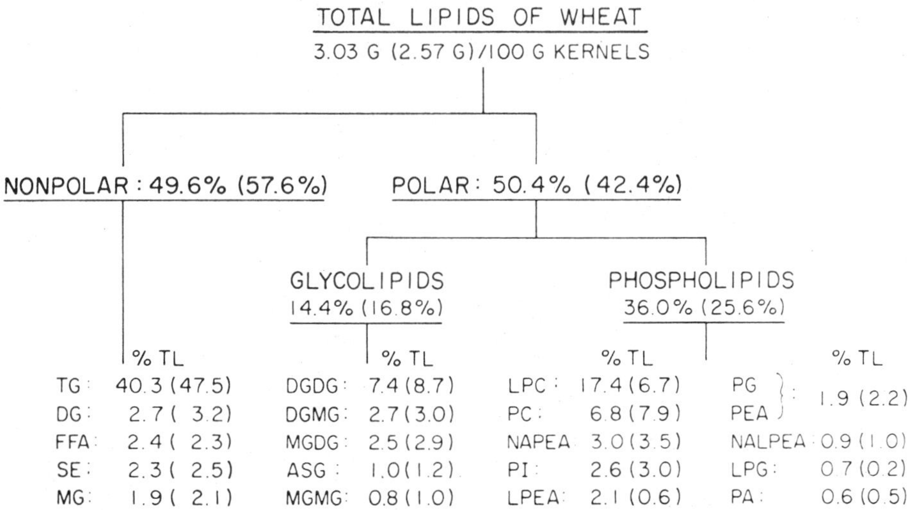


Fig. 13-1. Lipids of wheat, expressed as percentage of total lipids (TL). Data in parentheses are expressed as percentage of nonstarch lipids. Other abbreviations: ASG, 6-O-acyl steryl glucoside; DG, Diglycerides; DGDG, digalactosyldiglyceride; DGMG, Digalactosylmonoglyceride; FFA, Free fatty acid; LPEA, Lysophosphatidylethanoloamine; LPC, Lysophosphatidylcholine; LPG, Lysophosphatidylglycerol; MG, Monoglycerides; MGDG, monoglactosyldiglyceride; MGMG, Monogalactosylmonoglyceride; NALPEA, N-acyl phosphatidylethanolamine; PA, Phosphatidic acid; PC, Phosphatidylcholine; PEA, Phosphatidylethanolamine; PI, Phosphatidylinositol; PG, Phosphatidylglycerol; SE, Steryl ester, and TG, Triglyceride. From Hargin and Morrison (182).

glycolipids, and 14 to 16% phospholipids (22). Lipids in purified starches are almost exclusively monoacyl lipids (86 to 91% lysophospholipids) and their composition is remarkably uniform.

According to Acker (24), wheat phosphatides include phosphatidyl– and lysophosphatidylcholine, ethanolamine (and its N-acyl derivative), serine and inositol phosphatides and phosphatidylglycerol, and phosphatidic acid. Up to 0.7% lysophosphatidylcholine may be bound to starch and is unavailable for interaction with gluten proteins (23). In addition, several sphingolipids are present. During fractionation of wheat flour, lysophosphatides are bound to starch and other phosphatides mainly to gluten proteins (24). Lysolecithin comprised 23.5 to 24.9% and lecithin 3.0 to 3.9% of total polar lipids in 3 wheat flours; and 3.0 to 3.9 and 1.2 to 1.4% of total polar lipids washed out from those flours. Differences in lipid content and composition, including those of phosphatides, could not account for genetic differences in breadmaking quality of European wheat cultivars (24).

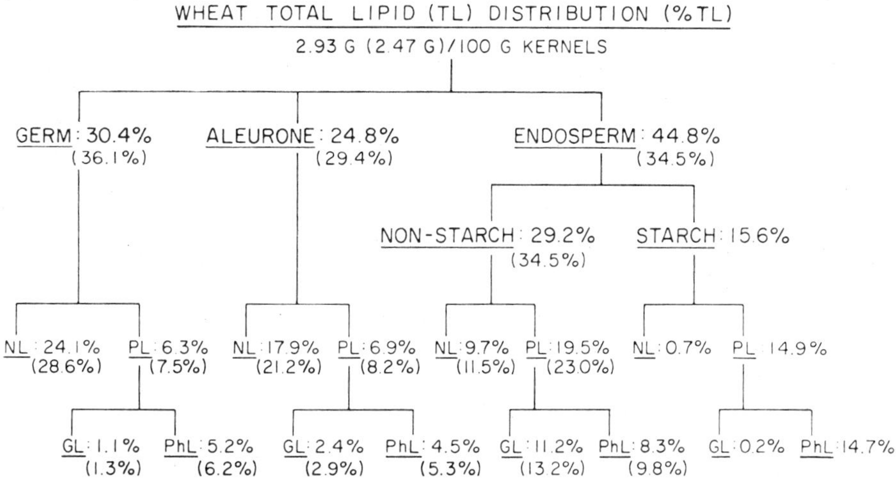


Fig. 13-2. Lipid distribution in wheat tissues expressed as total lipids. Data in parentheses are expressed as percentage nonstarch lipids. Abbreviations: GL, Glycolipid; NL, Nonpolar lipid; PhL, Phospholipid; and PL, Polar lipid. From Hargin and Morrison (182).

Auerman et al (30) described surface-active properties of phosphatide concentrates in interaction with flour, starch and gluten. Putschkova and Netschajev (31) reviewed the interaction of lipids with wheat flour components. Sidorova and Chukhlebova (32) studied the effects of surface active agents on properties of wheat flour gluten. Zimmerman (33), who followed noncovalent interaction of lipoids in wheat dough, compared the effects of phosphatides on decreased water absorption and dough consistency with the effects of fatty acids, triglycerides, and several emulsifiers.

## Types, Properties, and Combinations

Soybeans contain 1.5 to 3.0% phosphatides (34). Crude commercial soybean lecithin has the following representative composition: 16% phosphatidylcholine; 14% phosphatidylethanolamine; 10% phosphatidylinositol; 35% soybean oil; and 25% miscellaneous compounds (water, sugars, sterols, other phosphatides) (1). The crude dried product from degumming soybean oil contains 28 to 33% oil. This product can be used to prepare a variety of grades of lecithin by adding bleaching agents (such as $H_2O_2$) to lighten the tan-to-brown color, by adding fatty acids or oil to increase fluidity, by removing the oil to increase the phosphatide content, or by separating the oil-free lecithin into alcohol-soluble and alcohol-insoluble fractions (35) (36). Different types of commercial lecithins are described in Table I. Separating acetone-solubles increases the amount of phosphatides in the acetone-insoluble fraction by decreasing the amount of triglycerides. Further fractionation can separate the alcohol-soluble phosphatidylcholine and phosphatidylethanolamine from the alcohol-insoluble inositol phosphatides. The alcohol-soluble fraction gives stable oil-in-water emulsions; the alcohol-insoluble fraction stable water-in-oil emulsions. The latter can be used to produce antispattering and pan-grease (release) agents. Composition and properties of commercial lecithin fractions are described in Table II. Hydrogen peroxide, in addition to bleaching, can also hydroxylate, which imparts hydrophilic properties, improves moisture retention and contributes to the formation of stable oil-in-water emulsions.

**TABLE 13-I**
**Different Grades of Commercial Lecithin—Bleached**

| | Properties of Different Grades[1] | | | |
|---|---|---|---|---|
| Characteristics | A | B | C | D |
| Consistency | Plastic | Fluid | Fluid | Fluid |
| Acetone insoluble (%)[2] | 67–72 | 62–64 | 54–60 | 62–64 |
| Moisture (%) | 1.1 | 0.75 | 0.75 | 0.75 |
| Benzene insoluble (max %) | 0.1 | 0.1 | 0. 1 | 0.05 |
| Acid value | 25 | 32 | 40 | 32 |
| Centipoises at 80°F | — | 15,000 | 7,000 | 15,000 |

Source: (1)
[1]Grades with A, B, C, and D can result from different amounts of bleaching with hydrogen peroxide.
[2]Actual percentage of phosphatides—remainder primarily oil or additives like fatty acids.

The National Soybean Processors' Association classifies lecithin into six grades, depending on viscosity and color (Table III). Phosphatide content is specified in terms of acetone-insoluble; product clarity and purity are specified in terms of benzene-insoluble. Plastic lecithins are converted to fluid forms by adding 2 to 5% fatty acids or carriers such as soybean oil (36). Sullivan and Szuhaj (37) distinguished between natural lecithins and those refined or chemically modified (Table IV). The refined ones are obtained by either custom blending or alcohol fractionation; the modified ones are hydroxylated or acetylated lecithins. Hydroxylated lecithins are prepared by treating the phosphatides with peracids or hydrogen peroxide in the presence of a water-soluble aliphatic carboxylic acid. Hydroxylated leci-

**TABLE 13-II**
**Composition and Properties of Commercial Lecithin Fractions**

| Property | Oil-free Lecithin | Phosphatidyl Choline Fraction | Phosphatidyl Inositol Fraction |
|---|---|---|---|
| Solubility in: | | | |
| Alcohol | Dispersible | Soluble | Insoluble |
| Water | Dispersible | Dispersible | Dispersible |
| Oil | Soluble | Soluble | Soluble |
| Choline fraction (%) | 30 | 60 | 4 |
| Cephalin fraction (%) | 30 | 30 | 29 |
| Inositol fraction (%) | 32 | 2 | 55 |
| Oil (%) | 3 | 4 | 4 |
| Type of emulsifier | O/W[1] or W/O[2] | O/W | W/O |

SOURCE: (1)
[1]Oil in water.
[2]Water in oil.

thins are readily dispersed in water and have excellent fat-emulsifying properties; they are approved for food application in the USA. The amino group of phosphatidylethanolamine acetylates when treated with acetic anhydride. Introducing a substituent on the positively charged portion of the zwitterionic phosphatidylcholine converts it to a negatively charged lecithin with improved solubility and oil-in-water emulsifying properties (36).

**TABLE 13-III**
**Specifications for Commercial Lecithins**

| Analysis | Natural Fluid | Natural Plastic | Single-bleached Fluid | Single-bleached Plastic | Double-bleached Fluid | Double-bleached Plastic |
|---|---|---|---|---|---|---|
| Acetone insoluble, %, min | 62.0 | 65.0 | 62.0 | 65.0 | 62.0 | 65.0 |
| Moisture, %, max[a] | 1.0 | 1.0 | 1.0 | 1.0 | 1.0 | 1.0 |
| Benzene insoluble, %, max | 0.3 | 0.3 | 0.3 | 0.3 | 0.3 | 0.3 |
| Acid value, max | 32.0 | 30.0 | 32.0 | 30.0 | 32.0 | 30.0 |
| Color, Gardner, max[b] | 10.0 | 10.0 | 7.0 | 7.0 | 4.0 | 4.0 |
| Viscosity, poises, @ 25°C, max[c] | 150.0 | — | 150.0 | — | 150.0 | — |
| Penetration, mm, max[d] | — | 22.0 | — | 22.0 | — | 22.0 |

SOURCE: National Soybean Processors Association (1974).
[a]By toluene distillation for 2 hr or less (AOCS Method Ja 2–56).
[b]On a 5% solution in mineral oil.
[c]By any appropriate conventional viscometer, or by AOCS Bubble Time Method Ka 6–59, assuming density to be unity. Fluid lecithin with a viscosity less than 75 poises may be considered a premium grade.
[d]Using Precision cone 73525, Penetrometer 73510; sample conditioned 24 hr at 25°C.

**TABLE 13-IV**
Classification of Soybean Lecithins

- I. Natural
  - A. Plastic
    - 1. Unbleached
    - 2. Single-bleached
    - 3. Double-bleached
  - B. Fluid
    - 1. Unbleached
    - 2. Single-bleached
    - 3. Double-bleached
- II. Refined
  - A. Custom blended natural
  - B. Oil-free phosphatides
    - 1. As is
    - 2. Custom blended
  - C. Fractionated oil-free phosphatides
    - 1. Alcohol-soluble
      - a. As is
      - b. Custom blended
    - 2. Alcohol-insoluble
      - a. As is
      - b. Custom blended
- III. Chemically modified

Source: Sullivan and Szuhaj, cited by Wolf and Sessa, (36)

Lecithins that contain more than 67% acetone insolubles disperse with difficulty in aqueous media. Solubility is imparted by reaction with polyhydric alcohols and epoxy compounds, hydroxylation of unsaturated fatty acid constituents, fractionation, or compounding with dispersing agents. Oil-free lecithin is more susceptible to oxidative deterioration than is natural lecithin. Possibly, the small amounts of tocopherols in natural lecithins act as antioxidants (36). According to Schaefer (2), antioxidant activity of commercial lecithin is probably due to phosphatidylethanolamine as an effective scavenger of metal traces. The granular form has appreciably more surface area than do the thick-fluid or plastic forms. Exposure of the increased surface to oxygen accelerates deterioration and causes rancid or bitter flavors. These flavors are also generated by heating above 40 C. Heat discoloration of phospholipids added to oils and fats to prevent spattering can be prevented by adding a compound that forms carbon dioxide in heating, such as $NaHCO_3$ (38).

Lysolecithins, which have a more hydrophilic character than does lecithin, show stronger oil-in-water emulsifying properties. Hydrogenated phosphatides have a lighter color, greater resistance to oxidative rancidity, and less odor and flavor. They are, however, more difficult to disperse in aqueous systems and have poorer emulsifying properties than do natural phosphatides (36).

The main function of phosphatides is to emulsify fats. The long-chain fatty acid moieties contribute hydrophobic properties; those properties are counterbalanced by the polar or hydrophilic character of the phosphate moiety. In an oil:water system the phosphatide components concentrate at the oil:water interface. The polar, hydrophilic parts of the molecules are directed toward the aqueous phase and the nonpolar, hydrophobic (or lipophilic) parts are directed toward the oil phase. Concentration of phosphatides at the oil:water interphase lowers the surface tension and makes it possible for emulsions to form. Once the emulsion is formed, the phosphatide molecules at the surface of oil or water droplets act as barriers that prevent the droplets from coalescing and thus they stabilize the emulsion (36).

Instant dispersibility in water generally requires chemical modification. It is possible to attain the desirable properties in a blend of soybean lecithin and ethoxylated monoglycerides (39). In baking formulations that call for soy flour, lecithinated (usually at a level of 15%) low-fat or high fat flours can be used. Often, lecithin is incorporated into shortenings (solid or fluid) that are used in baking (40–43).

Many workers have demonstrated the synergistic effects in breadmaking of lecithin in combination with mono- and diglycerides (2) (44–46) and other surface active agents (47–50). According to Hampl and Tvrznik (51), the use of lecithin in combination with monoglycerides (a) improves quality characteristics of the raw materials, (b) optimizes technical processing, (c) reduces shortening requirements, and (d) improves overall quality of final products, including freshness retention and nutritive value. It has been suggested (1) that hydroxylated lecithins are particularly valuable in bakery products because of an apparent synergy with mono- and diglycerides, in addition to their high dispersibility in water systems in contrast to oil solubility of most lecithins.

The following product modifications and combinations were described in several publications from Lucas Meyer Co., Hamburg, Germany (52): (a) improved water dispersibility through formation of N-acylphospholipids, (b) improved water dispersibility through use of high levels of lysophosphatides, (c) interaction with amylose as co-emulsifier for monoglycerides, (d) defatted soy lecithin, 95% powder as emulsifier and dispersant, (e) spray-dried combination of standard soya lecithin and milk solids (mainly lactose) as emulsifier, dispersant, wetting agent, and (f) spray-dried combination of modified partial glycerides and milk solids for flour treatment and assurance of uniformity and high quality of breadmaking wheat flours.

## Wheat Phosphatides and Functional Properties

As early as 1924 Working (53) reported that phosphatides added to flour in small quantities injured gluten. Prolonged washing of soft gluten from low-grade flour removed phosphatides and gradually increased tenacity until the gluten was comparable to that from patent flour. Working (54) concluded that ordinary flours contained more phosphatides than necessary to develop dough properly. Sinclair and McCalla (55) suggested that polar lipids were essential for forming and maintaining characteristic wheat flour dough and gluten structure. Traub et al (56) reported, on the basis of X-ray diffraction patterns, that endosperm sections of wheat grains were uniquely spaced at 47 Å. The spacing was due to phospholipids associated with gluten proteins. Daniels (57) found that the spacing depended on the presence of lecithin. It is difficult to interpret those results in as much as there is basically no gluten in wheat. Wheat contains storage proteins, gliadin and glutelin, that can interact to form gluten when water and some energy input are added, such as in dough mixing, washing out gluten or kneading a dough. Olcott and Mecham (58) found that the amount of free lipids (ether-extractable) was reduced two-thirds during dough mixing; phospholipids were bound preferentially and were associated with the glutenin fraction in gluten. Hess (59) made extensive physical observations on fractions separated from wheat flours and concluded that polar flour lipids play an important role in determining the structure of doughs and baking properties of flour. Electron micrographs and x-ray studies reported by Grosskreutz (60) indicated that proteins existed in platelet form. Extracting phospholipids did not affect the basic platelet but seriously impaired its ability to bond into sheets capable of sustaining large plastic deformations. It was postulated that protein chains, as part of the lipoprotein structure, were bound to the outer edge of a phospholipid bimolecular array, probably by salt-like linkages between acidic groups of the phospholipid and the basic protein group. Zentner (61) suggested that phosophoric acid residues of flour lipids may be bound to proteins and may be involved in stabilizing dough. Coppock et al (62) reported that adding as little as 0.003% of an acetone-insoluble phosphatide fraction of flour oil substantially improved loaf volume of bread baked from defatted flour, though the response for some flours was inconsistent (63).

Pomeranz and co-workers conducted a series of studies on the binding, action, and role of wheat flour lipids (17). Chiu and Pomeranz (64) extracted free lipids with petroleum ether and total lipids with a chloroform-methanol mixture from flour, dry milk solids, yeast, dough, fermented dough, bread crumb and bread crust. Dough formulations used in breadmaking included —in addition to a basic formula of flour, water, yeast and sodium chloride— sugar, commercial vegetable shortening, dry milk solids and their combinations. The extracted lipids were fractionated by thin-layer chromtography (TLC). Petroleum-ether-soluble flour lipds were reduced to one-third during dough mixing or fermentation; subsequent baking lowered the residual free lipids to half. Only small amounts of hydrogenated vegetable shortening were bound during dough-mixing, but about one-third to one-half of the added shortening lipids became bound during baking. Processing flour into bread had no effect on the amounts of total lipids extractable by the chloroform-methanol mixture. Fractionation of extracted lipids by thin layer chromatography showed that much more polar wheat flour (glycolipids and phospholipids) than nonpolar components were bound during dough mixing.

In a subsequent study, Chiu et al (65) fractionated a composite hard red winter wheat flour containing 12.9% protein into gluten and a mixture of starch and water solubles. The original flour, the mixture of water solubles and starch and the reconstituted flours containing respectively 10, 13, and 16% protein were baked by a formula that included 3% shortening. In addition, 16 flours milled from eight hard red winter, five hard red spring, and one each soft red winter, club, and durum wheat varieties were baked. Free lipids were extracted with petroleum ether and bound lipids with water-saturated butanol following petroleum ether.

The gluten contained up to five times as much protein and lipids as the original flour. The mixture of starch and water solubles contained little lipids, and distribution of its nonpolar and polar lipid components varied from that of lipids in gluten. Most phosphatidylcholine and digalactosyl glyceride were bound by the gluten proteins. Lipid binding during dough mixing of flours that vary in breadmaking potential increased with protein content and mixing time and decreased in flours of poor protein quality.

Pomeranz et al (66–67) studied the effects of lipid fractions, mixing time and dough ingredients on lipid binding. Increasing length of dough mixing up to the point of minimum mobility increased binding of free flour lipids; binding decreased slightly during prolonged overmixing. Adding 2 to 4% NaCl to the dough had little effect on the binding of wheat flour polar lipids (phospholipids and glycolipids added to an extracted flour), but significantly reduced binding of nonpolar lipids. The results suggested that the action of salt was on the gluten proteins and not on salt-like linkages. Presence of nonpolar lipids reduced binding of polar lipids. More bound-flour lipids were released by remixing a dough with an NaCl solution than by remixing with water. Practically no added nonpolar lipids were bound by dry-mixing with flour, but substantial amounts of free polar lipids were bound during dry-mixing with a flour containing 4.4% moisture. The binding increased with increases in moisture content and was highest in dough.

Wheat flour produced by conventional roller milling contains particles of different sizes: large endosperm chunks, small particles of free protein, free starch granules and small chunks of protein with starch granules. Fine grinding followed by air classification can separate such a mixture into three main fractions: (a) a high-protein fraction with particle size under 15 $\mu$m; (b) a low-protein fraction with particle size of 15–40 $\mu$m; and (c) a coarse residue size of 40–120 $\mu$m with protein content approximating that of the original flour. Repeated grinding of the coarse residue to free more protein and starch particles, followed by air classification, increases the protein content of some fractions and the starch content of other fractions, and is the basis for numerous patents for the production of "tailor-made" flours from a single parent flour. Air classification is relatively inexpensive, and its advantages are numerous (68): (a) manufacture of more uniform flours from various wheats, (b) increase of protein content of bread flours and decrease of protein in cake and cookie flours, (c) controlled particle size and chemical composition and (d) production of specialty flours for specific uses.

Lipids were extracted with petroleum ether, water-saturated butanol following petroleum ether, or directly with water-saturated butanol from a composite hard red winter wheat flour and from five flours separated from the original flour by air classification (69). The lipids were fractionated by silicic acid column and thin-layer chromatography. Lipid contents increased as the protein contents of the flour fractions obtained by air classification increased. The phosphorus content of lipids, extracted by either petroleum ether or water-saturated butanol following petroleum ether from air-classified flour fractions, was fairly constant, indicating a rather uniform proportion of phospholipids in the total extracted lipids.

Pomeranz et al (70) studied the effects of polar and nonpolar wheat-flour lipids on the quality of bread baked from untreated wheat flour. Adding up to 3 g of vegetable shortening per 100 g of a hard winter wheat-flour composite improved crumb grain and increased loaf volume from 802 to 948 ml. Higher levels of shortening had no additional significant improving effect. When the bread formula included 3 g of vegetable shortening, adding 0.5 g of nonpolar, polar or unfractionated lipids from wheat flour had no significant effect on crumb grain or loaf volume. However, the loaf volume of bread baked without vegetable shortening was increased strikingly, from 802 to 958 ml, by adding 0.5 g polar-flour lipids. Adding 0.5 g of nonpolar lipids did not increase loaf volume significantly. Adding 0.5 g unfractionated flour lipids (approximately 1:1 ratio of polar and nonpolar) gave an intermediate volume of 893 ml expected from the contribution of lipid components.

In a subsequent study, free (petroleum-ether-soluble) and bound (soluble in water-saturated-butanol, following petroleum ether) lipids were extracted from a composite hard red winter flour, two other hard red winter flours varying widely in breadmaking potentialities, and a hard red spring flour (71). The lipids were fractionated into polar and nonpolar fractions by silicic acid column chromatography and by acetone precipitation. Free polar lipids substantially increased loaf volume; the increase was smaller when bound polar lipids were added. Lipid fractions isolated from various flours indicated no varietal differences.

Total free lipids containing a mixture of nonpolar and polar components (in a ratio of 3:1) improved bread quality less than did polar lipids alone. Nonpolar lipids decreased loaf volume and impaired crumb grain of baked bread from petroleum-ether-extracted flours; the deleterious effects were counteracted by polar lipids. The effects on bread depended on the levels and ratios of the nonpolar to polar lipids. Galactosyl glycerides increased loaf volume of bread baked from petroleum-ether-extracted flours substantially more than phospholipids did.

## Rheological Properties

Whereas views on the effects of lecithins on bread quality (loaf volume and crumb grain) vary, there is general concensus about the beneficial effects on dough-handling properties (72–76). Sullivan (77) reported in 1940 that adding small amounts of lecithin improved dough consistency but had little effect on baking properties. Pratt (78), who conducted a series of experiments with lecithin in bread, found that the presence in dough of 0.25% lecithin (flour basis), the optimum level, affected a slight mellowing of gluten and an increase in the mixing tolerance of the flour. The most noticeable result of adding lecithin to the dough was improved handling properties at make-up. The lecithin-supplemented doughs were more elastic and lively, with a distinctly drier feel, than were the control doughs. Dusting-flour requirements were reduced and the skin formed in the molding operation showed fewer signs of tenderness. There was no measurable effect of lecithin on absorption, fermentation, proof time or bake time (78). At constant water absorption, 0.3% lecithin was most beneficial in dough mixing. Lecithin increased extensibility of short doughs and shortened extensibility of long doughs. The lecithin effect was particularly noticeable and useful in bucky doughs. Lecithin levels above 0.5% impaired handling properties of straight and sponge doughs.

According to Schaefer (2), lecithin interacts with dough components through (a) physical enrichment on the surface of flour particles, (b) combination with the surface particles, and (c) formation of interaction products, lipoproteins. Those interactions improve water-binding capacity, rheological-dough-handling properties, and overall bread quality. Adding 0.2 to 0.5% lecithin improves dough consistency by softening it and making it more pliable; higher levels are desirable in bucky doughs.

A recent pamphlet (52) reported a series of rheological studies on the effects of lecithins on dough-handling properties. Alveograms, extensograms, and maturograms and oven-spring curves were included. Several products (spray-dried combinations of lecithins, partially modified glycerides, and milk solids) are available to control rheological properties. Those products were shown to (a) stabilize gluten quality, (b) stabilize low-gluten flours of low gassing power, (c) modify gluten in short and brittle doughs, (d) modify glutens in doughs of extreme extensibility, and (e) shorten gluten in oxidant-containing flours (i.e., potassium bromate).

The rheological effects were confirmed in actual performance tests. The lecithin formulations were effective either in flours that varied in end-use properties or in end products.

Pomeranz et al (66) reported effects of commercially available natural and modified soybean lecithins on rheological properties and breadmaking potentialities of untreated and petroleum ether extracted flours. Hydroxylated phosphatides increased Farinograph dough-development times, valorimeter values, and mixogram times. Alcohol-soluble phosphatides lowered the temperature and the hot-paste viscosity peak in the Amylograph. No correlation could be established, however, between bread quality and the effects of phospholipids on rheological properties.

## Functions in Breadmaking

The general uses of lecithin in bakery products and mechanism of their action and their effectiveness in improving quality of such products were described by Cook (79), Whillans (80), Aylward (81), Iveson (82), Stewart (83), Glabe et al (84), Zieglitz (85), Jakubczyk et al (86), Church (87), Anon. (88), Bade (89), Knyazhanska and Garmash (90), Kuntze and Schneeweiss (91–93), Nickerson and Ronsivalli (94), Morrison (19–20), Wasserman (95) and Rusch (96).

According to Goebbels (97), additives in breadmaking should improve baking properties, eating properties and consumer acceptance through the contributions of the additives to uniform browning, desirable intensity of browning, increased crispness of the crust, increased loaf volume, uniform crumb grain, crumb elasticity, adequate slicing properties, rounded-off taste, freshness retention, and resistance to microbial deterioration. Baking materials enhance and facilitate production of baked goods and reduce or eliminate faults in those goods through regulation of baking potential of flours, improvement in dough development and dough properties through adaptation of dough properties to processing conditions, improvement of fermentation stability and tolerance, regulation of enzymic activity, enhancement of dough aeration, stabilization of dough acidity, and regulation of swelling properties.

Knightly (98) presented the rationale for the use of surface-active agents in baked goods. In yeast-raised goods, the primary role of a surfactant is to retard the rate of staling; secondarily, a surfactant provides improved loaf volume and dough-handling properties. In continuous dough processing, certain surfactants may strengthen the gluten structure of the weaker doughs resulting from such processing.

Lecithin can be used in baking (99) as (a) an emulsifier (alone or in conjunction with other emulsifiers) to reduce emulsifier cost, to stabilize emulsions and mixing of fats and water, to improve moisture tolerance and to ensure uniform suspension of ingredients; (b) a wetting agent to provide instant wetting of powdered formulations to reduce mixing time; (c) a parting agent to provide cleaner and faster release from a mold; and (d) an antioxidant to stabilize vegetable and animal fats and oils, particularly as a synergist with other oxidants.

Pratt (78) reported that adding lecithin did not affect bread's volume and crumb color. Notable improvements were a tenderer crust, a finer grain and smoother texture, a more symmetrical appearance, and longer keeping quality. A loaf containing 0.3% lecithin with 1% shortening had keeping qualities equivalent to those of a lecithin-free bread containing 4% shortening. But a loaf with 3% shortening and 0.25% lecithin exhibited superior keeping qualities. Pratt (100) concluded that lecithin increased the effect of shortening in bread. Loaf scores were 93.0, 97.5, 97.5, 96.0, and 93.0 for 0, 0.1, 0.25, 0.50, 0.75, and 1.0% lecithin additions, respectively. All above lecithin levels are expressed on a flour weight basis.

Walrod (101) found that one to five ounces of lecithin per 100 lb of flour permitted reduction in mixing time and a slightly increased absorption; the added lecithin yielded doughs that felt drier, possessed improved machinability and showed greater elasticity and smoothness with almost complete absence of buckiness, regardless of the type of flour used. Lecithin-containing doughs produced more symmetrical loaves with a more uniform grain and a finer texture; lecithin contributed to freshness retention in bread. Sunflower phosphatides, added at 0.4%, were found by Radev and Kalichkov (102) to improve loaf volume, crumb grain, and keeping quality of bread. Rohrlich and Bruckner (103) reported that as little as 0.1% pure lecithin increased dough and bread yields, improved crumb grain and texture, and retarded staling. Excessive amounts of lecithin could have adverse effects on bread taste. In the baking industry in England, soya "lecithin" is used as a dough improver (104). Used in the proportion of 0.5 lb per sack (280 lb), it gives the dough a silky feel and makes the crumb of the resultant loaf softer.

According to Ponte (105) lecithin has practically no effect on dough handling or tolerance, on loaf volume or bread score, and a very minor effect on firmness inhibition. Bread effects are similar to those produced by lard. Very little lecithin is used commercially in bread in the USA (106–107). According to Pyler (74), however, hydroxylated lecithin significantly increases gluten extensibility. In products containing less than 8% shortening, the recommended lecithin level is 2 to 3 oz/100 lb of flour; in high-fat products 1 to 2% shortening basis. The use of lecithin in bread doughs increases fermentation tolerance, yields breads of more uniform crust color, gives a tenderer crust and a smoother crumb texture with a more uniform grain, and produces a softer and less rapidly staling loaf. The beneficial effects of phosphatides in dough fermentation were described by Popadic (108). Lehmann (109) discussed the use of lecithin in improving the baking quality of weak wheats. The use of phosphatides in regulating foam formation in liquid ferments was studied by Kazanskaja and Gurina (110). According to Kuntze (76), lecithins and mono-diglycerides are the most commonly and widely used surface active agents in East Germany. When she added eight surface-active agents at levels of 0.25 to 0.50%, flour basis, in baking buns from a low-protein flour with 3% yeast and 1.8% salt, she found that lecithin improved dough-handling properties, including dough stability, bun volume and bun quality. Starceus and Altaev (111), reported that adding vegetable phosphatides to a dough as an emulsion improved loaf volume, crumb grain, crumb elasticity, and overall loaf appearance; retarded staling; and increased the nutritional value of the product. Acker (25) isolated lysolecithins from wheat starch and showed that loaf volume was increased 10% and 28%, respectively, by adding 0.1% and 0.4% of the phosphatide. According to Wolf and Sessa (36), from 0.1 to 0.3% lecithin (flour basis) in bread modifies the characteristics of flour gluten and increases the shortening effect of the fat; improved flavor, shelf-life, texture, and moisture retention are claimed for lecithin-containing products.

In another publication (52), a powdered stabilizer and emulsifier to improve the standardization of wheat flours was described. In pan bread, adding up to 0.5% lecithin was claimed to (a) improve flour quality by stabilizing-providing uniform performance of flours of varying quality, including those from freshly harvested wheats or freshly milled wheats; (b) reduce the amount of strong wheats required to produce acceptable breads; (c) counteract to some extent the adverse effects of sprout damage; (d) improve gas retention, fermentation tolerance, viscosity, elasticity, and extensibility of dough; (e) increase loaf volume and improve crumb grain; (f) retard staling;

and (g) complement or act synergistically with dough improvers in the production of bread and other yeast-leavened products. The beneficial effects of lecithin to counteract the adverse effects of sprout damage were also claimed by Osadchaya et al (112).

The improver, it is believed, acts through improved water distribution in dough and formation of lipoproteins on the surface of starch granules; the lipoproteins presumably swell and improve the viscosity, expansion, and elasticity of doughs. Lecithin has been used in formulations that do not include shortening. It has been stressed that adequate diastatic activity is essential for producing acceptable bread (52).

Effects of commercially available natural and modified lecithins on breadmaking potentialities of untreated and petroleum-ether-extracted flours were reported by Pomeranz et al (66,67), who found that hydrogenated lecithin added to shortening-containing doughs reduced loaf volume to, or materially below, that of bread baked from doughs containing no shortening (Table V). Hydrogenated lecithin did not substitute for free flour lipids as did eight other soybean phospholipids when added to petroleum-ether-extracted flours baked with shortening. Alcohol-soluble phosphatides containing a 2:1 mixture of phosphatidylcholine and phosphatidylethanolamine were the most effective in improving loaf volume and crumb grain of bread baked from untreated and petroleum-ether-extracted flour without shortening. It is noteworthy that 0.5% alcohol-soluble phospholipids, when added to petroleum-ether-extracted flours, substituted for both 0.8% free flour lipids and 3% shortening. Softness retention of bread was generally correlated with loaf volume.

**TABLE 13-V**
**Effects of Adding 0.5 g Phospholipids per 100 g Composite (RBS-66) Flour on Loaf Volume and Crumb Grain of Bread Baked with Three Levels of Vegetable Shortening**

| | No shortening | | 0.5 g shortening | | 3.0 g shortening | |
|---|---|---|---|---|---|---|
| Phospholipid description | Loaf volume | Crumb grain | Loaf volume | Crumb grain | Loaf volume | Crumb grain |
| | (cc) | | (cc) | | (cc) | |
| None | 783 | Q–U[a] | 868 | Q–S | 925 | S |
| Natural | 885 | Q–S | 910 | S | 908 | S |
| Hydroxylated | 878 | Q–U | 928 | S | 930 | S |
| Alcohol soluble | 905 | Q–S | 920 | S | 913 | S |
| Alcohol insoluble | 875 | Q | 903 | S | 912 | S |
| Hydrogenated | 763 | Q–S | 773 | S | 785 | S |

[a]S = satisfactory, U = unsatisfactory, Q = questionable.
(from Pomeranz *et al.* (173).

Arai et al (50) and Kihara and Kato (113) received patents for including lecithin in combination with other surfactants to improve appearance, texture, and flavor of bread. Similar effects were reported in a patent granted to Shinka (114). Lecithin can be used in bread production in combination with polylactates (115).

## Specialty Baked Products

Knjaginicev (116) reviewed a series of compounds that can retard staling of rye bread (as determined by crumb compressibility, distribution and binding of water, crumbliness, crumb toughness and crumb dispersibility). Phosphatides were quite effective.

The use of lecithin in producing consumer-acceptable bread nutritionally enriched by soy flour was first described by Adler and Pomeranz (117). The formulations used in those baking tests included no shortening. In test bakes in which 3% shortening was included, lecithin still had a small improving effect on loaf volume, crumb color and crumb grain (118).

The improving effect of soy lecithin in the production of bread enriched by isolated soybean proteins was described by Mizrahi et al (119). Among numerous other reports on producing soy bread and improving its loaf volume, crumb grain and freshness retention are studies by Menger and Bretschneider (120), Sipos and Turro (121), and Seibel (122).

Pomeranz et al (123–124) studied the use of surfactants in producing consumer-acceptable bread nutritionally enriched by high levels of protein concentrates. They found that adding 0.5 g of sucroglycerides per 100 g of flour increased water absorption and maximum hot-paste viscosity; it did not significantly affect mixing time, mixing tolerance or extensibility characteristics. The sucroglycerides counteracted the deleterious effects of up to 16% of soy products (wheat flour basis) on loaf volume. The improving effect of 0.5% sucroglycerides was equal to, or better than, the effect of 3.0% commercial vegetable shortening. The effect of sucroglycerides increased with increase in hydrophilic-lipophilic balance (i.e., with decrease in number and chain lengths of fatty acids attached to the sucrose molecule). In addition to increasing loaf volume, sucroglycerides substantially improved crumb grain and softness. Largest improvement was obtained by adding 0.5% free polar-flour lipids, rich in glycolipids. Improvement of free polar flour lipids was observed both in the control and in soy-containing breads; the sucroglycerides rendered soy proteins functional in breadmaking. Lecithin had a highly significant and consistent improving effect, though not so large as the effect of sucroglycerides. In one series of experiments, the improving effects of sucroglycerides on loaf volume and crumb grain of bread containing 8 g (per 100 g of wheat flour) of seven commercial soy products were determined. Bread baked with a soy flour that contained 1.75% lecithin consistently had the largest and generally the best loaves.

A new soy product that has no beany flavor, which limits the use of soybean in foods, has been reported (125). It is claimed that 15% to 30% of the bland product can be used in baked products; the recommended formulations include lecithin.

Rao and Vakil (126) reported that replacing wheat flour with 5% to 10% defatted soybean or peanut flour altered water absorption and extensibility of the dough and impaired bread quality. Adding 0.5% soy lecithin improved sensory characteristics and acceptability of breads, supplemented with up to 15% oilseed flours. A dry protein mixture containing several additives, including lecithin, improved the organoleptic and physico-chemical characteristics of bread produced in the USSR (127).

The importance of dietary fiber in nutrition is well documented and has been the subject of many publications. One problem related to high-fiber bread is how to produce, by conventional and available technology, a reasonably inexpensive, high-fiber bread that will meet with consumer accep-

**TABLE 13-VI**
**Characteristics of Bread Baked from Blends of 81.8 g of Flour, 3.2 g of Vital Gluten, and 15 g of Wheat Bran With and Without Improvers**

| Improvers[a] | Loaf Volume (cc) | Crumb grain[b] | Crumb Compressibility at 1 hr | 3 days |
|---|---|---|---|---|
| None | 699 | UU | 142 | 412 |
| Shortening (3.0 g) | 934 | Q–U | 86 | 206 |
| Surfactant (1.0 g) | | | | |
| LEC | 932 | Q | 81 | 226 |
| SSL | 915 | Q–S | 83 | 166 |
| SMP | 1,024 | Q–S | 60 | 123 |
| EMG | 992 | S | 69 | 168 |
| DAT | 993 | S | 69 | 147 |
| Shortening (3.0 g and surfactant (0.5 g) | | | | |
| LEC | 945 | Q | 81 | 206 |
| SSL | 950 | Q | 86 | 148 |
| SMP | 1,005 | Q–S | 65 | 152 |
| EMG | 965 | Q–S | 71 | 155 |
| DAT | 983 | Q | 77 | 170 |

[a]LEC = lecithin, SSL = sodium stearoyl-2-lactylate, SMP = sucrose monopalmitate, EMG = ethoxylated monoglycerides, DAT = diacetyl tartaric acid esters of monoglycerides and diglycerides.
[b]Visually evaluated: S = satisfactory, Q = questionable, U = unsatisfactory.
(from Shogren *et al.,* 178).

tance. The deleterious effects of up to 15 parts of wheat bran per 85 parts wheat flour could be largely counteracted by the adding of vital gluten and one of several surfactants (128). Commercial lecithin (at the 0.5% flour level) was as effective as 3% shortening (flour basis) and slightly less effective than 0.5% ethoxylated monoglycerides, sucrose monopalmitate, or diacetyl tartaric acid esters (Table VI).

The usefulness of soy phosphatides in producing Zwieback—a toast with long shelf life—was described by Makarjan et al (129). Arya et al (130) studied the role of wheat-flour polar lipids (glycolipids and phospholipids) on texture, hardness and cohesiveness of chapaties; wheat varieties having higher percentages of polar lipids gave softer chapaties.

Pomeranz and coworkers conducted a series of studies on production of consumer-acceptable bread that contains large amounts of wheat germ (131). Raw germ decreased loaf volume of wheat bread, whereas up to 15% glutathione-inactivated germ or equiprotein amounts of dialyzed sodium chloride extracts of germ had no deleterious effect (131, 132). The addition of free-polar wheat-flour lipids increased loaf volume of bread nutritionally enriched with germ or germ proteins. Phosphatidylcholine had no effect on loaf volume of control bread, but substantially increased loaf volume of bread enriched with up to 30% germ or 9% germ proteins. Results were best if the lecithin-to-germ ratio was between 1:10 and 1.5:10. Several lipids varied in their effects on loaf volume and crumb grain. Inositol phosphatide, synthetic DL alpha-lecithin, and phosphatidylethanolamine increased loaf volume most; phosphatidylserine had little effect. In a subsequent study, bread was baked with 15% or 30% heat-treated defatted wheat germ and 2% to 4% commercial soybean phosphatides (132) (Table VII). A hydrogenated phosphatide had no improving effect. Phosphatides fractionated according to solubility in alcohol and hydroxylated phosphatides were most effective in counteracting the undesirable effects of wheat germ on loaf volume and crumb grain. Subsequently, Geoffrey (133) and Gasiorowski et al (134) described related effects.

Scerbatenko et al (135) were granted a patent on the production of whole wheat bread of acceptable quality, in a formulation that included 0.5% lecithin (flour basis).

De Ruiter (136) reported that full-fat soy flour is better than defatted flour in producing bread from composite flours containing wheat flour. Again, it is difficult to attribute the improving effect to oil, to phosphatides, or to both in the bread from composite flour that was produced from formulations that contained little, if any, shortening. The beneficial effect of phosphatides was demonstrated clearly, however, in a study of bread baked from non-wheat flours (137). Adding 1% or 4% lecithin (alone or especially in combination with 2% alginate) improved substantially loaf volume and crumb grain.

**TABLE 13-VII**
**Effects of Adding 1.5% Soybean Phosphatides on Loaf Volume and Crumb Grain of Bread Baked with 15% Heat-treated Germ**

| Phosphatide | Loaf volume (cc) | Crumb grain |
|---|---|---|
| None | 885 | Q–S[a] |
| Natural | 950 | S |
| Modified, emulsifier | 985 | Q–S |
| Fluidized, low viscosity | 975 | S |
| Alcohol soluble | 1005 | S |
| Alcohol insoluble | 1000 | S |
| Hydroxylated | 995 | S |
| Hydroxylated-granulated | 1010 | Q–S |
| Hydrogenated | 895 | S |

[a]Visually evaluated: S = satisfactory, Q = questionable, U = unsatisfactory.
(from Pomeranz, et al 131).

## Phosphatides and Bread Staling

According to Osman and Dix (138), the interaction of lecithins with starch, as measured with a Brabender Amylograph, was smaller than that of other surface active agents. There have been numerous reports on the contribution of lecithin to freshness retention in bread (101, 102, 135, 139, 140, 141, 142). Seibel and Menger (142) indicated that freshness retention from emulsifiers, added alone or in combination with shortening, is a combined effect-result relation of dough consistency, fermentation tolerance, loaf volume, crumb elasticity and overall bread quality. Kvetnyj and Scerbatenko (143) studied, by differential thermoanalysis, changes in stale bread that had been "refreshed" by heat treatment. The transfer from the crystalline to the amorphous phase was followed. Aging of bread biopolymers was accompanied by an increase in specific heat capacity and related enthalpy. Adding plant (vegetable) oils and surface active agents interferred with the formation of crystalline structures and lowered the temperature of phase transfer. Kroll et al (144) measured the hydrophilic-hydrophobic balance and amylose-binding index of several emulsifiers. The amylose binding index of lecithin is substantially lower than that of commercial surfactants, which might explain lecithin's limited capacity to slow down staling.

According to Acker (25), the lipids of wheat starch have a special technological function. They include lysophosphatides, rarely found in nature, and the main compound, lysophosphatidylcholine, is found only in the form of its l-acyl isomer. Lipids that interact with starch must have a special configuration. Lipids that meet the configuration requirement form inclusion compounds with amylose and exert a tremendous influence at very low concentrations on viscosity of starch pastes and on overall rheological properties of starch gels. Free fatty acids reduce swelling power, solubility and iodine-binding capacity of starches. Similar effects are related to retardation of staling (25); lysophosphatides retard staling through their interaction with starch (24). According to Morrison (22), the experimental evidence for the formation of lysolipid-amylose complexes is not conclusive.

## Cookies, Crackers, Extrusion Products

According to Matz (145), lecithins are relatively inexpensive and fairly potent emulsifiers in cookie and cracker technology. They are easily mixed into cracker or cookie doughs, modify the consistency and make machining easier by reducing stickiness. In the finished baked product, the effect on flakiness, tenderness and other shortening-related factors is not so pronounced as in the dough. Greasiness of cookies with high shortening content is often reduced by adding small amounts of lecithin to the dough.

Lecithin may reduce the shortening requirements; alternatively, it may increase the shortening effect of fats (145). It promotes a tendency for fat to cover or spread among slightly moist particles of sugar, flour and other ingredients. Such particles would otherwise repel the fat. Because of the emulsifying effect, lecithin makes rich, sweet doughs seem drier and improves their machinability. Better release from the die, especially on rotary pieces, makes a better impression and reduces the number of cripples. Lecithin also speeds the dispersion of the fatty and aqueous components of the dough so that mixing time can be reduced. That is especially important in relatively strong flours, because shorter mixing time reduces dough development, and the resultant cookies and crackers are tenderer (145). According to Adrian (146), lecithin may give crackers, cookies, and similar products a new lease on life. Knightly (147) reviewed the role of emulsifiers in improving the volume, shelf-life, crumb grain, crumb structure and uniformity of chemically leavened products, including cookies. The emulsifiers also can significantly improve eating quality and consumer acceptance of other products. Prokopenko et al (148) studied the effects of phosphatide concentrates on the properties of cookie fat emulsions. Replacing 0.5% to 5.0% of the fat by phosphatide concentrates in formulations that do not include milk or egg solids substantially increased dispersibility and stability of emulsions and improved hard-type cookies.

Phosphatidylcholine was slightly more effective than digalactosyl-diglyceride (both at 0.025% level, flour basis) in restoring cookie baking properties of defatted soft wheat flour (W. T. Yamazaki, private communication). Interestingly, however, in non-defatted flour, adding 1% to 10% bran lipids (flour basis) that are low in polar components (phospholipids and galactolipids) was as effective in improving cookie quality as adding 1% to 10% flour lipids that are rich in polar components (149). Lipid extracts from wheat flour, lecithins from soy, and a commercial safflower lecithin were evaluated at low concentrations as improvers for cookie spread in conjunction with protein-fortified wheat flour blends. Soy lecithins were most effective (150). Soy lecithin somewhat improved the quality of cookies made from sorghum and millet (rather than wheat) flour (151); soy lecithin was an effective improver in cookies made from formulations that included triticale flours (152). The use of lecithin in snack foods was discussed by Smith (153); Viguerie (154) described a patent on the use of lecithin in drying of foods.

As stated previously, lecithin-shortening mixtures reduce shortening requirements in regular yeast leavened bread, toast bread, and Zwieback (2). In egg-containing doughs, Brioches, lecithin lowers egg requirements; in "Blatterteig"-puff pastry (thin layer doughs) it improves layering and texture; in cookie production it yields products with better shelf life. Knightly and Lynch (155) pointed to the role of surface-active agents in providing the optimum hydrophilic-lipophilic balance in the production of yeast-leavened and chemically leavened pastry.

Schaefer (2) reported that adding lecithin produced doughs that handle better and that the texture and spread of cookies was improved. Waffles made from lecithin-containing doughs showed better iron-grid release and easier handling. The weight of the waffle could be reduced; the waffles were stronger, crisper, did not become soggy, and retained freshness better than did control waffles. Similar effects were recorded in omlette-type doughs (ready made or house made); the level of lecithin should be higher in weak than in strong baking flours. Lecithins produce in cookies or biscuits a drier product with improved machinability and reduced stickiness (39). They produce a uniform crust color and reduce checking because fats are more evenly distributed. Lecithin antioxidant properties reduce rancidity and prolong freshness. The presence of 0.5% to 1.0% lecithin (shortening basis) acts as a lubricant for rotary-die cutters for biscuits, crackers, and cookies; approximately 0.3% lecithin in ice cream cones (formula weight) promotes clean release from mold in products that contain little or no shortening.

## Cakes, Pastry, Doughnuts, Confectionary

According to Elsesser and Bogyo (156), incorporating lecithin in a formulation substantially improved the quality of cakes. Lecithins from various sources were effective. Soy-hydroxylated lecithins that have an intermediate degree of saturation were best. Obenauf and Tatter (157) recommended the use of lecithin as an emulsifying agent for cake mixes to assure easy pan release and to prevent the cakes from falling or dipping in the center. Weiss et al (158) described an emulsifier-liquid oil preblend; the preblend worked well in cake mix packages; the recommended emulsifier was lecithin. According to Del Vecchio (159), however, adding 0.1% lecithin produced tunnels and a slight dip in the tops of yellow cakes. MacDonald and Bly (160) reviewed the need for optimal levels of several emulsifiers in cake-mix shortenings. Wootton et al (161) described the complex role of emulsifiers in incorporating air into layer cake-batter systems. In a more recent patent-publication, Tomita and Mizukoshi (162) mentioned an emulsified fat composition that is useful for baking sponge cakes and that contains lecithin in addition to several surfactants. Differences in total lipid and fatty acid composition of doughnuts, as influenced by lecithin, leavening agent, and frying fat were noted by McComber and Miller (163). Bernotavicz (164) reported on soya lecithin for food and confectionary use and Prokopenko and Roiter (165) discussed the effect of phosphatide concentrates on the properties of emulsions for sweet pastry products. According to Kuntze and Luft (166, 167), emulsifiers can improve processing and quality of sweet pastry goods by contributing to freshness retention, and they can reduce requirements for eggs and/or shortening. The improving effects of lecithin in the production of cakes from formulations that contain triticale flours was described by Kissell and Lorenz (152).

Soy lecithin, 95% powder, is recommended as an emulsifier and dispersant for waffles; a spray-dried combination of standard soy lecithin and milk solids (mainly lactose) is claimed an effective emulsifier, dispersant, and wetting agent in the production of cookies, waffles and long shelf life products (52).

According to Knightly (98), both yeast-raised goods and chemically leavened goods are protein foams. The ability to entrap leavening gases is important from the standpoint of volume, symmetry, and grain of cakes. Chemically leavened batters have a relatively weak protein foam structure. Yet, they are required to carry a great proportion of antifoaming ingredients (fat, eggs, etc). The value of certain surfactants, including lecithin, is in protecting the batter foam structure, resulting in greater volume and improved crumb grain. Another important factor is the dispersion of surfactants. If the surfactant is not dispersed adequately, its efficiency per unit weight is reduced (98). A combination of lipophilic and hydrophilic emulsifiers generally provides optimum dispersibility and improved emulsion stability.

Many natural emulsifiers in cake batters (i.e., lecithins and lipoproteins) are from flour, milk and eggs; they impart to the surface of fat in cakes a predominantly negative electrical charge. This charge serves a very useful function in keeping the globules dispersed and in reducing the tendency of fats to clump. Similarly charged fat globules repel each other. In emulsion systems where the levels of such anionic surfactants are inadequate, fat clumping may build up thixotropic viscosity in the batter, particularly if the intrinsic viscosity of the aqueous phase is low (98). The lecithins of egg yolk contain both saturated and unsaturated fatty acids; the saturated ones are in the alpha position, whereas the unsaturated ones are in the beta position on the glycerol molecule (168). A replaceable hydrogen in the phosphatidylcholine molecule is in the phosphoric acid part, whereas the choline acts as a base. However, the basic properties are stronger. Because ethanolamine and serine are less basic than choline, the phosphatides containing them are more acid in character than in those containing choline.

In cake products, 1% to 2% lecithin (shortening basis) gave free-flowing batters, more uniform color, smoother texture, more uniform grain, better keeping quality and greater flavor stability (169). According to Wolf and Cowan (34), the emulsifying properties of phosphatides find extensive use in cake mixes and instant foods, in which variable amounts are used depending on the product: 0.5% to 1.0% in pies, based on shortening; 1.0% to 3.0% in cake mixes, based on shortening; and 0.5% to 3.0% in instant foods. Wolf and Sessa (36) stated that adding 0.5% to 1.0% lecithin promotes wetting, thereby speeding up mixing of cake-doughnut mixes. Prolonged mixing of cake doughnut batters causes toughness and poor eating quality in the finished product. Lecithin softens the batter so that it handles better. At levels of 0.025% to 0.5%, lecithin enables cake mixes to wet rapidly with water or milk. Lecithin is essential in white cakes because adding eggs would impart a yellow color. The presence of lecithin reduces the danger of collapse and loss of volume during the baking of the cake. In pancake mixes, 0.25% lecithin assists in wetting to provide rapid mixing and acts as a grill-release agent during frying. Lecithin is the classical emulsifier for margarine, an oil-in-water emulsion, and is added at the 0.15% to 0.5% level. It is commonly used in conjunction with mono- and diglycerides. Lecithin in margarine serves as an antispattering agent, promotes browning during frying, and acts as antioxidant (170). In margarine used for baking, lecithin has a shortening-saving effect. More complete and uniform blending of shortening occurs when 0.5% to 1.0% lecithin is added (36). Lecithins in which soybean oil has been replaced by hard butters are useful in producing cake and cookie icings according to a study reported by Lucas Meyer Co. (52). In that study, a mixture of 3% lecithin blended with mono- and diglycerides gave a volume equivalent to 8% mono- and diglycerides in cakemaking. Adding lecithin improved keeping qualities, grain and texture, and also produced "shorter" sweet-dough products (coffee cakes, sweet rolls, etc). Incorporating lecithin into pie crusts reduced mixing time, produced flakier doughs, enhanced release, contributed to uniform browning and aided as a moisture barrier to protect the crust (1).

Pritula et al (171) reported that oil-lecithin mixtures can be used in producing Lebkuchen-type cakes. The use of phosphatide concentrates to improve physicomechanical properties of emulsions for long pastry sweet loaves was reported by Prokopenko et al (172).

## Legal Aspects

The U.S. Standards of Identity for Bakery Products and Food Additive Regulations, published in the Code of Federal Regulations (Title 21), allow the use of lecithin and hydroxylated lecithin in breads without level limitations (106, 107). Similarly, the use of phosphatides is permitted, as additives and they are considered natural improvers in most countries. According to Seibel (173), use of additives in West Germany requires that they be acceptable from a nutritional standpoint, be functional (i.e., perform a defined technological function), and be technologically required. To meet those criteria, additives in baking should perform some of the following functions: increase loaf volume, improve appearance, contribute to bread-crumb elasticity, taste, and freshness retention. Some of those criteria are met by several emulsifiers and surface-active agents including lecithins. Use of lipids (emulsifiers) in the bread and bakery industry, with particular reference to the concept of "technological requirements," was reviewed by Jodlbauer (174). Benedickt (175) emphasized that in West Germany lecithin is considered as a generally permitted additive; the amount that can be added, basically, is not limited. Bruemmer (176) and Weith (177) discussed coordination of legal requirements for bread in the European Economic Community. Calvel (178) provided information on technological requirements for the manufacture of bread and bakery products in France. Schaefer (179) reviewed the uses of emulsifiers (including lecithin) that are allowed in the production of various types of breads, buns, rolls, pastry, cakes and fillings in West Germany.

## Final Comments

This review shows clearly that there are many things that lecithin can do; and that by chemical modification and combination with other functional ingredients lecithin can be made to do many additional things to improve processing, production, quality and consumer acceptance of baked products. This review also attempted to document the limitations of lecithin effects. Despite all that it can do, lecithin cannot perform wonders—such as make a feed wheat into a breadmaking wheat; it cannot overcome advanced damage to wheat during sprouting or storage under adverse conditions. It can, however, contribute to uniformity and improve processing and quality of final products, provided it is used at optimum or near-optimum level; and is an optional ingredient in a formula that includes sound and balanced ingredients.

There has been renewed interest in lecithin as a natural and versatile improver, even though its limitations in terms of potency are recognized. As described in a review in Food Engineering (180), the emulsifying properties of a lecithin product were improved by treating crude lecithin with fatty acids. "Born-again" lecithin is solving some tough, long-standing problems in food formulations (181). The "old" ingredient has been challenged to offer new solutions, and according to increasing numbers of reports, it is standing up to the challenge reasonably well.

## References

1. Anonymous, Lecithin for Central Soya, Central Soya, Ft. Wayne, IN, 1978.
2. Schaefer, W., Getreide, Mehl, Brot 27:132 (1973).
3. Sartoretto, P., Lecithin, in Kirk-Othmer Encyclopedia of Chemical Technology, 2nd edn., Interscience Publishers, Division of John Wiley and Sons Inc., New York and London, Vol. 12, 1967, pp. 343–361.
4. Kleinert, J., Zucker-Susswaren-Wirtsch. 29(6):179 (1976).
5. Eichberg, J., Lecithin, in Kirk-Othmer Encyclopedia of Chemical Technology, 3rd edn., Vol. 14, Wiley and Sons Inc., New York and London, 1981, pp. 250–262.
6. Mason, L. H., and A. E. Johnson, Cereal Chem. 35:435 (1958).
7. McKillican, M. E., JAOCS 44:200 (1967).
8. Wren, J. J., and D. S. Merryfield, J. Sci. Food Agric. 21:254 (1970).
9. Youngs, V. L., D. G. Medcalf and K. A. Gilles, Cereal Chem. 47:640 (1970).
10. Arunga, R. O., and W. R. Morrison, Lipids 6:768 (1971).
11. Yamaguchi, F., J. Food Sci. Technol. (Nipon Shokuhin Kogyo Gakkai-Shi) 19(7):327 (1972).
12. Ruckenbauer, P., Wiss. Veroff. Deut. Gesell. Ernahrung 24(7):48 (1973).
13. Kosmina, N. P., Biochemistry of Breadmaking (German translation from Russian), VEB Fachbuchverlag, Leipzig, 1977, p. 339.
14. Moruzzi, G., R. Viviani, A. M. Sechi and G. Lenaz, J. Food Sci. 34:581 (1969).
15. Pomeranz, Y., Food Technol. 24:927 (1970).
16. Pomeranz, Y., Baker's Dig. 44(6):30 (1970).
17. Pomeranz, Y., Baker's Dig. 41(5):48 (1967).
18. Mecham, D. K., Lipids, in Wheat Chemistry & Technology, edited by

Y. Pomeranz, American Association of Cereal Chemists, St. Paul, MN, 1971, pp. 393–451.
19. Morrison, W. R., Proceedings of the 9th Congress of the International Association of Cereal Chemists, Vol. II, ICC, Vienna, Austria, 1976.
20. Morrison, W. R., Getreide, Mehl, Brot 30:244 (1976).
21. Morrison, W. R., Adv. Cereal Sci. & Technol. 2:221 (1978).
22. Morrison, W. R., Lipids in Wheat and Their Importance in Wheat Products, in Recent Advances in the Biochemistry of Cereals, edited by D. L. Laidman and R. G. Wyn Jones, Academic Press, London, 1979, pp. 313–335.
23. Acker, L., Getreide, Mehl, Brot. 7:181 (1974).
24. Acker, L., Z. Ernahrungswiss 126–135, Supplement 17 (1976).
25. Acker, L., Fette, Seifen, Anstrichm. 79(1):1 (1977).
26. Acker, L., and G. Becker, Gordian 72(7/8):275 (1972).
27. Becker, G., and L. Acker, Staerke 23:339 (1971).
28. Becker, G., and L. Acker, Fette, Seifen, Anstrichm. 76:464 (1974).
29. Chung, O. K., and Y. Pomeranz, Baker's Dig. 55(5):38 (1981).
30. Auerman, L. Y., L. I. Puchkova et al., Vysshikh Uchebnykh Zavedenii, Pishchevaya Tekhnol. 4:75, Chem. Abstr. 55:2768d (1961).
31. Puchkowa, L. I., and A. P. Netschajev, Proceedings of the 7th International Congress on the Problems of Modern Grain Processing and Grain Chemistry, Sept. 22–25, 1975, Bergholz-Rehbrucke, 1977, pp. 123–127.
32. Sidorova, O. G., and L. N. Chukhlebova, Vys. Uchebnykh Zav., Pishchevaya Technolog. 2:150, Food Sci. Technol. Abstr. 8:11M1333 (1976).
33. Zimmerman, R., Nahrung 23(3):289 (1979).
34. Wolf, W. J., and J. C. Cowan, CRC Critical Rev. Food Technol. 2:81 (1971).
35. Cowan, J. C., and W. J. Wolf, Soybeans, in Encyclopedia of Food Technology, edited by A. H. Johnson and M. S. Peterson, Avi Publishing Co. Inc., Westport, CT, 1974, pp. 818–828.
36. Wolf, W. J., and D. J. Sessa, in Encyclopedia of Food Science, edited by M. S. Peterson and A. H. Johnson Avi Publishing Co., Westport, CT, 1978, pp. 461–467.
37. Sullivan, D. R., and B. F. Szuhaj, Commercial Lecithin: Types, Properties and Uses, in Encyclopedia of Food Science, edited by M. S. Peterson and A. H. Johnson, Avi Publishing Co., Westport, CT, 1975, p. 476.
38. Matsueda, Y., S. Matsubara, M. Komoda, O. Suzuki and S. Fujiwara, Jpn. Kokai Tokkyo Koho 79, 107,530, Chem. Abstr. 92:20901x (1979).
39. Anonymous, Cereal Foods World 24:52 (1979).
40. Kuramschin, J. N., and L. S. Surdina, Bread and Confectionery Ind. (Moscow) 7(10):1 (1963).

41. Coduro, E., Z. Lebensmittel-Unters. Forsch. 126(5):167 (1965).
42. Unilever, N. V., Pourable Margarine and Method for Its Manufacture, Netherlands Patent Application 7,112,119 (1972).
43. Crosby, T. G., UK Patent Application 2006,810 (Cl. A 23 Df/00) US Application 841,405, Chem. Abstr. 91:209684z (1977).
44. Anderson, G. E., Proceedings of the 55th Annual Meeting, American Society of Bakery Engineers, Chicago, IL, 1974, pp. 56–61.
45. Eder, N., Deut. Muller-Ztg. 15:261 (1976).
46. Gorjaceva, A. F., et al., Chlebopek. Kondit. Prom., Moskva 23(4):19 (1979).
47. De Renzo, D. J., Doughs and Baked Goods; Chemical, Air and Non-leavened, Noyes Data Corp., Park Ridge, NJ, 1975.
48. Korjackina, S. J., and L. I. Puckova, Chlebopek. Kondit. Prom. Moskva 21:19 (1977).
49. Franzke, Cl., J. Kroll and W. R. Stenzel, Proceedings of the 7th International Congress for Problems of Modern Grain Processing and Cereal Chemistry, Bergholz-Rehbrucke, 1977, pp. 49–53.
50. Arai, K., S. Tanaka and F. Umeda, Jpn. Kokai Tokkyo Koho 79 41,343, to Oriental Yeast Co. Ltd., Chem. Abstr. 91:54945j (1979).
51. Hampl, J., and K. Tvrznik, Proceedings of the 7th International Congress on the Problems of Modern Grain Processing and Cereal Chemistry, Bergholz-Rehbrucke, 1977, pp. 59–64.
52. Anonymous, Lucas Meyer Co., Hamburg, Germany (1978).
53. Working, E. B., Cereal Chem. 1:153 (1924).
54. Working, E. B., Cereal Chem. 5:223 (1928).
55. Sinclair, A. T., and A. G. McCalla, Can. J. Res. C 15:187 (1937).
56. Traub, W., J. B. Hutchinson and D. G. H. Daniels, Nature (London) 179:769 (1957).
57. Daniels, D. G. H., Chem. Ind. (London):633 (1958).
58. Olcott, H. S., and D. K. Mecham, Cereal Chem. 24:407 (1947).
59. Hess, K., Kolloid Z. 136:84 (1954).
60. Grosskreutz, J. C., Cereal Chem. 38:336 (1961).
61. Zentner, H., J. Sci. Food Agr. 15:629 (1964).
62. Coppock, J. B. M., M. A. Cookson, D. H. Laney and D. W. E. Axford, J. Sci. Food Agr. 5:19 (1954).
63. Cookson, M. A., M. L. Ritchie and J. B. M. Coppock, J. Sci. Food Agric. 8:105 (1957).
64. Chiu, C. M., and Y. Pomeranz, J. Food Sci. 31:753 (1966).
65. Chiu, C. M., Y. Pomeranz, M. D. Shogren and K. F. Finney, Food Technol. 33:1157 (1968).
66. Pomeranz, Y., M. D. Shogren and K. F. Finney, Food Technol. 22:897 (1968).

67. Pomeranz, Y., R. P–C. Tao, R. C. Hoseney, M. D. Shogren and K. F. Finney, Agr. Food Chem. 16:974 (1968).
68. Griffin, E. L., Proceedings of the First National Conference on Wheat Utilization Research, Lincoln, NE, 1962, pp. 109–120.
69. Daftary, R. D., A. B. Ward and Y. Pomeranz, J. Food Sci. 31:897 (1966).
70. Pomeranz, Y., G. L. Rubenthaler and K. F. Finney, Food Technol. 19:1724 (1965).
71. Daftary, R. D., Y. Pomeranz, M. D. Shogren and K. F. Finney, Food Technol. 22(3):79 (1968).
72. Zapryagaeva, A. V., L. I. Puchkova, G. S. Fedorova and L. E. Chernenko, Izv. Vysshikh Uchebnykh Zav. Pishchevaya Technolog. 1:53 (1970).
73. Aberham, H., Deutsche Mueller-Z., 70(11):228 (1972).
74. Pyler, E. J., Baking Science and Technology, Seibel Publishing Co., Chicago, IL, 1973, pp. 464–485.
75. Puchkova, L. I., S. Y. Koryachkina, et al., Izv. Vysshikh Uchebnykh Zaved. Piscevaya Technolog. 2:68 (1975). Food Sci. Technol. Abstr. 8:2M272, (1976).
76. Kuntze, R., Proceedings of the 7th International Congress on the Problems of Modern Grain Processing and Cereal Chemistry, Bergholz-Rehbruke, 1977, pp. 73–82.
77. Sullivan, B., Cereal Chem. 27:661 (1940).
78. Pratt, D. B., Jr., Bakers' Helper 84 (1053–1054):38, 64, 66, 99 (1945).
79. Cook, L. R., Bakers' Helper 77(1093):56,72,82,84,87 (1947).
80. Whillans, D. A., Canadian Baker 62:27, 30.
81. Aylward, F., Food Manufacture 27 (7):285, (8):311, (9):355 and (10):395 (1952).
82. Iveson, H., Bakers Rev. 110(1):26,28,29,36. (1955).
83. Stewart, J., Food Manufacturer 30(9):370 (1955).
84. Glabe, E. F., P. W. Anderson and E. C. Jertson, Cereal Sci. Today 9:300 (1964).
85. Zieglitz, R., Susswaren 14(10):490 (1970).
86. Jakubczyk, T., A. Horubalowa et al., Przeglad Piekarski Cukiernieczy 20(9):195 (1972).
87. Church, J. A., Process Biochem., May:1 (1973).
88. Anonymous, Bakery Prod. Market. 8:122 (1973).
89. Bade, V., Getreide, Mehl, Brot. 28:296 (1974).
90. Knyazhanska, B. A., and G. S. Garmash, Karchova Promislovist' (Naukovavirobnichii Zbornik) 3:40, Food Sci. Technol. Abstr. 8:3M387 (1976).
91. Kuntze, R., and R. Schneeweiss, Baeker Konditor 24(3):72, Chem. Abstr. 89:105944b (1976).

92. Kuntze, R., and R. Schneeweiss, Baeker Konditor 24(4):117, Chem. Abstr. 89:105943a (1976).
93. Kuntze, R., and R. Schneeweiss, Baeker Konditor 24(6):166, Chem. Abstr. 89:105946d (1976).
94. Nickerson, J. T. R., and L. J. Ronsivalli, Elementary Food Science, Avi Publishing Co., Westport, CT, p. 67.
95. Wasserman, L., Gordian 79(10):248 (1979).
96. Rusch, D. T., Cereal Foods World 26:110 (1981).
97. Goebels, F. J., Gordian 77(9):242 (1977).
98. Knightly, W. H., Monograph 32, Surface-active Lipids in Foods. pp. 131–157.
99. Anonymous, ADM soy products for the baking industry, ADM, Decatur, IL (1975).
100. Pratt, D. B., Jr., Food Ind. 18:16 (1946).
101. Walrod, F. E., Proceedings of the 23rd Annual Meeting of the American Society of Bakery Engineers, p. 76 (1947).
102. Radev, A., and M. Kalichkov, Khranitelna Prom. 12:13 Chem., Abstr. 59:6898f (1963).
103. Rohrlich, M., and G. Bruckner, Cereal Grains I: Cereal Grains and Their Processing, 2nd revised edn., Verlag Paul Parey, Berlin, 1966, pp. 181, 198, 206.
104. Kent-Jones, D. W., and A. J. Amos, Modern Cereal Chemistry, 6th edn., Food Trade Press Ltd., London, 1967, p. 143.
105. Ponte, J. G. Jr., Bread, in Wheat Chemistry and Technology, edited by Y. Pomeranz, American Association of Cereal Chemists, St. Paul, MN, 1971, p. 718.
106. Marnett, L. F., Getreide, Mehl, Brot 31:244 (1977).
107. Tenney, R. J., Baker's Dig. 52(4):24 (1978).
108. Popadic, I. A., Bread and Confectionery Ind. (Moscow) 18(4):9 (1974).
109. Lehmann, G., Muehle & Mischfuttertechnik 112(22):291 (1975).
110. Kazanskaja, L. N., and O. F. Gurina, Chlebopek. Kondit. Prom., Moskva 20(4):24 (1976).
111. Starceus, P. A., and S. A. Altaev, Proceedings of the 7th International Congress on the Problems of Modern Grain Processing and Cereal Chemistry, Bergholz-Rehbrucke, 1977, pp. 65–71.
112. Osadchaya, N. T., I. I. Puchkova, R. G. Rakhmankulova and I. S. Shub, Izv. Vyssh. Uchebn. Zaved., Pishch. Tekhnol., 1979(6):108, Chem. Abstr. 92:127201w (1979).
113. Kihara, H., and T. Kato, Jpn. Kokai Tokkyo Koho 80 26,804 to Nippon Oils and Fats Co. Ltd., Chem. Abstr. 93:24810k (1980).

114. Shinka Shokuhin, K. K., Jpn. Kokai Tokkyo Koho 80:138, 352, Chem. Abstr. 94:64090c (1980).
115. Belohlawek, L., and W. Weber, Ger. Offen. 2,854,291 to Diamalt A.-G., Chem. Abstr. 93:130965z (1980).
116. Knjaginicev, M. I., J. Chem. All-Union Mendeleev Soc. (Moscow) 10:277 (1965).
117. Adler, L., and Y. Pomeranz, J. Sci. Food Agric. 8:449 (1959).
118. Finney, K. F., G. Rubenthaler and Y. Pomeranz, Cereal Sci. Today 8:166 (1963).
119. Mizrahi, S., Z. Berk and U. Kogan, Cereal Sci. Today 10:145 (1965).
120. Menger, A., and F. Bretschneider, Getreide, Mehl, Brot 27:154 (1973).
121. Sipos, E. F., and E. Turro, Baker's Dig. 48:29, 34, 67, (1974).
122. Seibel, W., Gordian 74(2):46; (3):95 (1974).
123. Pomeranz, Y., M. D. Shogren and K. F. Finney, Cereal Chem. 46:503 (1969).
124. Pomeranz, Y., M. D. Shogren and K. F. Finney. Ibid. 46:512 (1969).
125. Anonymous, Milling & Baking News, July, 1975, pp. 55–58.
126. Rao, U. S., U. K. Vakil, J. Food Sci. Technol. 17:259 (1980).
127. Chumachenko, N. A., A. P. Demchuk, L. M. Markianova, G. N. Kovalenko and V. B. Verbrii, Khlebopek. Konditer. Prom-st. 1981(2):31, Chem. Abstr. 94:207322t (1981).
128. Shogren, M. D., Y. Pomeranz and K. F. Finney, Cereal Chem. 58:142 (1981).
129. Makarjan, E. S., et al., Bread and Confectionery Ind. (Moscow) 17(8):10 (1973).
130. Arya, S. S., N. Vanaja and D. B. Parihar, Role of lipids in chapati baking, Fats and Oils Related Food Products and Their Preparation, Symposium, 1976, edited by P. B. Rama Rao, Association of Food Science and Technology, Mysore, India, pp. 146–148, Chem. Abstr. 89:128 (1976).
131. Pomeranz, Y., M. J. Carvajal, R. C. Hoseney and A. B. Ward, Cereal Chem. 47:373 (1970).
132. Pomeranz, Y., M. J. Carvajal, M. D. Shogren, R. C. Hoseney and K. F. Finney, Cereal Chem. 47:429 (1970).
133. Geoffrey, R., Techniques Ind. Cerealieres 145:3 (1974).
134. Gasiorowski, H., S. Jankowski and W. Obuchowski, Przeglad Piekarski Cukierniczy 24(9):164 (1976).
135. Scerbatenko, V. V., et al., USSR Patent No. 460041, IPK A 21 d. 2/00 (from IGV Schnellinformationen 1-78-12) (1975).
136. De Ruiter, D., Adv. Cereal Science & Technol. 2:349 (1978).
137. Kim, J. C., and D. De Ruiter, Food Technol. 22:867 (1968).

138. Osman, E. M., and M. R. Dix, Cereal Chem. 37:464 (1960).
139. Vogel, R., Kakao Zucker Monchengladbach 19(9):454 (1967).
140. Szalai, L., Zagadnienia Piekarstwa, ZBBP, Warsaw 13(2):17 (1968).
141. Siebenwirth, W., R. Schneeweiss and S. Anders, Baecker Konditor 26(2):37, Chem. Abstr. 89:105945c (1978).
142. Seibel, W., and A. Menger, Brot Gebäck 22:193 (1968).
143. Kvetnyj, F. M., and V. V. Scerbatenko, Bread and Confectionery Ind. (Moscow) 20(3):21 (1977).
144. Kroll, J., W. R. Stenzel and C. Franzke, Proceedings of the 7th International Congress on the Problems of Modern Grain Processing and Cereal Chemistry, Bergholz-Rehbrucke, 1977, pp. 55–57.
145. Matz, S. A., Cookie and Cracker Technology, Avi Publishing Co., Westport, CT, 1968, pp. 54–55, 121–122.
146. Adrian, J., Biscuit Cracker Baker 50(3):27 (1960).
147. Knightly, W. H., Proceedings of the 45th Annual Meeting of the American Society of Bakery Engineers, 1969, pp. 128–140.
148. Prokopenko, A. D., et al., Izv. Vyss. Uchebn. Zaved. Piscevaja Technolog. Krasnodar 1:28 (1975).
149. Yamazaki, W. T., J. R. Donelson and R. L. Clements, Cereal Chem. 56:584 (1979).
150. Kissell, L. T., and W. T. Yamazaki, Cereal Chem. 52:638 (1975).
151. Badi, S. M., and R. C. Hoseney, Cereal Chem. 53:733 (1976).
152. Kissell, L. T., and K. Lorenz, Cereal Chem. 53:233 (1976).
153. Smith, W. H., Snack Food 60(6):54 (1971).
154. Viguerie, M. M., Process for Obtaining Dried Foodstuff, French Patent 1 599 543 (1980).
155. Knightly, W. H., and M. L. Lynch, Baker's Dig. 40(1):28 (1966).
156. Elsesser, C. C., and S. W. Bogyo, Powdered shortening composites, U.S. Patent 2,954,297 (1960).
157. Obenauf, C. F., and C. W. Tatter, US Patent 3,060,030 (1962).
158. Weiss, R. M., J. M. Sinner and W. F. Bloch, Emulsifier-liquid Oil Preblend, U.S. Patent 3,222,184 (1965).
159. Del Vecchio, A. J., Baker's Dig. 49(4):28, 52 (1975).
160. MacDonald, I. A., and D. A. Bly, Cereal Chem. 43:571 (1966).
161. Wootton, J. C., N. B. Howard, J. B. Martin, D. E. McOsker and J. Holme, Baker's Dig. 41(5):179 (1967).
162. Tomita, T., and M. Mizukoshi, Emulsified Fat Composition, U.S. Patent 3,782,970 (1974).
163. McComber, D., and E. M. Miller, Cereal Chem. 53:101 (1976).
164. Bernotavicz, J. A., Biscuit Maker Plant Baker 14(7):596, 610 (1963).
165. Prokopenko, A. D., and I. M. Roiter, Food Sci. Technol. Abstr.

1:12M912 (1969).
166. Kuntze, R., and C. Luft, Baeker Konditor 27(7):218, Chem. Abstr. 93:14188h (1979).
167. Kuntze, R., and C. Luft, Baeker Konditor (Leipzig) 27(8):249 (German), Chem. Abstr. 93:148189j (1979).
168. Lee, F. A., Basic Food Chemistry, Avi Publishing Co., Westport, CT, 1975, pp. 96–98.
169. Bradley, W. B., Proceedings of the American Society of Bakery Engineers, 1948, p. 53.
170. Fujikawa, T., A. Miyahara and M. Hamashima, Jpn. Kokai Tokkyo Koho 83:910, Chem. Abstr. 91:173696p (1979).
171. Pritula, et al., Chlebopek. Kondit. Prom. (Moscow) 10:20 (Russian: from IGU Schnellinformationen 1-80-3) (1979).
172. Prokopenko, A. D., L. V. Grishchenko and M. V. Dyakunovskaya, Pishch. Prom.-st. Kiev. 24:51–53, Chem. Abstr. 92:196618z (1978).
173. Seibel, W., Development in Baking Technology, Processing Technology, New Nutrients, Additives, in Cereals '78: Better Nutrition for the World's Millions, edited by Y. Pomeranz, American Association of Cereal Chemists, St. Paul, MN, pp. 187, 192–194 (1978).
174. Jodlbauer, H. D., Lebensmittelrecht 4(1):27 (1977).
175. Benedickt, G., Getreide, Mehl, Brot. 32:181 (1978).
176. Bruemmer, J. M., Getreide, Mehl, Brot. 25:7 (1971).
177. Weith, L., Lebensm. Ernahr. 39:159 (1976).
178. Calvel, R., Getreide, Mehl, Brot. 26:75 (1972).
179. Schaefer, W., Mühle und Mischfuttertechnik 116:542 (1979).
180. Anonymous, Food Eng. 52(8):122 (1980).
181. Worrall, C. T., Food Eng. 53(11):88 (1981).
182. Hargin, K. D., and W. R. Morrison, J. Sci. Food Agric. 31:877 (1980).

# Chapter Fourteen
## Lecithin in Health and Disease

**Steven H. Zeisel**
Nutrition-Pathology Unit
Department of Pathology
and
Department of Pediatrics
Boston University School of Medicine
85 East Newton Street
Room M1002
Boston, MA 02118

### Introduction

This chapter reviews current knowledge about the sources of lecithin in the diet, the absorption of lecithin from the gut, the metabolism of lecithin, and the use of lecithin as a therapeutic agent. The use of choline (a component of lecithin) and lecithin as treatments for certain diseases of the nervous system has aroused public interest in using lecithin as a dietary supplement. Although it is clear that lecithin administration can be beneficial for humans with tardive dyskinesia (a neurological disorder), and although there is reason to believe that normal memory may be influenced by the choline content of the diet, evidence available at this time does not justify the widespread use of lecithin for improved memory by the healthy general public.

In commercial usage, lecithin is a term which refers to a complex mixture of phosphatides. In scientific usage, it refers to the compound phosphatidylcholine (1). In this chapter, lecithin means pure phosphatidylcholine.

### The Diet as a Source of Lecithin (and Other Choline-containing Compounds)

Lecithin is a part of almost all biological membranes, therefore we consume a great deal of this phospholipid as part of our diet (2–4). Foods such as eggs, liver, soybeans and peanuts are rich in lecithin (Table I) (4–6). It is often added to processed foods because it acts as an emulsifying agent or as an antioxidant (1, 7). Healthy humans in the USA probably ingest ca. 6g of lecithin/day (100 mg/day of this amount is added to foods during processing) (1,2). As will be discussed later, choline is an important constituent of lecithin. Free choline is found in some foods such as liver, oatmeal, soybeans, cauliflower, kale and cabbage (4–6,8). Total choline intake (as free choline and the choline in lecithin) amounts to 600–1000 mg/day in the adult (2–4).

**TABLE 14-I**
**The Choline and Lecithin Content of Common Foods[a]**

| Food (per 100 g) | Choline (mg) | Lecithin (mg) |
|---|---|---|
| Calf liver | 650 | 850 |
| Lamb chop | — | 753 |
| Beef | — | 453 |
| Ham | — | 800 |
| Trout | — | 580 |
| Cheese | — | 100 |
| Egg | 0.4 | 394 |
| Oatmeal | 131 | 650 |
| Soybeans | 237 | 1480 |
| Wheat germ | — | 2820 |
| Peanuts | — | 1113 |
| Cauliflower | 78 | 2 |
| Kale | 89 | 2 |
| Potato | 40 | 1 |

[a]Data from references 4 and 5.

Lecithin is also available to the public as a retail dietary supplement. The capsules or granules sold in this manner are usually impure (<35% phosphatidylcholine) (9). It is possible that several grams of lecithin could be taken, doubling the consumer's intake from common foods. As will be described later, physicians are using 20–30 g of lecithin for the treatment of neurological diseases (2). The administration of such large doses as part of common foods increases the caloric load. For this reason, lecithins have been purified so that preparations are available which contain as much as 95% phosphatidylcholine. A list of the manufacturers of such lecithins is available from the author.

## Is Lecithin an Essential Nutrient?

Mammals do not have to eat lecithin molecules in order to survive; however, they may have to eat some source of choline molecules. Three enzymatic pathways catalyze lecithin biosynthesis, yet only one generates new choline molecules. The cytidine diphosphocholine (CDP) and base exchange pathways do not cause a net synthesis of choline but only redistribute preexisting molecules (10–13). The methylation pathway, catalyzed by the enzyme(s) phosphatidylethanolamine-N-methyltransferase (PEMT), makes new choline molecules by sequentially methylating phosphatidylethanolamine, using S-adenosylmethionine (SAM) as the methyl donor (14–18). This enzyme has its highest activity in the liver, but it is also found in kidney, testes, heart, adrenal, lung, erythrocyte, brain and spleen (18–20). PEMT is probably composed of several enzyme activities (18,20–22). In the liver and brain, the activity of the first enzyme is limiting, and it has a high affinity for SAM. It is located on the inner (cytoplasmic) side of the plasma membrane. In the membrane of the endoplasmic reticulum this enzyme is also on the cytoplasmic surface (23). As monomethyl-phosphatidylethanolamine is formed it probably flips into the middle of the membrane's phospholipid bilayer, where it is methylated by the second PEMT enzyme (20,21). This enzyme has a low affinity for SAM. A third methyl group is then added (either by the second enzyme or by a third enzyme) and the phosphatidylcholine formed flips into the outer (extracellular) leaflet of the phospholipid bilayer. This pathway is one of the major mechanisms whereby the asymmetry of the membrane bilayer is maintained (i.e., more phosphatidylethanolamine on the cytoplasmic side, more phosphatidylcholine on the extracellular side) (24,25). There are no accurate estimates of the activity of PEMT in vivo; in vitro data suggest that 15–40% of the lecithin present in liver is synthesized via this pathway, with the remainder coming from the CDP pathway (17,26). Choline deficiency and exposure to ethanol increase the activity of the methylation pathway in liver (27–29). Female rats have more hepatic PEMT activity than do males (14,17). In brain and red blood cells, stimulation of noradrenergic receptors acts to increase the rate of PE methylation (30–32). It has been suggested that the accumulation of the methylated products alters membrane fluidity and thereby modulates the activity of enzymes which are embedded within the membrane (31,32).

The methylation pathway makes new free choline molecules as the lecithins produced are hydrolyzed (34). We do not know whether this source of choline is sufficient to meet all of the needs of mammals. Early experiments with insulin first demonstrated choline's importance as a nutrient. Depancreatized dogs developed fatty liver and died (35). Choline administration prevented this (36–38). Chronic ingestion of a diet deficient in choline has major consequences: these include hepatic, renal, memory and growth disorders. In the rat, hamster, pig, dog and chicken, choline deficiency results in fatty infiltration of the liver (36,39–48). This is probably due to a disturbance in the synthesis of lecithin which is needed to export triglycerides as part of lipoproteins (see section on lecithin metabolism). Renal function is also compromised, with abnormal concentrating ability, free water reabsorption, sodium excretion, glomerular filtration rate, renal plasma flow and gross renal hemorrhaging (49–53). Infertility, growth impairment, bony abnormalities, decreased hematopoiesis and hypertension have also been reported to be associated with diets low in choline content (54–58). In mice, choline deficiency is associated in an impairment of memory (59). Decreased serum choline concentrations have been reported in humans being fed by total parenteral nutrition (using solutions low in choline content). It is possible that the liver disease associated with parenteral nutrition is related to choline deficiency (60).

In view of all the pathology associated with choline-deficient diets, why are we not able to say definitively that humans must eat some choline-containing compounds in their diet? It is because we are unsure to what extent methionine and other labile methyl donors can spare choline requirements. Methionine or vitamin B12 deficiency exacerbates the hepatic and renal damage associated with choline deficiency (28,61,62). Choline plays a part in a labile methyl group cycle. As will be discussed later, choline can be degraded to betaine which can donate a methyl group to S-adenosylhomocysteine, forming SAM. This universal methyl donor can then use the methyl group in one of several pathways (e.g., the formation of lecithin) (63–66). These complex interrelationships between choline, SAM and lecithin make it difficult to be sure how much choline must be included in the diet. It is possible that exogenous choline is only needed during periods of rapid tissue growth (i.e., during infancy), for it is at these times that the sequelae of choline deficiency are the easiest to elicit (61). We have studied PEMT-catalyzed lecithin synthesis in newborn rats and have found the enzyme to be relatively inactive.

## Absorption of Lecithin and Other Choline-containing Compounds from the Gut

The ultimate source of most dietary choline is lecithin (5). Both pancreatic secretions and intestinal mucosal cells contain enzymes capable of hydrolyzing lecithin. Phospholipase A2 (which cleaves the $\beta$-fatty acid moiety) is found in pancreatic juice and in the intestinal brush border (67–73). It is secreted as a zymogen, which is activated in the gut by trypsin, calcium and bile salts

(67,68,70). Within the gut mucosal cell, phospholipase A1 cleaves the $\alpha$-fatty acid, and phospholipase B cleaves both fatty acids (74). These enzymes degrade much less lecithin than the pancreatic lipase does. The net result is that most ingested lecithin is absorbed as lysolecithin (deacylated in the $\beta$-position) (69). Within the enterocyte, lysolecithin can be deacylated to form glycerophosphorylcholine (GPC), or it can be acylated to reform lecithin (69,71,72,74,75). The latter reaction is catalyzed by a dismutase found in both the microsomal and soluble fractions of the mucosal cell (74). The dismutase is maximally active when high concentrations of lysolecithin are present (Km = 3.6 mM). It uses a second lysolecithin molecule as the fatty acid donor, thus two lysolecithin molecules are converted to a lecithin and a GPC molecule. For this reason, approximately twice as many lecithin molecules are absorbed from the gut as are reconstituted and secreted from the mucosal cell into the lymphatic circulation (69,71,76,77).

Free choline is absorbed from the upper small intestine, chiefly in the jejunum (78–81). This transport is mediated by a carrier system at choline concentrations less than 4 mM, and by passive diffusion at higher concentrations (80). This carrier is located in the brush border of jejunal mucosal cells and it is not sodium- or energy-dependent (80,81). Some of the ingested choline is metabolized before it can be absorbed from the gut. Enterocytes degrade it to form betaine, and gut bacteria use it to make methylamines (9,79,83). The formation of methylamines does not occur when choline is injected intravenously, or when it is fed to germ-free or antibiotic-treated animals (82,84–86). Ingested lecithin should not liberate free choline within the gut lumen, therefore bacteria should not produce methylamines. However, when humans ingest commercially prepared lecithins they excrete large amounts of methylamines in their urine. We believe that this occurs because such lecithin preparations contain significant quantities of methylamines (see Table II) (9).

GPC is present in small quantities in the diet, and, as discussed above, is formed from ingested lecithin. Within the gut mucosal cell, GPC diesterase (L-3-glycerophosphorylcholine glycerophosphohydrolase, EC 3.1.4.2) catalyzes the conversion of GPC to glycerylphosphate and free choline. This free choline enters the portal circulation of the liver (70,72). Phosphorylcholine is also present in small amounts in the diet. It is rapidly degraded by intestinal alkaline phosphatases, liberating free choline and inorganic phosphate (87).

**TABLE 14-II**
**Methylamine Contaminants in Preparations of Choline-Containing Compounds Used to Treat Humans**

| Preparation | micromoles amine per gram preparation DMA | TMA | mg lecithin per gram preparation |
|---|---|---|---|
| Choline chloride | nd | nd | — |
| Choline stearate | — | 500.00 | — |
| Soy lecithin (Bolec-soup) | 0.75 | 11.95 | 857 |
| Soy lecithin (Bolec, 1979) | nd | 14.62 | 796 |
| Soy lecithin (Bolec, 1980) | nd | 11.68 | 858 |
| Egg lecithin (Bolec, 1979) | 0.55 | 6.83 | 597 |
| Soy lecithin (Phospholipon 100) | nd | 5.38 | 687 |
| Egg lecithin (Nattermann, 1978) | 0.12 | 2.86 | 223 |
| Soy lecithin (Alcolec 640) | nd | 2.84 | 258 |
| Soy lecithin (Alcolec granules) | nd | 1.47 | 142 |
| Soy lecithin (Pharmacaps capsules) | nd | .33 | 190 |
| Soy lecithin (Food-Tritional) | nd | 0.78 | 162 |
| Soy lecithin (Formula 92 granules) | nd | 0.76 | 163 |
| Soy lecithin (Fearn Liquid) | nd | 0.87 | 97 |
| Soy lecithin (Schiff) | nd | 0.93 | 152 |
| Soy lecithin (Three Star Soya) | nd | 0.94 | 140 |
| Soy lecithin (Lecitol) | 1.26 | 4.51 | 69 |
| Soy lecithin (Leci-Fiber) | 1.78 | 1.57 | 47 |
| Soy lecithin (Golden Cal Powder) | 1.68 | 4.20 | 68 |

Choline chloride was dissolved in water and 1 micromole was injected onto the gas chromatograph. Choline stearate and lecithins were dissolved in chloroform and water was then added. An aliquot of the aqueous phase was then acidified and prepared for injection onto the gas chromatograph using the methods described in the text. Lecithin purity was determined using an assay for choline which is described in the methods.
nd = not detected.
Data taken from reference 9.

## Metabolism of Lecithin

The lecithin synthesized within gut mucosal cells enters the lymphatic circulation; that made in the liver enters the blood. Transport lipoproteins, which are rich in lecithin content, are formed in both the intestine and the liver, and occur in four major classes that differ in particle size and density (88, 89). They are differentiated by their relative flotation rates after ultracentrifugation. Within each class, proteins and lipids exist in a relatively fixed ratio. Chylomicrons contain large amounts of triglycerides within a lipoprotein envelope, they have a particle size of 1000 nm and a low density. Very low density lipoproteins (VLDL) contain more phospholipids than do chylomicrons and are smaller (50 nm) and denser. Low density lipoproteins (LDL) contain more cholesterol and more phospholipid than do VLDL, they are also smaller (20 nm) and denser. High density lipoproteins (HDL) contain esterified cholesterol and little triglyceride. These particles are smaller than 10 nm, and are the most dense lipoproteins (89).

After a lipid-rich meal is eaten, chylomicrons are released into the lymph from the small intestine. The liver synthesizes triglyceride from glucose and fatty acids, and releases it into the blood packaged as VLDL. Circulating HDL transfers apoprotein C to the chylomicrons and VLDL, this apoprotein activates lipoprotein lipases which are in capillary endothelial cells. This results in the hydrolysis of triglycerides and the release of free fatty acids into the blood. As triglycerides are removed from VLDL, a transport lipoprotein of intermediate size is formed, and the excess surface components (phospholipids, proteins and unesterified cholesterol) are released and enter HDL. The intermediate size lipoproteins are converted to LDL by the liver. LDL transports cholesterol and its esters to tissues, where they bind to specific receptors and are internalized and catabolized (89). HDL are produced by the liver and the small intestine (88). It takes tissue cholesterol and transfers it back to the liver and to VLDL (90). The enzyme lecithin-cholesterol acyl transferase (LCAT) is secreted by the liver and acts on HDL cholesterol. It transfers an acyl group from lecithin to cholesterol, forming esterified cholesterol (91). In terms of lecithin metabolism, the net result is that lysolecithins are formed which are transferred to albumin for transport within blood.

Many tissues possess enzymes which are capable of degrading lecithins and lysolecithins. Phospholipases A1 and A2 remove lecithin's fatty acid moieties (92). Lysophospholipase performs the same function for lysolecithin. GPC diesterase then acts to release free choline and glycerol phosphate (94,95). Lecithins can also be degraded by phospholipase C activity which produces diacylglycerol and phosphorylcholine (93). Phospholipase D activity has recently been identified in mammalian tissues: it forms free choline and diacylglycerolphosphate from lecithin (96). Lysophospholipase D acts in a similar manner on lysolecithin (97). Although we are sure that blood choline concentration increases after humans eat lecithin (2), we do not know which organ or which enzyme activity is responsible for the liberation of most of the free choline seen.

## Metabolism of Choline

In mammalian tissue free choline participates in four enzyme-catalyzed pathways: oxidation, phosphorylation, acetylation and base exchange.

Choline is oxidized to betaine aldehyde which is then converted to betaine. These reactions are catalyzed by the enzyme system choline oxidase (actually two enzyme activities—choline dehydrogenase and betaine aldehyde dehydrogenase). Choline oxidase is present in several mammalian tissues, including liver and kidney. It is not present in blood, brain or muscle (98–103). Betaine formation is probably the predominant fate of most excess choline in the liver (104). In this organ the maximal rate of betaine formation from choline is more than ten times faster than choline incorporation into phosphorylcholine (the first step in lecithin synthesis) (101,105). Once betaine is formed it cannot be reduced to reform choline; however, it can donate a methyl group, thereby forming dimethylglycine and methionine. Dimethylglycine is converted to sarcosine and then to glycine. The oxidation pathway thus acts to remove free choline from tissues while scavenging methyl groups.

Choline kinase catalyzes the phosphorylation of the hydroxyl group of choline, using ATP as the phosphate donor (106). It is widely distributed in mammalian tissues including the liver, brain, kidney and lung (13,106–109). Phosphorylation of choline is the first step in the CDP pathway for lecithin synthesis. (10) In this pathway cytidine triphosphate and phosphorylcholine are combined, generating cytidine diphosphocholine (CDP-choline). The CDP-choline is then combined with diacylglycerol, forming lecithin.

The acetylation of choline, catalyzed by choline acetyltransferase (ChAT), is important because of the role that acetylcholine plays as a neurotransmitter. Only a small fraction of ingested choline is acetylated (110–112). The enzyme ChAT is highly concentrated in the terminals of cholinergic neurons; however, it is also present in such nonnervous tissues as the placenta (113–115). Several factors influence in vivo ChAT activity, including the availability of choline and acetyl-CoA (116). In brain it is unlikely that ChAT is saturated with either of its substrates. (116) Thus choline (and possibly acetyl-CoA) availability determine the rate of acetyl-choline synthesis (110,111).

A carrier mechanism, within the blood-brain barrier, transports free choline into the brain against a concentration and charge gradient. It has a Km for choline of 0.44 mM, and a Vmax of 10 nmol/g brain per min. At physiologic concentrations of choline in serum (10 $\mu$M), this carrier is unsaturated and is able to carry choline into brain at a rate that is proportional to serum choline concentration (117). Another carrier system transports choline out of the brain. Choline efflux is proportional to the gradient of choline that exists between brain and blood (118). When large doses of choline and lecithin are ingested blood choline concentrations increase (2), thereby accelerating choline's entry into and inhibiting its efflux from brain. Some choline molecules may also be transported into the brain as part of phospholipids such as lecithin or lysolecithin (119). These molecules could cross membranes without the help of a carrier, as they are lipid-soluble. It has not been conclusively demonstrated that such lipids do enter the brain intact (117).

As ChAT is not saturated with choline, more acetylcholine will be synthesized when brain choline concentrations increase. This occurs when large doses of choline or lecithin are fed to animals (110,111,120). The increase in brain acetylcholine content is associated with an augmented release into the synapse of this neurotransmitter. Postsynaptic activation of tryosine hydroxylase, a phenomenon mediated by acetylcholine complexing with its receptor (121–123), occurs whenever choline is administered (124,125). This activation can be blocked by cutting the presynaptic nerve or by blocking the acetylcholine receptor (124,125). This suggests that choline is being converted to acetylcholine within the presynaptic neuron, that acetylcholine is released and that a postsynaptic receptor is then activated. It has been possible actually to measure an increase in the number of acetylcholine molecules released by certain in vitro preparations of nerves exposed to higher than normal levels of choline in their perfusion media (126–128).

A base-exchange pathway has been described in several tissues including liver and brain (11,12,129). It involves the substitution of choline for serine, inositol or ethanolamine head groups on endogenous phospholipids. The reaction is reversible and requires the presence of calcium.

## Side Effects of Lecithin and Other Choline-containing Compounds

Much of the clinical use of choline and lecithin has occurred recently. Little attention has been given to the pharmacology and side effects of these compounds. In the treatment of neurologic diseases, large doses of choline and lecithin have been used. Such doses of choline have been associated with "fishy" body odor, vomiting, salivation, sweating and gastrointestinal distress (1). These side effects are probably dose-dependent, and all but the "fishy" body odor can be prevented by prior administration of a muscarinic receptor blocker such as methscopolamine (130). Rarely, humans have become depressed after ingesting large amounts of choline (131).

Large doses of lecithin are also associated with "cholinergic" side effects (sweating, gastrointestinal distress, vomiting and diarrhea) (1). Again, these side effects are dose-dependent: most humans tolerate 10 g of lecithin, many have no problems with 30 g/day. When lecithin is consumed as a component of common foods, as much as 35 g/day have been ingested for several days without difficulty (132). Lecithins contain fatty acids, and therefore have a relatively high caloric density. Weight gain has been a side effect of lecithin ingestion.

The high linoleic acid content of soy lecithin could be a potential problem, as polyunsaturated fatty acids have been reported to be associated with the formation of gallstones in men (133). This has not been seen in clinical studies to date, and in three patients, lecithin administration (in combination with cholic acid) has been associated with a decrease in the size of gallstones (134).

Large doses of lecithin have not been associated with "fishy" body odor (2). This is probably because little free choline is released from lecithin within the gut lumen, and therefore gut bacteria cannot act upon the choline to form trimethylamine. We have found that most commercially available lecithins contain some trimethylamine (see Table II) (9). Probably these amounts of trimethylamine are not sufficient to be detected by our noses. The formation of methylamine, dimethylamine and trimethylamine from choline and lecithin could have the potential for causing yet another side effect of these compounds. The methylamines formed after choline is ingested could be substrates for the formation of nitrosamines (9). Nitrosamines have marked carcinogenic activity in a wide variety of animal species (135). Nitrite reacts with amines most rapidly at acid pH such as that present in the human stomach, but even under conditions present in the human intestine (pH 8) bacteria can catalyze the endogenous formation of nitrosodimethylamine (NDMA) (136,137). The major role of bacteria in the nitrosation of dimethylamine is the reduction of nitrate to nitrite and the lowering of gut pH. Once ingested, or formed within the gut, methylamines are rapidly absorbed and excreted in the urine. NDMA can also be formed in urine contaminated with bacteria.

A diet of rapeseed oil has been associated with cardiac pathology in rats. The erucic acid (C22:1) content of rapeseed oil has been implicated (138). In view of the toxicity of erucic acid, care must be taken when administering lecithins derived from rapeseed. A recent report to the FDA has recommended that the content of eicosa- or docosamonoenoic fatty acid in lecithin used in humans be less than 6% of the total fatty acids (1).

Humans with neurological diseases have been ingesting large doses of lecithin (15–30 g/day) for as long as 2 years. Although we have listed many potential side effects, these doses of lecithin have been remarkably benign.

The investigation of these potential problems should be encouraged, but at the present time lecithin, as used in the treatment of human diseases, seems to be nontoxic.

## The Use of Lecithin and Other Choline-containing Compounds in the Treatment of Human Diseases

Choline and lecithin have been used as treatments for human illnesses for decades. The major sites of action proposed for these compounds have been the nervous system, the cardiovascular system, the immune system, and the organs which store and transport lipids within the body.

## Diseases of the Nervous System

Choline and lecithin therapy benefits patients with tardive dyskinesia (139–144). This disorder is characterized by involuntary choreic movements of the tongue, lips and jaw. It is thought to be a side effect of many of the antipsychotic drugs currently being marketed in the USA. These neuroleptic drugs are dopamine receptor blockers, and their chronic use presumably causes a postsynaptic receptor to become supersensitive to dopamine (145,146). An acetylcholine neuron, located in the caudate nucleus of the brain, normally acts to inhibit unwanted movements. This neuron receives input from a dopamine neuron which inhibits its activity. In healthy humans the normal rates of firing of these two neurons result in the prevention of twitching of the face. After supersensitivity to dopamine develops, the acetylcholine neuron receives extra inhibitory input, and it in turn sends fewer messages to the area of motor cortex that it is supposed to inhibit. The net result is that unwanted facial movements appear (147). Within months after the first demonstrations that choline could elevate the levels of acetylcholine in rat brain, humans with tardive dyskinesia were being treated with choline with some success (139, 140). Choline therapy had one bothersome side effect—subjects developed a body odor reminiscent of rotten fish as a result of the formation of trimethylamine. This made careful, double-blind studies difficult (140). For this reason the choline-containing compound lecithin was tried and several studies, including some with a double-blind crossover protocol, have demonstrated lecithin administration to be an effective therapy for tardive dyskinesia, and it has not been associated with body odor (141,142,144).

Cholinergic neurons are vital for normal memory in rats and humans. Memory function was markedly impaired after humans were treated with scopolamine, an acetylcholine receptor blocking agent, and memory was restored by treatment with physostigmine, a drug which increases cholinergic transmission (148). Humans with surgical or traumatic lesions of the hippocampus are unable to remember recent events. Within this area of the brain there are many cholinergic neurons, which are thought to be the site of action of scopolamine and physostigmine. Choline administration has been reported to improve the memory of humans who are poor initial learners in serial-learning and selective-reminding tasks (149, 150). Mice exhibit an age-related diminution of memory which is exacerbated by a diet low in choline content and is alleviated by a diet high in choline content (59). It may be that such diets are also associated with anatomical differences in the brains of these mice. Diminished memory in old age is not peculiar to mice; benign senescent forgetfulness is a problem in humans as well.

Some humans are afflicted with rapid loss of memory at a relatively young age, a syndrome called Alzheimer's disease. This disease is characterized by the loss of acetylcholine neurons in brain (151, 152). This loss of neurons is much greater than that seen in normal elderly brain. Central cholinergic receptors are apparently intact in Alzheimer's disease (152). A number of reports have described memory function in Alzheimer's patients treated with choline or lecithin. Most have described open-label studies, but several have been double-blind (189). In most studies a portion of the subjects improve significantly, whereas others are not helped at all (153–155). For example, in one study 7 of 11 patients showed a 50–200% improvement on memory tests after 2–8 weeks of lecithin (155). There are serious problems facing investigators who wish to study memory in humans. There is little agreement as to what constitutes an objective memory test, and clinicians are not able to identify a homogeneous population to study. It is analagous to trying to treat all cancers with one drug; of course, only a specific subpopulation will respond (i.e., leukemia might respond to a drug that might leave brain tumors unaffected). There are probably several subclasses of Alzheimer's; some patients may have selective loss of some of their cholinergic neurons, others may have lost several types of neurons, and yet others may have lost so many cholinergic neurons that few or none are left. We would expect that only the first class would be responsive to the administration of precursor, as the second group would require treatments capable of correcting several types of deficits, and the third group is left with no neurons capable of synthesizing acetylcholine from precursor. Once a homogeneous group of patients can be identified, it will be important to determine the dosage and route of administration of precursor that is most effective. It may be that multidrug pharmacy is required; preliminary reports indicate that choline or lecithin plus piracetam (a drug thought to increase the firing rate of acetylcholine neurons) has more effect than precursor alone (189). In order to accelerate progress in the search for a treatment for memory disorders, the "International Study Group on the Pharmacology of Memory Disorders Associated with Aging" has been formed, with members drawn from the international community interested in memory (189). Definitive evidence of the efficacy of lecithin treatment in especially selected patients with Alzheimer's disease awaits further investigation.

It has been proposed that mania is the result of increased activation of adrenergic receptors in brain, whereas depression is the result of increased cholinergic stimulation (150). Lithium therapy is commonly used to treat mania, its mechanism of action is unknown. Lithium is a toxic drug, and psychiatrists desire to use the smallest doses possible. Physostigmine, a cholinergic agonist, induces depression when administered to persons with affective disorders, and it decreases manic symptoms when administered to manic patients (157). This action of physostigmine suggested that the administration of choline or lecithin to manic patients might be beneficial. When lecithin was administered along with lithium the improvement in patient's behavior was marked (158). Three to seven days after cessation of lecithin treatment the manic symptoms returned. Chronic lithium treatment retarded choline efflux from rat brain, while treatment with lecithin accelerated choline influx into rat brain (118). This may explain why a lecithin-lithium combination therapy was so effective in humans with mania. One manic patient was treated with large doses of choline chloride alone and demonstrated less psychopathology.

Gille's de la Tourette's disease is characterized by abrupt involuntary movements of the muscles of the face, neck, torso and extremities. Involuntary vocalization also occurs (159). Red blood cell choline appears to be markedly elevated in these patients (160). This may reflect a genetic defect, as biological relatives also had similar elevations. The significance of this finding is unknown, however it may reflect an abnormality in transport systems for choline. There have been reports that administration of choline or lecithin helped patients with Tourette's; however, these studies have not always been replicated. These differences in clinical response may be related to the use of different lecithin preparations (161–163).

There are several diseases of the nervous system which are characterized by ataxia (balance problems). Friedreich's Ataxia is a recessive hereditary condition in which there is chronic degeneration of the cerebellum, spinal cord and peripheral nerves. These lesions result in balance problems and muscle weakness. In Friederich's ataxia there is a defect in the enzyme pyruvate dehydrogenase (164) which could impair acetyl-CoA synthesis, thereby altering acetylcholine production. Choline and lecithin therapies have been tried in this condition with some encouraging results (165). Unfortunately, the clinical response to these treatments was insufficient to improve the lifestyle of the patients, and the design of the experiments was not adequate to conclude that choline or lecithin are efficacious. Another ataxic disorder, spinocerebellar degeneration was not improved by lecithin administration (165). Oral choline produced sustained benefit for one patient with cerebellar ataxia (166). In another study of ataxic disorders, only 1 of 13 patients demonstrated reproducible improvement after treatment with choline (167).

The myasthenic syndrome is a disease characterized by weakness and easy fatigability. It is caused by deficient release of acetylcholine at the myoneural junction. This rare syndrome is sometimes associated with bronchiogenic carcinoma. Choline administration resulted in significant improvement in one patient (168).

### The Cardiovascular System

In both humans and animals, intravenous choline administration lowers blood pressure (169–171). This effect is prevented by pretreatment with hemicholinium-3, which suggests that choline must be transported via a carrier into some cell, possibly a neuron, to have this effect (172). Oral administration of choline has a very slight, but significant hypotensive effect in humans (153). Choline could be acting by increasing vagal tone to the heart, or by dilating arterioles. Although we know that choline administration increases acetylcholine release in in vitro preparations of heart (127), we have not observed changes in cardiac rate in healthy humans treated with choline. Rats exposed to transient periods of dietary choline deficiency develop irreversible hypertension (49, 58). These changes may be related to changes in renal as well as in cardiovascular function.

### Mechanisms of Lipid Transport

Lecithin has been used as a treatment to prevent the development of atherosclerosis. The accumulation of cholesterol and its esters in the arterial wall are thought to result in athersclerosis. The importance of cholesterol intake and the resulting hypercholesterolemia has been controversial. There has been agreement that lowering plasma LDL cholesterol is beneficial. As described earlier, one of the important functions of lecithin is to contribute to the formation of transport lipoproteins. Administration of large amounts of lecithin changes the plasma lipoprotein pattern temporarily (173–175). Lecithin rich in linoleic acid alters the size and density of chylomicrons and VLDL by altering the coat (phospholipid) to core (triglyceride) ratio (176). Esterification of cholesterol is mediated by the enzyme lecithin cholesterol acyltransferase. This enzyme prefers lecithins which have a linoleic or arachidonic acid in the 2 position (1). In rabbits fed an atherogenic diet, lecithin administration decreases blood cholesterol and arteriosclerotic plaques (177). Although choline itself appears to lower serum cholesterol in rats, it does not have this effect in rabbits (1). In humans with hypercholesterolemia lecithin treatment also lowers blood cholesterol. Six months after lecithin therapy was discontinued the blood cholesterol in these patients had returned to high levels. These studies have been reproduced several times using soybean lecithins of various purities (178–181). Results have not been as encouraging in studies on patients with type II hyperlipoproteinemia treated with Lipostabile, an impure lecithin preparation (175, 182–184). This may be because less lecithin was used. Both corn oil and lecithin decrease LDL-cholesterol in serum, suggesting a mechanism mediated by the fatty acid moieties. However, only lecithin results in an increase in HDL-cholesterol (185). As noted earlier, HDL-cholesterol might be beneficial as it is necessary for the removal of cholesterol from tissues. Lecithin may also suppress intestinal cholesterol absorption (186).

Disaturated lecithins adsorb to surfaces more readily than do lecithins with unsaturated fatty acids. In humans with a past history of athersclerotic heart disease (myocardial infarction and premature atherosclerosis without hyperlipemia) there is much more disaturated lecithin in plasma than is normally seen in healthy humans. It has been postulated that high levels of disaturated lecithin in plasma is a risk factor for development of heart disease (187).

### The Immune System

Lecithin is often purchased and consumed by the public because they believe that it decreases their susceptibility to disease. There is little evidence to support this claim. In one brief report lecithin was demonstrated to decrease the mortality associated with administering bacteria (*Staphylococcus aureus*) to mice (188). The investigators were able to demonstrate a five-fold reduction in mortality in a very large number of mice, making the results highly significant. They found that the protective effect was present when lecithin was administered intraperitoneally from 1 to 7 days prior to the bacterial challenge. They postulate that the immune response of monocytes and macrophages was augmented by the lecithin treatment.

## Acknowledgments

Dr. Zeisel is a John A. and George L. Hartford fellow of the John A. Hartford Foundation. This work was supported in part by a grant from the National Institutes of Health (1-P01-CA26731-01).

## References

1. Wood, J. L., and R. G. Allison, Effects of Consumption of Choline and Lecithin on Neurologic and Cardiovascular Systems, Life Sciences Research Office of the Federation of American Societies for Experimental Biology Technical Report to the Bureau of Foods, Food and Drug Administration, Washington, DC, 1981, pp. 1–105.
2. Zeisel, S. H., J. H. Growdon, R. J. Wurtman, S. G. Magil and M. Logue, Neurology 30:1226 (1980).
3. Zeisel, S. H., P. J. Garry, M. Brigida, S. G. Magil, J. S. Goodwin and N. Alvarez in Alzheimer's Disease: A Report of Progress in Research, edited by S. Corkin, K. Davis, J. Growdon, E. Usdin and R. Wurtman, Raven Press, New York, 1982, pp. 45–47.
4. Wurtman, J. J., in Nutrition and the Brain, edited by A. Barbeau, J. Growdon and R. J. Wurtman, Raven Press, New York, 1979, pp. 73–81.
5. Zeisel, S. H., Ann. Rev. Nutr. 1:95 (1981).
6. Engel, R. W., J. Nutr. 25:441 (1943).
7. Food and Nutrition Board, National Academy of Sciences USA, Comprehensive GRAS Survey, Usage Levels Reported for NAS Appendix A Substances (Group 1) Used in Regular Foods, Washington, DC, 1973.
8. McIntire, J. M., B. S. Schweigert and C. A. Elvehjem, J. Nutr. 28:219 (1944).
9. Zeisel, S. H., P. M. Michel and J. K. Blusztajn, J. Pharmacol. Exp. Ther., 225:320–324 (1983).
10. Kennedy, E. P., and S. B. Weiss, J. Biol. Chem. 222:193 (1956).
11. Orlando, P., G. Arienti, F. Cerrito, P. Massari and G. Porcellati, Neurochem. Res. 2:191 (1977).
12. Salerno, D. M., and D. A. Beeler, Biochim. Biophys. Acta. 326:325 (1973).
13. Sundler, R., G. Arvidson and B. Akesson, Ibid. 280:599 (1972).
14. Bjornstad, P., and J. Bremer, J. Lipid Res. 7:38 (1966).
15. Bremer, J., P. Figard and D. Greenberg, Biochim. Biophys. Acta 43:477 (1960).
16. Bremer, J., and D. Greenberg, Ibid. 46:205 (1961).
17. Linblad, L., and T. Schersten, Scand. J. Gastroenterol. 11:587 (1976).

18. Blusztajn, J. K., S. H. Zeisel and R. J. Wurtman, Brain Res. 179:319 (1979).
19. Hirata, F., and J. Axelrod, Proc. Natl. Acad. Sci. USA 75:2348 (1978).
20. Hirata, F., O. H. Viveros, E. J. Dilberto and J. Axelrod, Ibid. 75:1718 (1978).
21. Schneider, W. J., and D. E. Vance, J. Biol. Chem. 254:3886 (1979).
22. Hoffman, D. R., and W. E. Cornatzer, Lipids 16:533 (1981).
23. Higgins, J. A., Biochim. Biophys. Acta 640:1 (1981).
24. Crews, F. T., F. Hirata and J. Axelrod, Neurochem. Res. 5:983 (1980).
25. Rothman, J. E., and E. P. Kennedy, Proc. Natl. Acad. Sci. USA 74:1821 (1977).
26. Sundler, R., and B. Akesson, J. Biol. Chem. 250:3359 (1975).
27. Fallon, H. J., P. M. Gertman and E. L. Kemp, Biochim. Biophys. Acta 187:94 (1969).
28. Hoffman, D. R., E. O. Uthus and W. E. Cornatzer, Lipids 15:439 (1980).
29. Uthus, E. O., D. H. Skurdal and W. E. Cornatzer, Ibid. 11:641 (1976).
30. Hirata, F., W. J. Strittmatter and J. Axelrod, Proc. Natl. Acad. Sci. USA 76:368 (1979).
31. Strittmatter, W. J., F. Hirata, J. Axelrod, P. Mallorga, J. F. Tallman and R. C. Henneberry, Nature 282:857 (1979).
32. Leprohan, C., personal communication, 1982.
33. Hirata, F., and J. Axelrod, Nature 275:219 (1978).
34. Blusztajn, J. K., and R. J. Wurtman, Ibid. 290:417 (1981).
35. Allan, F. N., D. J. Bowie, J. J. R. Macleod and W. L. Robinson, Br. J. Exp. Pathol. 5:75 (1924).
36. Best, C. H., H. J. Channon and J. H. Ridout, J. Physiol. 78:409 (1934).
37. Best, C. H., and M. E. Huntsman, Ibid. 75:405 (1932).
38. Hershey, J. M., Am. J. Physiol. 93:657 (1930).
39. Ketola, H. G., and M. C. Nesheim, J. Nutr. 104:1484 (1974).
40. Deeb, S. S., and P. A. Thornton, Poultry Sci. 38:1198 (1959).
41. Fairbanks, B. W., and J. L. Krider, North Am. Vet. 26:18 (1945).
42. Lombardi, B., Fed. Proc. 30:139 (1971).
43. Dutra, F. R., and J. M. McKibben, J. Lab. Clin. Med. 30:301 (1945).
44. Fouts, P. F., J. Nutr. 24:217 (1943).
45. Handler, P., and F. Bernheim, Proc. Soc. Exp. Med. 72:569 (1949).
46. Aoyama, H. J., H. Yasui and K. Ashida, J. Nutr. 101:730 (1971).
47. Atsushi, I., E. E. Hellerstein and D. M. Hegsted, Ibid. 79:488 (1963).
48. Daft, F. S., W. H. Sebrell and R. D. Lillie, Proc. Soc. Exp. Biol. Med. 48:288 (1941).
49. Michael, U. F., S. L. Cookson, R. Chavez and V. Pardo, Ibid. 150:672 (1975).

50. Baxter, J. H., J. Nutr. 34:333 (1947).
51. Best, C. H., and W. S. Hartroft, Fed. Proc. 8:610 (1949).
52. Griffith, W. H., and N. J. Wade, J. Biol. Chem. 131:567 (1939).
53. Patterson, J. M., and E. W. McHenry, Ibid. 156:265 (1944).
54. Chang, C. H., and L. S. Jensen, Poultry Sci. 54:1718 (1975).
55. Ensminger, M. E., R. W. Colby and T. J. Canha, Wash. Agric. Exp. Stn. Circ. 134:1 (1951).
56. Jukes, T. H., J. Biol. Chem. 134:789 (1940).
57. Caniggia, A., Haematologica 34:625 (1950).
58. Kratzing, C. C., and J. J. Perry, J. Nutr. 101:1657 (1971).
59. Bartus, R. T., R. L. Dean, A. J. Goas and A. S. Lippas, Science 209:301 (1980).
60. Burt, M. E., I. Hanin and M. F. Brennan, Lancet 2:638 (1980).
61. Sidransky, H., V. B. Mitbander and S. Clark, J. Nutr. 80:117 (1963).
62. Akesson, B., C. Fehling and M. Jagerstad, Br. J. Nutr. 40:521 (1978).
63. Mudd, S. H., and J. R. Poole, Metabolism 24:721 (1975).
64. Jukes, T. H., Fed. Proc. 30:155 (1971).
65. Finkelstein, J. D., W. E. Kyle and B. J. Harris, Arch. Biochem. Biophys. 165:774 (1974).
66. Finkelstein, J. D., B. J. Harris and W. E. Kyle, Ibid. 153:320 (1972).
67. DeHaas, G. H., N. M. Postema, W. Nieuwenhuizen and L. L. M. Van Deenen, Biochim. Biophys. Acta 159:103 (1968).
68. DeHaas, G. H., N. M. Postema, W. Nieuwenhuizen and L. L. M. Van Deenen, Ibid. 159:118 (1968).
69. Fox, J. M., H. Betzing and D. Lekim, in Nutrition and the Brain, edited by A. Barbeau, J. Growdon and R. Wurtman, Raven Press, New York, 1979, pp. 95–108.
70. Houtsmuller, U.M.T., in Nutrition and the Brain, edited by A. Barbeau, J. Growdon and R. Wurtman, Raven Press, New York, 1979, pp. 83–93.
71. Lekim, D., and H. Betzing, Hoppe-Seyler's Z. Physiol. Chem. 357:1321 (1976).
72. Parthasarathy, S., P. V. Subbaiah and J. Ganguly, Biochem. J. 140:503 (1974).
73. Subbaiah, P. V., and J. Ganguly, Ibid. 118:233 (1970).
74. Subbaiah, P. V., and J. Ganguly, Ind. J. Biochem. Biophys. 8:197 (1971).
75. Epstein, B., and B. Shapiro, Biochem. J. 71:615 (1959).
76. Scow, R. O., Y. Stein and O. Stein, J. Biol. Chem. 242:4919 (1967).
77. Bloom, B., J. Y. Kiyasu, E. D. Reinhardt and I. L. Chaikoff, Am. J. Physiol. 177:84 (1954).

78. Budowski, P., I. Kafri and D. Sklan, Poultry Sci. 56:754 (1977).
79. Flower, R. J., R. J. Pollitt, P. A. Sanford and D. H. Smyth, J. Physiol. 226:473 (1972).
80. Kuczler, F. J., D. L. Nahrwold and R. C. Rose, Biochim. Biophys. Acta 465:131 (1977).
81. Sanford, P. A., and D. H. Smyth, J. Physiol. 215:769 (1971).
82. Asatoor, A., and M. Simenhoff, Biochim. Biophys. Acta. 111:384 (1965).
83. De La Huerga, J., and H. Popper, J. Clin. Invest. 31:598 (1952).
84. De La Huerga, J., and H. Popper, Ibid. 30:463 (1951).
85. De La Huerga, J., H. Popper and F. Stiegmann, J. Lab. Clin. Med. 38:904 (1951).
86. Prentiss, P. G., H. Rosen, N. Brown, R. E. Horowitz, O. J. Malm and S. M. Levenson, Arch. Biochem. Biophys. 94:424 (1961).
87. McFarlane, M. G., L. M. B. Petterson and R. Robison, Biochem. J. 28:720 (1934).
88. Tall, A. R., and D. M. Small, N. Engl. J. Med. 299:1232 (1978).
89. Havel, R. J., Ann. N.Y. Acad. Sci. 348:16 (1980).
90. Carew, T. E., T. Koschinsky, S. B. Hayes and D. Steinberg, Lancet 1:1315 (1976).
91. Aron, L., S. Jones and C. J. Fielding, J. Biol. Chem. 253:7220 (1978).
92. Goracci, G., G. Porcellati and H. Woelk, Adv. Prostaglandin Thromboxane Res. 3:55 (1978).
93. Hostetler, K. Y., and L. B. Hall, Biochem. Biophys. Res. Commun. 96:388 (1980).
94. Webster, G. R., E. A. Marples and R. H. S. Thompson, Biochem. J. 65:374 (1957).
95. Baldwin, J. J., and W. E. Cornatzer, Biochim. Biophys. Acta 176:193 (1969).
96. Taki, T., and J. N. Kanfer, Methods Enzymol 71:746 (1981).
97. Wykle, R. L., and J. M. Schremmer, J. Biol. Chem. 249:1742 (1974).
98. Haubrich, D. R., and N. H. Gerber, Biochem. Pharmacol. 30:2993 (1981).
99. Bernheim, F., and M. L. C. Bernheim, Am. J. Physiol. 104:438 (1933).
100. Bernheim, F., and M. L. C. Bernheim, Ibid. 121:55 (1938).
101. Hatefi, Y., and D. Stiggall, in The Enzymes, edited by P. Boyer, Academic Press, New York, 1976, pp. 261–263.
102. Mann, P. J., and J. H. Quastel, Biochem. J. 31:869 (1937).
103. Wilken, D. R., Anal. Biochem. 36:323 (1979).
104. Zeisel, S. H., D. L. Story, R. J. Wurtman and H. Brunengraber, Proc. Natl. Acad. Sci. USA 77:4417 (1980).
105. Weinhold, P. A., and R. Sanders, Life Sci. 13:621 (1973).

106. Brophy, P., P. Choy, J. Toone and D. Vance, Eur. J. Biochem. 78:491 (1977).
107. Haubrich, D. R., J. Neurochem. 21:315 (1973).
108. McCaman, R. E., J. Biol. Chem. 237:672 (1962).
109. Farrell, P. M., M. F. Epstein, A. R. Fleischman, G. K. Oakes and R. A. Chez, Biol. Neonate 29:238 (1976).
110. Cohen, E. L., and R. J. Wurtman, Life Sci. 17:1095 (1975).
111. Haubrich, D. R., P. F. L. Wang, D. E. Clody and P. W. Wedeking, Ibid. 17:975 (1975).
112. Haubrich, D. R., P. F. L. Wang and P. W. Wedeking, J. Pharmacol. Exp. Therap. 193:246 (1975).
113. Fonnum, F., Brain Res. 62:497 (1973).
114. Wajda, I. J., I. Manigault, J. P. Hudick and A. Lajtha, J. Neurochem. 21:1385 (1973).
115. Rama Sastry, B. V., J. Olubadewo, R. D. Harbison and D. E. Schmidt, Biochem. Pharmacol. 25:425 (1976).
116. White, H. L., and J. C. Wu, J. Neurochem. 20:297 (1973).
117. Pardridge, W. M., E. M. Cornford, L. D. Braun and W. M. Oldendorf, in Nutrition and the Brain, edited by A. Barbeau, J. Growdon and R. Wurtman, Raven Press, New York, 1979, pp. 25–34.
118. Millington, W. R., A. L. McCall and R. J. Wurtman, in Nutrition and the Brain, edited by A. Barbeau, J. Growdon and R. Wurtman, Raven Press, New York, 1979, pp. 417–424.
119. Illingworth, D. R., and O. W. Portman, Biochem. J. 130:557 (1972).
120. Magil, S. G., S. H. Zeisel and R. J. Wurtman, J. Nutr. 111:166 (1981).
121. Iggo, A., and M. Vogt, J. Physiol. 150:114 (1969).
122. Mueller, R. A., H. Thoenen and J. Axelrod, Mol. Pharmacol. 5:463 (1969).
123. Thoenen, H., R. A. Mueller and J. Axelrod, Nature 221:1264 (1969).
124. Ulus, I. H., M. J. Hirsch and R. J. Wurtman, Proc. Natl. Acad. Sci. USA 74:798 (1977).
125. Ulus, I. H., and R. J. Wurtman, Science 194:1060 (1976).
126. Bierkamper, G. G., and A. M. Goldberg, Brain Res. 202:234 (1980).
127. Loffelholz, K., R. Lindmar and W. Weide, in Nutrition and the Brain, edited by A. Barbeau, J. Growdon and R. Wurtman, Raven Press, New York, 1979, pp. 233–241.
128. Werter, M. H., U. Misgeld, I. J. Bak and D. J. Jenden, Abstr. Soc. Neurosci. 5:81 (1979).
129. Porcellati, G., G. Arienti, M. Pirotta and D. Giorgini, J. Neurochem. 18:1395 (1971).
130. Jenden, D. J., in Brain acetylcholine and neuropsychiatric disease,

edited by K. Davis and P. Berger, Plenum Press, New York, 1979, pp. 483–513.
131. Tamminga, C. A., R. C. Smith, S. Chang, J. S. Haraszti and J. M. Davis, Lancet 2:905 (1976).
132. Hirsch, M. J., J. H. Growdon and R. J. Wurtman, Metabolism 27:953 (1978).
133. Sturdevant, R. A., M. L. Pearce and S. Dayton, N. Engl. J. Med. 288:24 (1973).
134. Toouli, J., P. Jablonski and J. M. Watts, Lancet 2:1124 (1975).
135. Lijinsky, W., and S. S. Epstein, Nature 225:21 (1970).
136. Kunisaki, N., and M. Hayashi, B. Appl. Environ. Microbiol. 37:279 (1979).
137. Ralt, D., and S. R. Tannenbaum, Am. Cancer Soc. Symposium on N-nitrosocompounds, Atlanta, GA (in press).
138. Beare-Rogers, J. L., in Modification of Lipid Metabolism, edited by E. G. Perkins, and L. Witting, Academic Press, New York, 1975, pp. 43–57.
139. Davis, K. L., P. A. Berger and L. E. Hollister, N. Engl. J. Med. 293:152 (1975).
140. Growdon, J. H., M. J. Hirsch, R. J. Wurtman and W. Weiner, Ibid. 297:524 (1977).
141. Growdon, J. H., A. J. Gelenberg, J. Doller, M. J. Hirsch and R. J. Wurtman, Ibid. 298:1029 (1978).
142. Gelenberg, A. J., J. C. Doller-Wojick and J. H. Growdon, Am. J. Psychiatry 136:772 (1979).
143. Tamminga, C. A., R. C. Smith, S. E. Ericksen, S. Chang and J. M. Davis, Ibid, 134:769 (1977).
144. Jackson, I.V., E. A. Nuttall, I. O. Ibe and J. Perez-Cruet, Ibid. 136:1458 (1979).
145. Ayd, F. J., Med. Sci. 18:32 (1967).
146. Gerlach, J., N. Reisby and A. Randrup, Psychopharmacologia 34:21 (1974).
147. Davis, K. L., and P. A. Berger, Biol. Psychol. 13:23 (1978).
148. Drachman, D. A., Neurology 27:783 (1977).
149. Sitaram, N., H. Weingartner, E. D. Caine and J. C. Gillin, Life Sci. 22:1555 (1978).
150. Sitaram, N., H. Weingartner and J. C. Gillin, Science 201:274 (1978).
151. Davies, P., and A. J. F. Maloney, Lancet 2:1903 (1976).
152. Perry, E. K., R. H. Perry, G. Blesses and B. E. Tomlinson, Ibid. 1:189 (1977).

153. Boyd, W. D., J. Graham-White, G. Blackwood, I. Glen and J. McQueen, Ibid. 2:711 (1977).
154. Fovall, P., M. W. Dysken, L. W. Lazarus, J. M. Davis, R. L. Kahn, R. Jope, S. Finkel and P. Rattan, Commun. Psychopharmacol. 4:141 (1980).
155. Hu, R., G. Vroulis and R. C. Smith, Abstr. Gerontology Soc. Am., 33rd scientific meeting, 1980, p. 128.
156. Janowsky, D. S., M. K. El-Yousef, J. M. Davis and H. J. Sekerke, Lancet 2:632 (1972).
157. Janowsky, D. S., M. K. El-Yousef, J. M. Davis and H. J. Sekerke, Arch. Gen. Psychiatry 28:542 (1973).
158. Cohen, B. M., A. L. Miller, J. F. Lipinski and H. G. Pope, Am. J. Psychiatry 137:242 (1980).
159. Golden, G. S., Ann. Neurol. 4:145 (1978).
160. Hanin, I., J. R. Merikangas, K. R. Merikangas and U. Kopp, Engl. J. Med. 301:661 (1979).
161. Barbeau, A., Ibid. 302:1310 (1980).
162. Polinsky, R. J., M. H. Ebert, E. D. Caine, C. Ludlow and C. J. Bassich, Ibid. 302:1310 (1980).
163. Stahl, S. M., and P. A. Berger, Ibid. 302:1311 (1980).
164. Barbeau, A., R. F. Butterworth, T. Ngo, G. Breton, S. Melancon, D. Shaplott, G. Geoffry and B. Leemieux, Can. J. Neurol. Sci. 3:379 (1976).
165. Barbeau, A., N. Engl. J. Med. 299:200 (1978).
166. Legg, N. J., Br. Med. J. 2:1403 (1978).
167. Lawrence, G. M., P. Millac, G. S. Stout and J. W. Ward, J. Neurol. Neurosurg. Psychiatry 43:452 (1980).
168. Kranz, H., D. J. Caddy, A. M. Williams and W. Gay, Ibid. 43:483 (1980).
169. Anton, V., Rev. Esp. Fisiol. 10:179–188, 1954.
170. Kapp, J., M. S. Mahaley and G. L. Odom, J. Neurosurg. 32:468 (1970).
171. Steigmann, F., R. Firestein and J. De La Huerga, Fed. Proc. 11:393 (1952).
172. Singh, G. S., Ind. J. Physiol. Pharmacol. 17:125 (1973).
173. Greten, H., H. Raetzer, A. Stiehl and G. Schettler, Athersclerosis 36:81 (1980).
174. Grundy, S. M., J. Lipid Res. 21:525 (1980).
175. Cobb, M., P. Turkki, W. Linscheer and K. Raheja, Nutr. Metab. 24:228 (1980).
176. Rosseneu, M., B. Declercq, D. Vandamme, R. Vercaemst, F. Soetewey, H. Peeters and V. Blaton, Atherosclerosis 32:141 (1979).
177. Kesten, H. D., and R. Silbowitz, Proc. Soc. Exp. Biol. Med. 49:71 (1942).

178. Adlersberg, D., and H. Sobotka, J. Mt. Sinai Hosp. NY 9:955 (1943).
179. Morrison, L. M., Geriatrics 13:12 (1958).
180. Steiner, A., and B. Domanski, Proc. Soc. Exp. Biol. Med. 55:236 (1944).
181. Davis, O., A. Levine, M. Bengal, N. Sloan and C. Beck, Ill. Med. J. 127:567 (1965).
182. Ter Welle, H. F., C. M. Van Gent, W. Dekker and A. F. Willebrands, Acta Med. Scand. 195:267 (1974).
183. Davies, L. G., and L. Murdoch, Br. Med. J. 2:619 (1959).
184. Svanberg, U., A. Gustafson and R. Ohlson, Nutr. Metab. 17:338 (1974).
185. Childs, M. T., J. A. Bowlin, J. T. Ogilvie, W. R. Hazzard and J. J. Albers, Atherosclerosis 38:217 (1981).
186. Rampone, A. J., and C. M. Machida, J. Lipid Res. 22:744 (1981).
187. Gershfeld, N. H., Science 204:506 (1979).
188. Sowar, M. C., Y. Tsuchiya and L. G. Nutini, IRCS Med. Sci. 9:459 (1981).
189. Zeisel, S. H., D. Reinstein, S. Corkin, J. H. Growdon and R. J. Wurtman, Nature 293:187 (1981).

# Chapter Fifteen

## Role of Phospholipids in Flavor Problems

**David J. Sessa**
Northern Regional Research Center
Agricultural Research Service
US Department of Agriculture
Peoria, IL 61604

### Role of Phospholipids in Flavor and Taste—General Introduction

Flavors of food systems may originate from the basic food ingredients—lipids, proteins and carbohydrates—and also from the minor vitamin and mineral constituents. Each food item, whether of plant or animal origin, has its own characteristic color and flavor as a result of the metabolic processes that occur in the live system and the deterioration and interactions of the constituents that occur during processing and storage. This chapter gives a brief overview of the contribution of phospholipids to flavor and taste in cereals, legumes and muscle meats; how the flavors are generated from these phospholipids; and methods for inhibiting their oxidation or methods for effecting removal of phospholipids and their oxidation products from foodstuffs. Phospholipids, minor constituents that are ubiquitous in all plant and muscle tissues, are distributed in membranes throughout the aqueous media of cells usually as very thin layers with relatively large surface area in close contact with either prooxidant or hydrolytic catalysts in or near the cellular membrane structures. These phospholipids are labile to lipid peroxidation because of their high content of polyunsaturated fatty acids (PUFA) and their propinquity to both oxidative and hydrolytic enzymes. Therefore, through oxidation, the phospholipids have the potential to participate greatly in the generation of off-colors, off-flavors and off-tastes.

The role of phospholipids in flavor problems is threefold: (a) phospholipids containing the ester-bound, oxidatively deteriorated fatty acids can possess off-flavors themselves; (b) phospholipids provide a source of PUFA for subsequent oxidative deterioration to produce off-flavors; (c) the amine base portion of the phospholipid can interact with the reactive carbonyl compounds resulting from PUFA oxidation to yield carbonyl-amine compounds with off-flavor.

## Flavor Development via Hydrolytic and Oxidative Rancidity

Unoxidized phospholipids have very little flavor; unoxidized soybean phosphatidylcholine (PC), for example, tastes only slightly bitter (1), with the weak bitter taste elicited by the choline base (2). Commercial soybean lecithin, a composite of 65–75% phospholipids and other constituents such as carbohydrates, pigments, sterols and triglycerides, tastes nutty, vegetable-shortening, paraffin-like (3). However, phospholipids containing oxygenated fatty acids are a source of rancid off-flavor and bitter taste (1,3,4).

Naturally occurring enzymes, termed phospholipases $A_1$ and $A_2$, directly hydrolyze the fatty acid ester bonds of phospholipids, producing free fatty acids. The scheme for various enzyme actions on a glycerophosphatide is illustrated in Figure 1. Flavors developed by this action depend on the composition of the fat, with short-chain fatty acids to $C_{10}$ having particularly disagreeable odors and flavors, and those above $C_{10}$ possessing waxy or, at alkaline pH, soapy flavors. Unless the food system contains short-chain fatty acids such as butyric, caproic or capric, the hydrolytic activity alone is of minor importance. Acids from acetic through $\eta$-caproic were identified in raw soybean meal (5). However, with the amounts present, these authors did not assign any practical effect on soybean flavor. The longer chain acids do not contribute substantially to off-flavors unless they are PUFA oxidized beyond the hydroperoxide stage. Secondary products arising from further oxidation of the hydroperoxides and the subsequent degradation of these secondary compounds can generate flavors and odors associated with rancidity. Strong soybean-like flavors can be generated from linoleic and linolenic acids via soybean lipoxygenase (6). The volatile compounds produced contribute to the rancid, grassy/beany, cereal/grain, musty/stale flavors, whereas the nonvolatile components cause bitter taste and astringency. All these flavor descriptions and tastes are associated in some way with the original soybean, solvent extracts of the bean, or its processed products. Oxidation of PUFA from phospholipids by virtue of the hydrolytic activity of the phospholipases can indeed result in the generation of some of the off-flavors in soybeans.

Phospholipases C or D (see scheme, Fig. 1) can hydrolyze the phosphate linkage, resulting either in a diglyceride with release of a phosphate base or in a phosphatidic acid with release of a free base. The released bases have no significant taste. Actions of these phospholipases do not contribute directly to off-flavors. However, these enzymes modify the physical characteristics of the phospholipid, thereby affecting solubility, surface activity, viscosity and emulsification properties. The impact of the changes in physical character

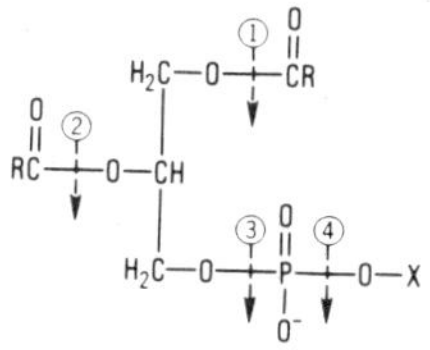

Fig. 15-1. The principal sites of hydrolytic attack on phospholipids by phospholipases: (1) $A_1$; (2) $A_2$; (3) C; (4) D.

on flavor or taste perception has not been clearly established. With regard to influence of surface activity on taste, Baur et al. (7) demonstrated that trihydroxyoctadecenoic acids dispersed with a surface-active sugar phosphate ester taste bitter.

## Interaction of Flavor Compounds with Amino Groups

Carbonyl-amine reactions in a simulated dry food system involving phosphatidylethanolamine (PE) and carbonyl compounds, such as 2-hexene-1-al, 2,4-hexadienal, malonaldehyde and nonanal, all representing oxidation products of PUFA, proceed readily at low moisture levels (2.5%) and either ambient or elevated temperatures (8). The resultant Schiff bases, polymers and scission products possessed off-odors described as burnt, beany, fishy; painty and rancid flavors; and bitter taste.

Nonvolatile oxygenated products derived from lipid hydroperoxide decomposition of PUFA contain hydroxy, epoxy and carbonyl functional groups that can react with active sites in proteins such as the $\epsilon$-amino group of lysine and thiol group of cysteine (9) to form lipoprotein complexes. Not all lipid decomposition products form stable lipoprotein complexes (10). For example, soybean proteins appear to have a much stronger affinity for binding aldehydes and ketones compared to alcohols (11). In fact, aldehydes, especially when unsaturated, react strongly with the protein (12–14), whereas ketones bind weakly and alcohols ($C_6$ and above) do not bind at all. According to Arai et al. (15), the amount of bound ligand increases with denaturation of the protein.

Beyeler and Solms (11) showed that soybean protein has a distinct affinity for certain flavor substances. Anderson and Warner (16) found that specific soybean proteins, termed acid-sensitive fraction, bind the grassy-beany, bitter and astringent soybean flavor components. Removal of this fraction from soybean protein preparations caused reduction in intensity of these flavors in soybean protein isolates. Yamauchi et al. (17) characterized the lipid components bound to the acid-sensitive fraction.

The ester-bound oxidized PUFA, liberated PUFA subsequently oxidized, carbonyl-amine reaction products and nonvolatile oxygenated PUFA that interact with proteins can all contribute to off-odor, off-flavor and off-taste in food systems.

## Assessment of Flavor in a Food Matrix

Individual flavor components extracted from a food matrix and tasted as dilute solutions in water or fat can differ greatly in intensity of flavor and also flavor description from those tasted in the original matrix. Lea and Swoboda (18) found that $C_8$-$C_{12}$ saturated aldehydes and $C_9$ 2-enals were all detectable in water at dilutions as high as one part in $10^8$ or $10^9$, the lowest concentration being reached with dodecanal at 0.9 part in $10^9$. However, when tasted in fat, in which the strongly flavored, less volatile, higher members of the aldehyde series are much more soluble than in water, the thresholds for these substances were ca. 100 times higher, presumably owing to reduced concentration achieved in the aqueous phase that is available at the olfactory area. Consequently, the proportion of lipid in plant or animal tissue need not be high in a more aqueous system before autoxidation or enzymic oxidation troubles become serious; oxidation of susceptible foods with low lipid content can be just as troublesome as in fatty foods. Meats with high fat content, for example, can actually reduce the intensity of rancid odors by trapping volatile decomposition products.

Individual flavor compounds can taste entirely different when evaluated as a composite in a food matrix. Linalool at 5 ppm in a beverage base tastes lemony, woody; 10 ppm *cis*-3-hexenal in beverage base has a green, sharp flavor. Yet the combination of the two at those concentrations gives the beverage base a blueberry flavor (19). Raw and cooked foods with acceptable flavors contain carbonyl compounds of the same general types as those produced in rancid-tasting, oxidatively deteriorated systems. The green-beany character of raw defatted soybean flakes is lost on extraction of the crude residual lipids with an azeotropic mixture of hexane/ethanol (20). The residual meal possessed very little taste and the crude lipid extract, rather than tasting green-beany, tasted oily or hydrocarbony with lingering aftertaste and possessed a characteristic described by our panel as throat-catching. The green-beany character apparently was not at a level high enough to give this characteristic to the crude lipid, mainly because the flavor constituents, even though they are concentrated, were now tasted in a predominantly phospholipid matrix (21) rather than in the more aqueous protein-carbohydrate matrix from which these constituents were extracted. Pokorný (22) reported the taste of methyl linolenate as varnish-like, but when reacted with a bland-tasting protein at 60 C the byproduct tasted hay-like. The relationship of individual flavor components to the entire food matrix must be clearly established before progress can be made for eliminating or masking off-flavors.

Physical state of the food may also affect flavor. In aqueous systems undergoing lipid oxidation, cardboard flavors predominate, whereas in dry systems, such as dry whole milk, oily, tallowy flavors prevail (23). Emulsions, e.g., butter, can possess an array of off-flavors described as cardboard, oily, tallowy, fishy, painty, metallic and mushroom-like.

Lipid composition of the food can also influence the susceptibility of phospholipids to oxidize, thereby generating off-flavors. In aqueous systems that contain both phospholipids and triglycerides, such as milk, the phospholipids are preferentially oxidized (23). However, in the absence of water, of the two lipids, the triglycerides are more susceptible to oxidation. With butter emulsions, which represent an aqueous concentrate of phospholipid dispersed in fat, the phospholipid is more readily oxidized.

## Warmed-Over Flavor (WOF) Development in Meat Systems

Rancidity in uncured cooked meats, termed WOF by Tims and Watts (24), occurs rapidly during short-term refrigerated storage. Phospholipids have been implicated as being major contributors to WOF in beef, chicken, mullet, mutton, pork and turkey (25–28). Less than 10% of the triglycerides but ca. 50% of the phospholipids in pork and beef intramuscular lipids are PUFA (29). These authors found that triglycerides heated in air produced aromas associated with pork and beef; the PE fraction produced strong fishy odors; the PC-sphingomyelin fraction was somewhat fishy with an odor of broiled liver predominating. They concluded that phospholipids do not contribute to desirable meat flavor; in lean meat, phospholipids contribute to poor flavor.

Igene and Pearson (30) investigated the role of: total lipids, phospholipids, both total and individual species, and triglycerides and their oxidation to WOF development in cooked beef and chicken systems. Their model meat system consisted of lipid-extracted muscle fiber as the matrix reconstituted with varying levels of the above lipids, then cooked and refrigerated at 4 C for 48 hr. These authors demonstrated that total phospholipids, especially PE, were the major contributors to development of WOF in cooked meat;

triglycerides enhanced development of WOF only when combined with total lipids, including the phospholipids; PC did not influence WOF. Quantitative changes in the $C_{18:2}$, $C_{20:4}$ and $C_{22:4}$ PUFA from the phospholipids were shown to be related to development of WOF in the cooked meat model systems.

## Bitter Taste of Lipids in Cereals, Grains and Legumes

Cereal products and legumes such as oats, millet, corn, buckwheat, wheat, rye and soybeans become bitter on prolonged storage. With wheat and rye flours, bitterness is of minor importance because it occurs only after long storage and is reduced on baking so that it is not noticed in the freshly baked product (31). With whole oats and oat products, however, the problem of bitter taste development is much greater. Under unfavorable storage conditions of increased temperature and high humidity, bitter taste develops within a short period and makes oat products inedible. Rothe (32) experimentally produced bitter taste in oats within a few hours by subjecting oat products to 95–100% relative humidity at 38 C. He attributed bitter taste to enzymic oxidation of PUFA. According to Popov and Cheleev (33), the PUFA resulting from hydrolytic action of lipases on triglycerides and phospholipids undergo oxidation, with subsequent generation of bitter taste.

Weiss and Diemair (34) established that the bitterness in lecithin is due to oxidation of ester-bound fatty acids, and their work suggested that the oxidation of lipids could also contribute to bitterness in oats (32). According to their findings (34), bound PUFA can indeed be the precursors of bitter substances; it is their work that was the basis of our research on the identification of a bitter principle in soybean meal (4).

Lipase action on phospholipids and triglycerides is not a necessity for developing bitterness in cereal products and legumes, even though oxidizing enzymes may act more readily on the free PUFA. Essential to the development of bitter taste is oxidation of the free or bound PUFA beyond the hydroperoxide stage, with subsequent generation of multioxygenated fatty acids such as the trihydroxyenoic acids (7, 35–37).

## Role of Phospholipids in Bitter Taste of Soybeans

### Isolation of Bitter Principles from Soybeans

Of the soybean flours, concentrates and isolates taste-evaluated by panel members, bitter taste and beany flavor predominated. Alcohol removed much of the flavor and lipid material from dehulled defatted soybean flakes (38). Teeter et al. (39) fractionated alcohol extracts of defatted soybean flakes in attempts to isolate and characterize bitter and beany fractions. An acetone-soluble bitter concentrate with a nitrogen:phosphorus ratio indicative of phospholipid was obtained from the alcohol concentrate by repeated fractionation with acetone. To minimize the carbohydrates, phenolic pigments and saponins extracted with the crude lipids via ethanol, Honig et al. (40) used a hexane/ethanol azeotropic mixture. A component, ethyl-$\alpha$-D-galactopyranoside, isolated from the hexane/ethanol azeotropic extract was identified as a bitter component. This compound, found at 0.2% in full-fat flakes and 0.03% in defatted flakes, was formed as an artifact during extraction. Since a 0.5% aqueous suspension was needed to elicit a bitter response, this compound was found at levels below its threshold value and is not likely to contribute to the bitter taste of alcohol-treated soybeans.

### Association of Bitter Taste with Soybean PC

The major components of the hexane/ethanol azeotropic extracts of soybean meals consisted of phospholipids, steryl glycosides, acylated steryl glycosides and nonpolar lipids, mainly tri-, di- and monoglycerides, sterols and some hydrocarbons (21). In an assessment of phospholipids as a contributor to off-flavors in soybean, Sessa et al. (3) found that soybean lipid extracts possessing bitter taste had similar composition to commercial soybean phosphatides. A 5-hr ultraviolet light irradiation of commercial soybean phosphatides generated intensely bitter-biting rancid flavors, the first definitive evidence showing that autoxidation of soybean phosphatides produces a bitter taste. This finding confirmed that of Weiss and Diemair (34) who established that bitter taste arose from oxidation of lupine and rape phospholipids. Sessa et al. (4) proved that autoxidation of purified soybean PC as an aqueous dispersion generates a bitter taste with a threshold calculated to be 60 ppm.

Three oxidized PC designated SPC–A, SPC–B and lyso SPC were isolated from the residual lipids of hexane-defatted flakes (1). A 0.05% aqueous dispersion of either SPC–B or lyso-SPC was strongly bitter, whereas the same concentration of SPC–A was rated weakly bitter. Defatted soybean flakes contain at least 0.08% oxidized PC. Based on the bitter threshold of oxidized soybean PC, 60 ppm, and the content of 0.08% oxidized soybean PC in defatted soybean flakes, it was concluded that oxidized soybean PC was largely responsible for the bitter taste of raw soybeans and products prepared from them.

Analysis of the fatty acids isolated from intact oxidized soybean PC revealed the presence of several oxygenated forms, including trihydroxyoctadecenoic acid (41). Baur and Grosch (35) reported that both tri- and tetrahydroxyoctadecenoic acids are bitter; the presence of a double bond in the trihydroxy acid enhanced the bitter taste threefold. The bitter threshold for the unsaturated trihydroxyoctadecenoic acid on emulsification with a sugar ester ranged from 200 to 1500 ppm. Yet, the bitter taste of soybean PC containing unsaturated trihydroxyoctadecenoic acid as an acyl residue is 60 ppm (1), which is lower than that of the free acid. The water solubility and surfactant properties of the PC moiety must play an important role and will be discussed later in the section "Bitter Taste as Related to Molecular Structure." Baur et al. (7) demonstrated that trihydroxyoctadecenoic acids were generated enzymically, with crude soybean extracts containing both lipoxygenase and peroxidase acting on linoleic acid. Storage of a soybean flour at 22 C increased the trihydroxy fatty acid content from 0.03 to 0.05% in 3 months (36). The amount found includes both free and bound forms. On flavor evaluation of the 3-month stored soybean flour, the predominant flavor was green-beany; however, the panel also noted a slight increase in bitter taste. Just how much trihydroxy fatty acid was bound to phospholipid was not determined. Conceivably, the amount of trihydroxy fatty acid that Moll et al. (36) found, even if only a small portion was bound to phospholipid, can definitely contribute to the bitter taste of soybean flour.

### Development of Bitter Taste at Last Stages of Maturity

The findings that oxidized PC contributes to the bitter taste of soybeans and that bitter-tasting trihydroxyoctadecenoic acids arise from endogenous enzymes in soybeans are both pertinent to explaining the observations of Rackis et al. (42), who established that bitter taste develops during the latter stages of maturation. They found a significant correlation at the 1% level ($r = 0.73$) between increase in the bitter intensity value and lipoxygenase activity during seed maturation. From their findings, they postulated that bitter taste of soy developed from the formation of nonvolatile oxidized lipids. Kalbrener et al. (6) found that action of soybean lipoxygenase on linoleic acid generates a bitter taste. The bitter-tasting compound is an oxidative decomposition product of linoleic acid hydroperoxide formed via a peroxidase (7). Both enzymes are active in soybeans (42–44) and both have a relatively low activation energy of 12–20 kJ $mol^{-1}$ (45) necessary to catalyze the oxidation. Soybean lipoxygenase isozymes can act on either free or ester-bound PUFA of triglycerides (46) or phospholipids (47–49). At the latter stages of soybean maturation, the membrane phospholipids, mainly PC and PE, are laid down (50). In the period from 60 days after flowering, PC increased from 9.8 to 46.0 mole % and PE from 8.6 to 25.0 mole %. Since PC and PE both possess a high degree of PUFA (51) and lipoxygenases can act on either the free or ester-bound PUFA, the phospholipids can either be a prime source of PUFA for oxidation or oxidation can occur on intact PUFA to produce bitter taste and other off-flavors.

## Bitter Taste of Lipids as Related to Molecular Structure

### Chemical and Physical Basis of Bitter Taste

A chemical basis for the human sense of bitter taste resulted from the research of Dastoli et al. (52). They isolated proteins from the back of pig tongues which form complexes with bitter substances in decreasing order: quinine, brucine, naringin and caffeine. The objective test for degree of bitterness of the compounds consisted of measurement of changes in the refractive index noted on complex formation between the test substance and receptor protein. Adsorption of a chemical stimulus to a given receptor site initiates a series of electrical and chemical events that, in vivo, lead to nerve excitation. The electrical impulses are carried to the brain, which in turn acts as a computer to determine whether a food is salty, sour, sweet or bitter (53). The magnitude of the taste response can be related to the chemical stimulus concentration mathematically, according to Beidler's fundamental equation (54). The strength of the complexes formed between the bitter substances and receptor protein, i.e., the porcine association constants, correlated with either the range or median of human bitter threshold values for the bitter substances (55). The porcine protein that complexes with bitter substances is a mucoprotein with neutral charge and a molecular weight of 166,000 (56), a hexose/hexoseamine content of 33%, and a sialic acid content of 20% (57).

A vast array of bitter substances exists in nature, ranging from simple salts to alkaloids, amino acids, glycosides, lipids, peptides, saponins and terpenoids. Thus, no single class of compounds can account for the bitter taste response. The relationship between molecular structure and taste sensation has intrigued many researchers. Within certain classes of compounds, such as the diterpenes from plants of the Isodor species, a correlation was established between bitterness and steric structure (58). For a compound of this series to taste bitter, the compound must contain a "unit" of bitter taste, where the unit consists of a proton-donor group and a proton-acceptor group with an interorbital distance of 1.5Å. This distance makes it possible to form an intramolecular hydrogen bond, which apparently is a requisite for the bitterness of diterpenes in the Isodor species.

Sugars are usually sweet, but some are bitter (59). Birch and Lee (60) explored the molecular characteristics of sugars and their analogues responsible for eliciting the bitter response. They found that increase in either the lipophilic or polar character of a sugar molecule by chemical modification usually increases the bitter characteristic; marked increases in both these molecular features will produce the most bitter analogues. Chemical modification to increase lipophilicity of the sugar molecule is done by esterification, etherification or glycosidation. Sugar acetates, some methyl ethers, and dideoxy glucopyranosyl structures are bitter. The position of the hydroxyl group relative to the anomeric oxygenation affects polarity pf the molecule. $\alpha$-D-Mannopyranose is sweet, whereas $\beta$-D-mannopyranose is bitter. The anomeric oxygen atom is necessary, since a cyclitol analogue of $\beta$-D-mannopyranose, epi-inositol, is not bitter. The second hydroxyl group is essential in the molecule since 1-deoxy-D-mannopyranose is sweet. The most polar molecule of the three, i.e., $\beta$-D-mannopyranose, epi-inositol and 1-deoxy-D-mannopyranose, is the $\beta$-D-mannopyranose structure.

The findings of Ney (61) to explain the bitterness of sugars support those of Birch and Lee (60). Ney concluded that sugars, normally hydrophilic, become bitter by reactions that increase hydrophobicity. His data also show that lipids, normally hydrophobic, become bitter when enzymically or chemically modified to increase their hydrophilic character. Based on his findings, he set forth a mathematical expression taking a ratio of the number of C atoms of a molecule to the number of hydroxyl groups to predict whether a lipid or carbohydrate was sweet, bitter or nonbitter. For sweet compounds, the ratio $R = 1.00–1.99$; bitter, $R = 2.00–6.99$; and for nonbitter, $R > 7.00$. Based on this ratio, trihydroxyoctadecenoic acid with $R = 6.00$ should taste bitter, whereas mono- and dihydroxystearic acid should not. This prediction was verified by Grosch and his coworkers (7,35,36). Since the degree of unsaturation and also the functional groups (carboxyl, hydroxyl and/or peroxyl) are requisite for bitter taste of autoxidized fatty acids (37), the mathematical expression of Ney (61) for predicting bitter taste may be oversimplified.

With phosphatides, Ney (61) made the assumption that PC is equivalent to 2 hydroxyl groups and lyso-PC is equivalent to 3. Sessa and coworkers established that oxidized soybean PC and lyso PC, both containing oxygenated fatty acids which include trihydroxyoctadecenoic acids, tasted bitter (1,41). Lyso PC with either mono-, di- or trihydroxyoctadecenoic, which according to Ney's mathematical expression gives R = 4.33, 5.20 and 6.50, respectively, should all be bitter. For soybean PC to taste bitter, an oxygenated species with 2 PUFA containing both a di- and a trihydroxyoctadecenoic acid would have to exist to give a ratio of 6.29. Such a species may be likely, since the di- 18:2 species is predominant in soy PC (62). However, according to Usuki and Kaneda (37), a monohydroxyoctadecenoic acid such as ricinoleic acid tastes strongly bitter. Ney's mathematical expression for predicting bitter taste falls short in several instances, including explaining why the $\beta$-anomer of D-mannose tastes bitter, whereas the $\alpha$-anomer tastes sweet (60). Perhaps a more valid expression to predict bitter taste of oxidized lipids should be based on the empirical hydrophilic-lipophile balance (HLB) value for nonionic surfactants (63). HLB, an expression of the relative simultaneous attraction of an emulsifier for water and for oil, is calculated from the formula H/5 where H represents the weight percentage of hydrophilic groupings in the molecule. Other methods for calculating HLB numbers can involve functional group, degree of unsaturation and interfacial tension measurements (64).

Oxidized PC tends to disperse in water better than unoxidized species; highly oxidized species will form optically clear dispersions with estimated HLB range 10–13 on sonication, whereas unoxidized species tend to be milky with estimated HLB range 6–8. The phosphocholine moiety lowers surface activity, promotes oil in water emulsions and lowers viscosity. How these properties relate to requirements for bitter taste is not known. However, it is known that many detergents are bitter, and the HLB for detergents is in the HLB range 13–15 (64). HLB should perhaps be investigated further for developing a means to predict bitter taste of oxidized lipids.

Physical properties of the taste-stimulating molecules such as water solubility, surface activity and emulsifying ability, as well as the physical properties of the solution, gel or dispersion in which the molecules are tasted with regard to density, viscosity and surface tension, must also be critically assessed to predict with some degree of confidence the taste of individual compounds. Relative sweetness of a sugar solution is, indeed, affected by the viscosity of the medium in which the sweet substance is tasted

(65). Response to aqueous sugar samples was much faster than to sugar samples tasted in selected gums. Viscous solutions containing sucrose were ranked more sweet; however, ranking between pairs of viscous and nonviscous solutions was not significantly different. Birch and Ogunmoyela (66) observed enhanced sweetness values of chocolate drinks with lecithin. They attributed this enhancement to lecithin's property of lowering surface tension and viscosity, whereby sucrose molecules or their complexes can diffuse more readily to taste receptors. They concluded from the results of their experiment that lipophilic character in sapid molecules will facilitate their accession to receptor sites located on the surface of taste cell membranes. Also, the increased persistence times of sweet sensation recorded for their chocolate solutions containing surfactant reflected increased localized concentrations of sucrose molecules at or near the receptor.

## Biogenesis of Flavor from Phospholipids

### Phospholipid Autoxidation

Since phospholipids contain phosphorus, a nitrogen-containing component such as choline, ethanolamine or serine, in addition to unsaturated and saturated fatty acids, autoxidation of this system is far more complicated than that of free PUFA or triglycerides. The nitrogenous and phosphorylated bases can affect the autoxidation of the PUFA constituents. Corliss and Dugan (51), who reported on the oxidation of aqueous emulsions of soy and egg PC and PE, found that oxidation is limited by physical factors including dispersibility of the phospholipids in water, effect of pH on autoxidation and fatty acid composition. Data in Figure 2 show that emulsified soy PE absorbed oxygen more rapidly than soy PC at pH 7.0 and 40 C. Oxygen consumption was greatest for PC at pH 7.0 and PE at pH 8.0. The rate of oxidation during the steady state was reported to be a function of the acyl PUFA components, with PE possessing 66.6% PUFA as compared to 56.1% for PC. Not only is the amount of PUFA higher in PE than in PC but also the degree of unsaturation is higher with 10.6% linolenic acid in PE vs 5.7% in PC. Egg PE, which contains the more highly unsaturated arachidonic acid, oxidizes even more readily than soy PE.

Tsai and Smith (67) explored the effects of nitrogenous bases on autoxidation of methyl linoleate in aqueous emulsion. They reported that ethanolamine and *o*-phosphoethanolamine both increased oxygen uptake of methyl linoleate at pH 7.9, whereas at pH 10.2 these amines each significantly

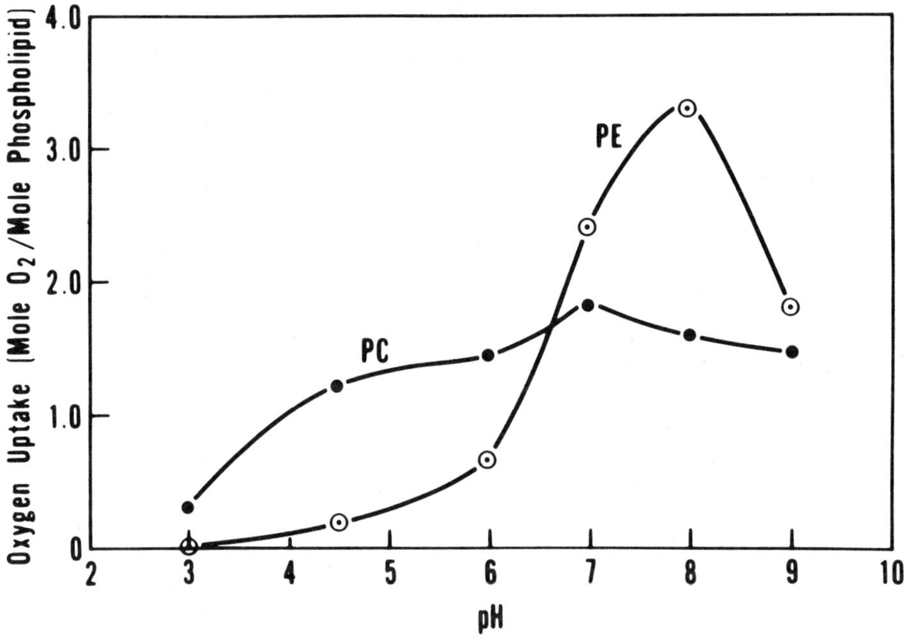


Fig. 15-2. Effect of pH on oxygen consumption of phosphatidylethanolamine (PE) and phosphatidylcholine (PC) (51).

decreased oxygen absorption. This effect of pH can be explained on the basis of pK for ethanolamine, which is 9.5. Ethanolamine ion at pH 7.5 must contribute to the prooxidant properties of ethanolamine and *o*-phosphoethanolamine. At either pH, neither choline or *o*-phosphocholine had much effect on rate of methyl linoleate oxidation, whereas serine and its derivatives decreased oxygen consumption of methyl linoleate. Yukami (68) demonstrated that not only ethanolamine but also 2-mercaptoethanol, histidine, arginine and bovine serum albumin promoted the autoxidation of sodium linoleate solutions. Their catalytic effect was dependent on concentration of these factors.

In view of these findings, PE should be particularly susceptible to oxidation in meat. This observation was substantiated by Keller and Kinsella (69), who found a more than 25% loss of arachidonic acid from PE in cooked ground beef. Love and Pearson (70) demonstrated a sixfold greater oxidation of an aqueous dispersion of PE at pH 7.0 than at pH 5.5. In early studies on WOF of cooked meats, the heme pigments were perceived as the catalysts for lipid oxidation (71–75). Others found that ferrous ion accelerated autoxidation (76–78). Igene et al. (79) solved the dispute by demonstrating that ferrous ion was released from myoglobin during cooking, and this ion catalyzed the autoxidation. Pokorný et al. (80), who separated the oxidation products of egg PE, presented evidence for conjugated unsaturated imines formed from interaction of the primary reaction product, the conjugated unsaturated hydroperoxide, with the amino group of ethanolamine and subsequent transformation to the imine. The amino-carbonyl reaction not only may decrease the oxidative activity of PE in off-flavor formation based on the antioxidant effect of browning compounds, but also can react with aldehydes that give rise to off-flavors.

In food systems, phospholipids are in contact with metal ions, metalloproteins and free amino groups in the presence of oxygen. Metals such as copper, iron and heme proteins can catalyze oxidation of the PUFA constituents of phospholipids. Morita and Fujimaki (81) found that 0.2–5 ppm copper ion definitely promoted oxygen uptake of PE, whereas ethylenediamine tetraacetic acid (EDTA) inhibited oxidation. The ability of EDTA to retard oxidation suggests that trace metals play a significant role in lipid oxidation. By complexing these metal impurities, EDTA can reduce the formation of free radicals and thereby provide a partial inactivation of oxidation. Effect of copper ion on oxygen uptake of soy PC is not as great as with soy PE. PUFA composition of the two phospholipids differs, and effect of copper ion on the more unsaturated PE should show increased oxygen uptake.

Corliss (82) demonstrated that the combination of cupric ion and ethanolamine acts as a prooxidant for soy PC. Although complex-forming (i.e., chelating) agents are frequently added to organic substrates to reduce the effect of metal catalysts, many chelating agents can actually increase rates of autoxidation. Morita and Fujimaki (81) found that compounds possessing a common functional group with that of PE act as prooxidants in phospholipid emulsion containing copper ion (Table I). Yet, iminodiacetic acid depresses the oxygen uptake of PE-copper ion emulsions. The affinity of copper ion to iminodiacetic acid is 30 times greater than the affinity of copper ion to ethanolamine. The most stable complex is that of copper ion and EDTA. The effect of chelating agent as a prooxidant may well depend on the stability of the complex formed with the least stable complex possessing the most activity. Other effects pertinent to prooxidant activity may involve steric effect of the metal in the complex and also a change in the redox potential on chelation.

**TABLE 15-I**
**Addition Effects of PE Constituents on the Copper-Catalyzed $O_2$ Uptake of Soy PE Emulsion (81)**

| Compounds added | Relative $O_2$ uptake (%) |
|---|---|
| Control; 1% PE emulsion ($1.3 \times 10^{-2}$ M) containing 1 ppm copper, pH 7.5, 37 C | 100 |
| + $2.7 \times 10^{-2}$ M monoethanolamine | 122 |
| + $2.7 \times 10^{-2}$ M O-phosphorylethanolamine | 110 |
| + $2.7 \times 10^{-2}$ M glyceryl phosphorylethanolamine | 130 |

## Membrane Lipid Peroxidation

Membrane lipid peroxidation has been studied extensively in the last 10–12 years. Basically, model systems have been developed with PC in neat films, aqueous emulsions or solutions in *tert*-butanol, chlorobenzene or benzene, with exposure to air or oxygen. When exposed to air (62,83) or singlet oxygen (84), the primary oxidation product is PC hydroperoxide. Liposomes of soy and egg yolk PE oxidized in the presence of copper yielded fatty acyl moieties with hydroperoxy and hydroxy groups (85). Lipid peroxidation products from the PC bilayer system of Mead (86) are mainly hydroperoxides when highly unsaturated soybean PC is used, but the products are epoxides, hydroperoxides and hydrolysis product-trihydroxyoctadecenoic acid when PC with increased proportion of saturated fatty acids is used. The autoxidation products from model systems with soybean PC (86) resembled those identified by Sessa et al. (41) from bitter-tasting soybean PC that they isolated from hexane-defatted soybean flakes.

Lipid peroxidation can be induced with chaotropic agents such as $SCN^-$, guanidine-HCl, $ClO_4^-$, $I^-$, $Br^-$, $NO_3^-$, urea (87); or heme compounds such as hematin, microperoxidase, methemoglobin, cytochrome C (88); or iron-sulfur proteins when destabilized in the presence of an iron chelator or chaotropic ion (88); or reduced triphosphopyridine nucleotide oxidase catalyzed systems (89).

Microsomal NADH-cytochrome reductase catalyzes an essential reaction involved in the initiation of lipid peroxidation with tissue homogenates (90). Svingen et al. suggest that the flavoprotein catalyzes reduction of ferric ions complexed to ADP, and the reduced iron nucleotide complex may be the key component responsible for both initiation and propagation of lipid peroxidation in the isolated membrane fragments. The free radical oxidation occurring in the phospholipids will alter the permeability of membranes in the cell by making them more polar. Hence, the protein components of the membrane are more readily accessible to proteolytic enzymes.

The liposomal model of Mead (86) involves a simplified lipid oxidation at a lipid/water interface. The mechanism for deterioration of phospholipids in a food matrix involves not only effects of other constituents such as carbohydrates, proteins and metals, but also other environmental factors such as water activity.

## Effect of Water Activity on Lipid Oxidation in Food Systems

The effect of moisture on the rate of oxygen uptake by lipids in foods and in model systems has been extensively studied by Karel, Labuza and coworkers (91). Labuza et al. (92) reviewed the kinetics of lipid oxidation as a function of water activity and showed that water was protective of dehydrated foods. As illustrated in Figure 3, deteriorations involving lipid oxidation, browning and enzyme activity can occur in intermediate moisture foods (93). Lipid oxidation rate decreased with increased water activity. However, in the capillary region of the isotherm equivalent to water activity in intermediate moisture foods, the oxidation rate increased. The protective effect of water at low water activity and low moisture content can be attributed to hydration of metal catalysts, thereby decreasing their effectiveness and hydrogen bonding of peroxide and slowing the chain reaction (91). According to Salwin (94), the beneficial effects of water on the flavor of stored food products are due to the moisture forming an effective barrier between atmospheric oxygen and lipid constituents of the food, thereby retarding lipid oxidation. Salwin feels that any food product is best stored at a water activity corresponding to the BET (95) monolayer moisture content of the product. The accelerative effect of water at higher water activity and moisture content, on the other hand, was postulated by Labuza (91) to be due to the increased mobility of the metal catalysts by lowering of liquid phase viscosity and through the swelling of the solid matrix with exposure of new catalyst sites. Haydar and Hadziyev (96) observed increased oxidation of pea mitochondrial lipids on mitochondrial swelling in the presence of transitional metal ions and ascorbic acid. They concluded that the coupled oxidation of mitochondrial lipids in the soaking and dehydration steps used in pea processing may well contribute to pea seed off-flavors. The effects of catalyst mobility and swelling can be distinguished based on the findings of Chou et al. (97), who designed

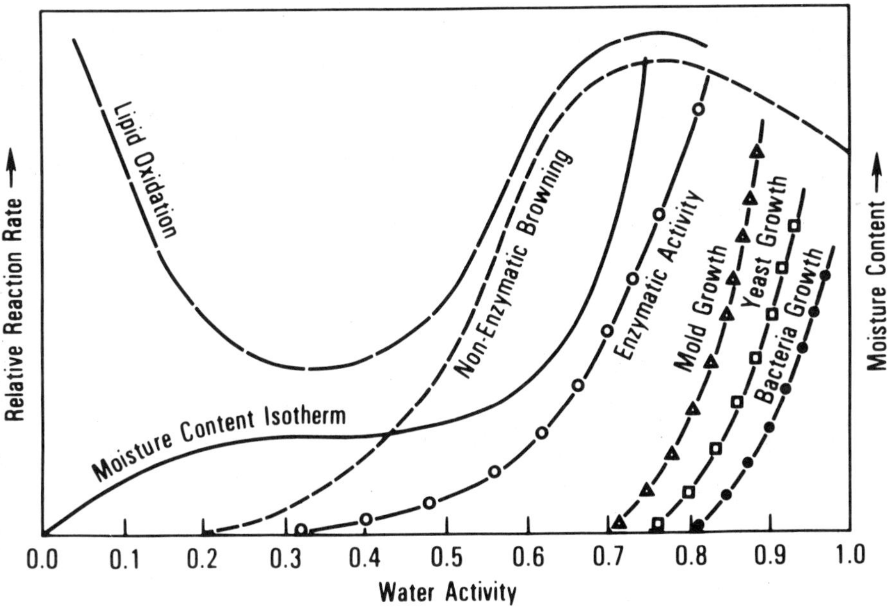


Fig. 15-3. Stability map of foods as a function of water activity (93).

model systems of methyl linoleate on either microcrystalline cellulose or amylopectin.

Blair studied the effects of water availability on enzyme activity using liquid model systems (98). When water was made less available by adding water-binding compounds such as propylene glycol, glycerol or sucrose to the substrate solution in these model systems, several hydrolytic and oxidative enzymes were inhibited. Likewise, Guardia and Haas (99) noted inhibition of hog pancreatic lipase activity in liquid oil emulsions at pH 9.0 on addition of water binders. At pH 7.0, however, addition of water binders significantly accelerated lipolytic activity. A lipase of fungal origin was completely inhibited by propylene glycol at concentrations which yielded maximum activation of pancreatic lipase. Enhancement or inhibition of enzyme activity in the presence of water binders appears to be both enzyme-specific and pH-dependent. Liquid model systems permit the evaluation of the relative importance of other factors, such as water-binding compounds, because these systems are not affected by the problems which are inherent to dry systems, such as lack of homogeneity.

Interaction of enzyme and substrate in dry systems is largely dependent on free capillary water (100). The state of water at these highly localized capillary sites rather than measurement of water activity of the whole tissue determines the enzyme activity. In general, activity of the endogenous enzymes of foods decreases as the free water in the tissue is reduced. However, Brockmann and Acker (101) report that in lipase-free systems, soybean lipoxygenase activity was evident at very low moisture levels. Therefore, presence of water is not a requisite for enzymic lipid oxidation. Both lipase and lipoxygenase activities, with subsequent formation of rancid products, occur in foods with moisture contents within the range in which water is rigidly bound (102). Phospholipases $A_1$ and $A_2$ and lipoxygenase acted on soy phospholipids under these conditions. Dry food systems containing these enzymes generated carbonyl compounds at water levels below the BET monolayer moisture content.

Lück (103) reviewed the influence of water activity on shelf-life, lipid oxidation, nonenzymatic browning reactions, enzyme activities, chemical deteriorations and microbiological stability as related to the properties, composition and manufacture of intermediate moisture foods. With regard to specific food systems, Hansen (104) developed a hydration model with soybean protein. By a combination of nuclear magnetic resonance and sorption isotherm measurements, he could distinguish three types of water. From his data, he speculated on which type of water corresponds to browning, enzyme activity and bacterial growth in foods. Love and Dugan (105) assessed physicochemical changes in instant navy bean powder during accelerated storage. Water activity was important in affecting the stability of this powder. At 5.23% moisture and water activity of 0.23, there was extensive protein insolubilization, lipid oxidation and browning during storage in air at 37 C for a month. Storage below the monolayer stage at 4.0% moisture and water activity of 0.11 resulted in lipid oxidation but less sugar amine browning. At both water activities, PC in air was oxidized to the greatest extent.

Once we understand the role that water activity and content of a food matrix plays on degradation processes, we can use hydration state as a means of food stability control. Water can act as an antioxidant at very low levels by decreasing the catalytic activity of metal catalysts, by promoting recombination of free radicals, and by promoting nonenzymatic browning which results in the production of active antioxidants (106).

## Processing Steps to Eliminate Lipid-Generated Off-Flavors

### Oxidation of Lipids During Processing

Whole soybeans are usually processed to a full-fat grit or flake for easier extraction of oil. Once the cell structure is broken during the cracking or crushing operation, metabolic reactions can run rampant because the cell's control mechanisms are no longer very effective. Not only is there an increased enzyme activity that acts on the ruptured oil globules and exposed membranes, but also the released oil and phospholipid membranes are exposed as surface films on flakes and grits and readily subjected to autoxidation catalyzed by metals, heat and light.

Oxidation during processing of whole soybeans to chips, full-fat flakes, and then to defatted flakes was assessed via thiobarbituric acid (TBA) analysis of the processed products (3,20,42). Data in Table II show that whole soybeans in acid medium possessed a TBA number of 1, and that it increased twofold on freshly cracked chips. When tempered to 11% moisture and flaked, the TBA number increased 11-fold, and finally, on defatting with pentane-hexane TBA, increased 12-fold that of the whole soybean. Based on these results, extensive oxidation of lipids occurred during the processing steps. Deterioration took place more rapidly after flaking because a relatively large surface area was presented to the air for possible oxidation and hydrolysis after the cells were ruptured. When the oxidizing enzymes are not inactivated by acid during homogenization, the TBA number of the whole soybean approximates that of defatted soybean flakes with either acid or no treatment. Apparently, the lipid oxidation potential of the whole bean is similar to that of the defatted flake by virtue of the fact that the cellular spherosome structure composing oil globules is not completely disrupted during the homogenization step. In contrast to intact seeds and defatted flakes, raw full-fat flakes, in which more than 99% of the oil can readily be extracted, had a TBA number of 84. In untreated aqueous medium of full-fat flakes, the oxidizing enzymes released oil from the disrupted cells, and air whipped in during the homogenization provides optimum conditions for lipid oxidation. The high TBA number attained represents the potential of lipids in full-fat flakes to oxidize, thereby generating strong, painty, rancid odors.

Since oxidation of PUFA, either free or bound, is the cause of objectionable flavors in legumes, vegetables and grains, and of the WOF in uncured, refrigerator-stored cooked meats, the obvious step to eliminate flavor generation is either to limit oxidation or effectively remove the oxidation products or their precursors.

**TABLE 15-II**
**Thiobarbituric Acid (TBA) Number Changes During Preparation of Defatted Soybean Flakes (3,20,42)**

| Sample | TBA number[a] | |
|---|---|---|
| | Without acid[b] | With acid[b] |
| Whole soybean | 12 | 1 |
| Chips | — | 2 |
| Full-fat flakes | 84 | 11 |
| Defatted flakes | 14 | 12 |

[a]TBA number expressed as mg malonaldehyde/kg sample.
[b]Treatment before blending for assay.

### Extraction Methods

Taste-panel evaluations of raw full-fat and defatted soy flours show flavor scores of ca. 4 based on a 10-point scale (107). The beany, bitter, astringent flavors of soybeans remain after extracting 97% of the lipids with pentane/hexane. Hydrogen bond-breaking solvents such as aqueous ethanol (108) or an azeotropic mixture of hexane/ethanol (107) can remove residual lipids, mainly phospholipids (21), and flavor constituents from defatted soybean flakes to yield products with flavor scores of 7 and above. Combinations of extraction with toasting procedures yield even blander tasting products with scores approaching 8 (109).

Significant improvement in flavor scores with azeotrope extractions is attributed to the large reduction in flavor intensity value of the green-beany component(s), whereas the intensities of bitter and astringent tastes are reduced to a lesser extent. Apparently solvent and heat treatments extract, distill or modify the beany flavor principles and also inhibit the oxidizing enzymes such as lipoxygenase and peroxidase that can generate off-flavors. The nonvolatile bitter and astringent principles are not completely extracted with hexane/ethanol azeotrope, and perhaps some are generated by autoxidation during processing with solvent and steam.

Aqueous extraction techniques designed for the simultaneous recovery of oil and protein concentrates or isolates were developed and applied to coconuts and peanuts (110); a combined aqueous extraction and membrane isolation technique was applied to soybeans (111). The aqueous process consists of an extraction of the comminuted material, followed by centrifugal separation of the slurry into oil, solid and aqueous phases. The proteins, either as concentrates or isolates, are recovered in either the solid or liquid phase, depending on the selected conditions. The concentrate procedure is a process in which acid-soluble proteins, carbohydrates, oil and other constituents are extracted, leaving the fiber and acid-insoluble substances in the protein concentrate. The isolate procedure separates alkaline-soluble substances from fiber and other insoluble materials prior to the recovery of oil and protein. A major disadvantage of the aqueous process is that only 95% of the oil is recovered and thus residual lipid is left in the protein. The lipids cause storage stability problems. Advantages of this method include low initial capital investment, greater safety since flammable solvents are not used, simplicity of operation and no residual solvent problems.

Another feasible alternative to extraction methods with water or organic solvents is extraction with supercritical fluids such as carbon dioxide. Friedrich and List (112,113) report on the efficient extraction of soybean oil with supercritical $CO_2$ to yield oils equivalent to a degummed, hexane-extracted crude oil. The phospholipids remain with the residual flakes. Storage stability of the extracted flakes may present problems, particularly if the oxidizing and hydrolytic enzymes are still viable. If the residual flakes are to be free of lipid, perhaps supercritical ammonia can be used rather than $CO_2$.

### Inhibition of Lipid Oxidation

In addition to reducing objectionable flavors by solvent extraction, flavor of food products can be improved by inhibiting in situ, lipid oxidizing enzymes. To prevent formation of flavor, inactivation steps may involve food treatment with steam, hot water, dilute acid, alcohol or antioxidants. Mustakas and coworkers (114) reported that extrusion cooking immediately after cracking and dehulling soybeans gave full-fat flour with flavor improvement over flour that was toasted after the tempering and flaking operations. Mustakas et al. (115) showed that a number of different heat treatments of dehulled beans gave rapid deactivation of lipoxygenase and relatively higher flavor scores for the resulting flours, but reduced protein solubility.

Flavor scores of legumes can also be improved significantly by steeping or wet milling whole legumes with aqueous ethanol (116). Over 99% of the lipoxygenase activity originally present in the raw legumes is destroyed by the alcohol treatment. Flavor improvement may be attributed either to enzyme inactivation by the alcohol or to irreversible binding of the objectionable flavors with alcohol-denatured proteins, which binding also can reduce intensity of flavor perception. Alcohol treatments, whether for purposes of extraction or enzyme inactivation, denature the protein and thereby cause undesirable changes in solubility and grainy texture and loss in certain functional characteristics.

Although extractions, weak acid and alcohol treatments prove effective in removing or inhibiting formation of off-flavors in grains and legumes, they are not applied to meat systems because processing conditions are very different. Lipid oxidation in cooked meat systems is usually controlled with antioxidants and chelating agents and also by oxygen exclusion. Some antioxidants and chelating agents used in meats are: propyl gallate, butylated hydroxyanisole, citric acid (117,118), polyphosphates (119,120) and sodium ascorbate (121).

Prolonged heat treatment stabilizes uncured canned meats against oxidative rancidity by producing antioxidative substances (122). Also, active antioxidative substances produced in retorted beef, pork and turkey meats were isolated in the diffusate fraction of aqueous meat extracts (77). The diffusates, possessing a pleasant meaty odor, can be used to prevent WOF in precooked or partially cooked meat products. Avoidance of WOF in meat is of great importance to the food industry because of the increased market for precooked convenience foods.

Alpha tocopherol and isoflavones, both naturally occurring in soybeans, are antioxidants that inhibit enzymic oxidation of linoleate with soy lipoxygenases (123,124). Hot water extracts from many vegetable sources retard the oxidation of lipids in cooked meats. Full-fat soybean flour proved to be an effective antioxidant in cooked pork (125). Soybeans heated to destroy lipoxygenase activity possess appreciable antioxidant activity (126,127), and adding various vegetable protein products such as textured soy flour to meat loaves provided protection against WOF development (77). Defatted soybean, cottonseed and peanut extracts were all highly effective in model systems for inhibiting linoleate peroxidation catalyzed by nonheme iron chelate, metmyoglobin and beef homogenates (128). Of the three concentrates tested, cottonseed concentrate showed a higher water-soluble antioxidant activity than did soy or peanut product in the three catalytic systems. With two protein isolates tested, cottonseed isolate had less antioxidant effect than soybean isolate in the three systems. Their effectiveness in meat systems was not tested.

Browning reaction products can function as inhibitors of lipid oxidation. Products obtained from interaction of sugars such as dextrose or lactose with amino acids glycine, leucine or lysine retarded WOF development in cooked ground beef (77). Absence of rancid odors and TBA-reactive materials indicated that possibly some of these compounds were free radical inhibitors that function by interfering or reacting with free radicals to produce harmless products. Various reducing compounds called reductones, such as amino-hexose reductones, are effective antioxidants in soybean, cottonseed and corn oils (129). Reductic acid retards fat oxidation in frozen, minced red salmon and also in herring (130). Effectiveness of reductic acid as an antioxidant may be attributed to the presence of the ene-diol portion of the molecule similar to that in ascorbic acid. Although many of the browning reaction products are excellent antioxidants, many of them are mutagenic (131) and their presence in food systems should be minimized.

All of the methods to limit flavor generation via lipid oxidation discussed so far involved extraction techniques and methods of inhibition. Proteins such as soy have a strong affinity for aldehydes (12,132). Flavor volatiles once irreversibly bound to proteins should lose their olfactory effect because they are no longer volatile. Because binding of aldehydes increases on protein denaturation (133), flavor of protein is improved with alcohol or heat treatments. Conversion of flavor-causing aldehydes to alcohols via alcohol dehydrogenases could cause diminished flavor because alcohols, in general, have higher flavor thresholds than corresponding aldehydes (134). In addition, alcohols produced by this reaction can act as enzyme inhibitors in the raw food matrix, or they can be more readily extracted from this matrix because alcohols do not bind as readily to proteins (12). Also, conversion of aldehydes to their respective acids via aldehyde dehydrogenase should markedly eliminate the flavor caused by aldehydes. Combination of aldehyde dehydrogenase and diaphorase incubated with defatted-soybean extracts yielded soy beverage with greatly reduced green-beany flavor (135).

Flavor chemistry spans many scientific fields; solutions to the problems are still in their infancy and much of our understanding is still empirical. Much more research has yet to be done.

## Acknowledgment

I dedicate this chapter to my very special friend, Kit Lowder.

## References

1. Sessa, D. J., K. Warner and J. J. Rackis, J. Agric. Food Chem. 24:16 (1976).
2. Ismail, F., M. Vaisey-Genser and B. Fyfe, J. Food Sci. 46:1241 (1981).
3. Sessa, D. J., D. H. Honig and J. J. Rackis, Cereal Chem. 46:675 (1969).
4. Sessa, D. J., K. Warner and D. H. Honig, J. Food Sci. 39:69 (1974).
5. Arai, S., H. Suzuki, M. Fujimaki and Y. Sakurai, Agric. Biol. Chem.

30:863 (1966).

6. Kalbrener, J. E., K. Warner and A. C. Eldridge, Cereal Chem. 51:406 (1976).
7. Baur, C., W. Grosch, H. Weiser and H. Jugel, Z. Lebensm. Unters. Forsch. 164:171 (1977).
8. Dugan, Jr., L. R., and G. V. Rao, Technical Report 72-27-FL, U.S. Army Natick Lab., 1972.
9. Pokorný, J., G. Janicék and J. Davídek, Zesz. Probl. Postepow Nauk Roln. 167:155 (1975).
10. Pokorný, J., S. Klein and M. Zelinková, Nahrung 10:125 (1967).
11. Beyeler, M., and J. Solms, Lebensm. Wiss. Technol. 7:217 (1974).
12. Gremli, H. A., JAOCS 51:95A (1974).
13. Franzen, K. L., and J. E. Kinsella, J. Agric. Food Chem. 22:675 (1974).
14. Solms, J. F., Osman-Ismail and M. Beyeler, Can. Inst. Food Sci. Technol. J. 6:A10 (1973).
15. Arai, S., M. Noguchi, M. Yamashita, H. Kato and M. Fujimaki, Agric. Biol. Chem. 34:1569 (1970).
16. Anderson, R. L., and K. Warner, J. Food Sci. 41:293 (1976).
17. Yamauchi, F., Y. Kurosawa, Y. Kamata and K. Shibasaki, Agric. Biol. Chem. 44:2455 (1980).
18. Lea, C. H., and P. A. T. Swoboda, Chem. Ind. (London) 1289 (1958).
19. Parliment, T. H., and R. Scarpellino, J. Agric. Food Chem. 25:97 (1977).
20. Rackis, J. J., D. H. Honig, D. J. Sessa and F. Steggerda, Ibid. 18:977 (1970).
21. Honig, D. H., D. J. Sessa, R. L. Hoffman and J. J. Rackis, Food Technol. 23:95 (1969).
22. Pokorný, J., in Lipids/Technology, edited by R. Paoletti, G. Jacini and G. Porcellati, Raven Press, New York, Vol. 2, 1976, p. 475.
23. Patton, S., in Symposium on Foods: Lipids and Their Oxidation, edited by H. W. Schultz, E. A. Day and R. O. Sinnhuber, Avi Publishing Co., Inc., Westport, CT, 1962, p. 190.
24. Tims, M. J., and B. M. Watts, Food Technol. 12:240 (1958).
25. Jacobsen, M., and H. H. Koehler, J. Agric. Food Chem. 18:1069 (1970).
26. Wilson, B. R., A. M. Pearson and F. B. Shorland, Ibid. 24:7 (1976).
27. Younathan, M. T., and B. M. Watts, J. Food Sci. 25:538 (1960).
28. Zipser, M. W., and B. M. Watts, Food Technol. 15:318 (1961).
29. Hornstein, I., P. F. Crowe and M. J. Heimberg, J. Food Sci. 26:581 (1961).
30. Igene, J. O., and A. M. Pearson, Ibid. 44:1285 (1979).
31. Thomas, B., and M. Rothe, Getreide Mehl 4:17 (1954).

32. Rothe, M., Fette Seifen 55:877 (1953).
33. Popov, M. P., and D. A. Cheleev, Biokhim. Zerna, Shornik 5:263 (1960).
34. Weiss, K., and W. Diemair, Biochem. Z. 302:173 (1939).
35. Baur, C., and W. Grosch, Ibid. 165:82 (1977).
36. Moll, C., U. Biermann and W. Grosch, J. Agric. Food Chem. 27:239 (1979).
37. Usuki, R., and T. Kaneda, Agric. Biol. Chem. 44:2477 (1980).
38. Beckel, A. C., and A. K. Smith, Food Ind. 16:616 (1944).
39. Teeter, H. M., L. E. Gast, E. W. Bell, W. J. Schneider and J. C. Cowan, JAOCS 32:390 (1955).
40. Honig, D. H., J. J. Rackis and D. J. Sessa, J. Agric. Food Chem. 19:543 (1971).
41. Sessa, D. J., H. W. Gardner, R. Kleiman and D. Weisleder, Lipids 12:613 (1977).
42. Rackis, J. J., D. H. Honig, D. J. Sessa and H. A. Moser, Cereal Chem. 49:586 (1972).
43. Pinsky, A., S. Grossman and M. Trop, J. Food Sci. 36:571 (1971).
44. Sessa, D. J., and R. L. Anderson, J. Agric. Food Chem. 29:960 (1981).
45. Eriksson, C. J., Ibid. 23:126 (1973).
46. Christopher, J., E. Pistorius and B. Axelrod, Biochim. Biophys. Acta 198:12 (1970).
47. Brockman, R., and L. Acker, Lebensm. Wiss. Technol. 10:24 (1977).
48. Morrison, W. R., and R. Panpaprai, J. Sci. Food Agric. 26:1225 (1975).
49. Sessa, D. J., J. Agric. Food Chem. 27:234 (1979).
50. Wilson, R. F., and R. W. Rinne, Plant Physiol. 54:744 (1974).
51. Corliss, G. A., and L. R. Dugan, Jr., Lipids 5:846 (1970).
52. Dastoli, F. R., D. V. Lopiekes and A. R. Doig, Nature 218:884 (1968).
53. Beidler, L. M., J. Food Sci. 31:275 (1966).
54. Beidler, L. M., J. Gen. Physiol. 38:133 (1954).
55. Dastoli, F. R., Nature 223:524 (1969).
56. Dastoli, F. R., and S. Price, U.S. Patent 3,494,741 (1970).
57. Dastoli, F. R., D. V. Lopiekes and A. Doig, Abstract, Amer. Chem. Sci. 154th Meeting, Sept. 1967.
58. Kubota, T., and I. Kubo, Nature 223:97 (1969).
59. Shallenberger, R. S., Agric. Sci. Rev. 2:1 (1964).
60. Birch, G. G., and C. K. Lee, J. Food Sci. 41:1403 (1976).
61. Ney, K. H., Fette Seifen Anstrichm. 81:467 (1979).
62. Crawford, C. G., Fed. Proc. 40:1632 (1980).
63. Griffin, W. C., J. Soc. Cosmet. Chem. 1:311 (1949).
64. Lynch, M. J., and W. C. Griffin, in Emulsions and Emulsion Technology,

Part 1, edited by K. J. Lissant, Marcel Dekker, Inc., New York, 1974, p. 264.
65. Stone, H., and S. Oliver, J. Food Sci. 31:129 (1966).
66. Birch, G. G., and G. Ogunmoyela, Ibid. 45:981 (1980).
67. Tsai, L.-S., and L. M. Smith, Lipids 6:196 (1971).
68. Yukami, S., Agric. Biol. Chem. 36:871 (1972).
69. Keller, J. D., and J. E. Kinsella, J. Food Sci. 38:1200 (1973).
70. Love, J. D., and A. M. Pearson, J. Agric. Food Chem. 24:494 (1976).
71. Hirano, Y., and H. S. Olcott, JAOCS 48:523 (1971).
72. Kendricks, J., and B. M. Watts, Lipids 4:454 (1969).
73. Lew, Y. T., and A. L. Tappel, Food Technol. 10:285 (1956).
74. Liu, H.-p., and B. M. Watts, J. Food Sci. 35:596 (1970).
75. Younathan, M. T., and B. M. Watts, Food Res. 24:728 (1959).
76. Sato, K., and G. R. Hegarty, J. Food Sci. 36:1098 (1971).
77. Sato, K., G. R. Hegarty and H. K. Herring, Ibid. 38:398 (1973).
78. Love, J. D., and A. M. Pearson, J. Agric. Food Chem. 22:1032 (1974).
79. Igene, J. O., J. A. King, A. M. Pearson and J. I. Gray, Ibid. 27:838 (1979).
80. Pokorný, J., P.-T. Tài and G. Janícek, Nahrung 20:149 (1976).
81. Morita, M., and M. Fujimaki, Agric. Biol. Chem. 36:1751 (1972).
82. Corliss, G. A., thesis, Michigan State University, Ann Arbor, MI 1968.
83. Porter, N. A., R. A. Wolf and H. Weenen, Lipids 15:163 (1980).
84. Terao, J., Y. Hirota, M. Kawakatsu and S. Matsushita, Lipids 16:427 (1981).
85. Morita, M., and N. Tsushimi, Agric. Biol. Chem. 45:1413 (1981).
86. Mead, J. F., JAOCS 57:393 (1980).
87. Hatefi, Y., and W. G. Hanstein, Arch. Biochem. Biophys. 138:73 (1970).
88. Kaschnitz, R. M., and Y. Hatefi, Ibid. 171:292 (1975).
89. Tam, B. K., and P. B. McCay, J. Biol. Chem. 245:2295 (1970).
90. Svingen, B. A., J. A. Buege, F. O. O Neal and S. D. Aust, Ibid. 254:5892 (1979).
91. Labuza, T. P., Crit. Rev. Food Technol. 2:355 (1971).
92. Labuza, T. P., H. Tsuyuki and M. Karel, JAOCS 46:409 (1969).
93. Labuza, T. P., L. McNally, D. Gallagher, J. Hawkes and F. Hurtado, J. Food Sci. 37:154 (1972).
94. Salwin, H., in Freeze Drying of Foods, edited by F. R. Fisher, Natl. Acad. Sci., Natl. Research Council, Washington, D.C., 1962.
95. Brunauer, S., P. H. Emmett and E. Teller, J. Am. Chem. Soc. 60:309 (1938).
96. Haydar, M., and D. Hadziyev, J. Sci. Food Agric. 25:1285 (1974).
97. Chou, H.-E., K. M. Acott and T. P. Labuza, J. Food Sci. 38:316 (1973).

98. Blain, J. A., Recent Adv. Food Sci. 2:41 (1962).
99. Guardia, E., and G. J. Haas, J. Agric. Food Chem. 15:412 (1967).
100. Acker, L., Adv. Food Res. 11:263 (1962).
101. Brockmann, R., and L. Acker, Ann. Technol. Agric. 26:167 (1977).
102. Purr, A., Fette Seifen Anstrichm. 72:725 (1970).
103. Lück, E., Z. Lebensm. Unters. Forsch. 153:42 (1973).
104. Hansen, J. R., J. Agric. Food Chem. 24:1136 (1976).
105. Love, M. H., and L. R. Dugan, Jr., J. Food Sci. 43:89 (1978).
106. Karel, M., Abstract: Autoxidation Processes in Food and Related Biological Systems, Natick, MA, 1979.
107. Rackis, J. J., A. C. Eldridge, J. E. Kalbrener, D. H. Honig and D. J. Sessa, AIChE Symp. Ser. 69:5 (1974).
108. Baker, E. C., G. C. Mustakas and K. A. Warner, J. Agric. Food Chem. 27:969 (1979).
109. Honig, D. H., K. Warner and J. J. Rackis, J. Food Sci. 41:642 (1976).
110. Cater, C. M., K. C. Rhee, R. D. Hagenmaier and K. F. Mattil, JAOCS 51:137 (1974).
111. Lawhon, J. T., L. J. Manak, K. C. Rhee, K. S. Rhee and E. W. Lusas, Ibid. 46:912 (1981).
112. Friedrich, J. P., G. R. List, and A. J. Heaken, JAOCS 59:288 (1982).
113. Friedrich, J. P., and G. R. List, J. Agric. Food Chem. 30:192 (1982).
114. Mustakas, G. C., E. L. Griffin, Jr., L. E. Allen and O. B. Smith, JAOCS 41:607 (1964).
115. Mustakas, G. C., W. J. Albrecht, J. E. McGhee, L. T. Black and G. N. Bookwalter, Ibid. 46:623 (1969).
116. Eldridge, A. C., K. Warner and W. J. Wolf, Cereal Chem. 54:1229 (1977).
117. Hanley, J. W., C. W. Everson, L. Q. Ashworth and R. E. Morse, Food Technol. 7:429 (1953).
118. Lineweaver, H., J. D. Anderson and H. L. Hanson, Ibid. 6:1 (1952).
119. Greene, B. E., J. Food Sci. 34:110 (1969).
120. Tims, M. J., and B. M. Watts, Food Technol. 12:240 (1958).
121. Greene, B. E., I-M. Hsin and M. W. Zipser, J. Food Sci. 36:940 (1971).
122. Zipser, M. W., and B. M. Watts, Food Technol. 15:445 (1961).
123. Siddiqi, A. M., and A. L. Tappel, Arch. Biochem. Biophys. 60:91 (1956).
124. Naim, M., B. Gestetner, A. Bondi and Y. Birk, J. Agric. Food Chem. 24:1174 (1976).
125. Neill, J., and L. Page, Food Technol. 10:315 (1956).
126. Pratt, D. E., J. Food Sci. 37:322 (1972).
127. Hayes, R. E., G. N. Bookwalter and E. B. Bagley, Ibid. 42:1527.

128. Rhee, K. S., Y. A. Ziprin and K. C. Rhee, Ibid. 44:1132 (1979).
129. Evans, C. D., H. A. Moser, P. M. Cooney and J. E. Hodge, JAOCS 35:84 (1958).
130. Tarr, H. L. A., and N. E. Cooke, J. Fish. Res. Board Can. 7:522 (1949).
131. Shibamoto, T., Food Technol. 36:59 (1982).
132. Kinsella, J. E., and S. Damodaran, in The Analysis and Control of Less Desirable Flavors in Foods and Beverages, edited by G. Charalambous, Academic Press, New York, 1980, p. 95.
133. Arai, S., M. Noguchi, M. Koji, H. Kato and M. Fujimaki, Agric. Biol. Chem. 34:1420 (1970).
134. Eriksson, C. J., J. Agric. Food Chem. 23:126 (1975).
135. Takahashi, N., N. Kitabatake, R. Sasaki and H. Chiba, Agric. Biol. Chem. 44:1669 (1980).

# Chapter Sixteen

## Effect of Phospholipids on the Flavor Stability of Soybean Oil

**David B. Min**
**and**
**Triantafyllos G. Stasinopoulos**
Ohio State University
2121 Fyffe Rd.
Columbus, OH 43210

### Introduction

Removal of phospholipids from crude vegetable oils has long been recognized as an important factor influencing the keeping quality of the finished product. Phospholipids are present in soybean oil at 1.5–2.5% and are generally satisfactorily removed during oil processing (1). These compounds are removed from the crude oil and are processed to produce the commercial "lecithin" which is mainly a mixture of phosphatidylcholine, phosphatidylethanolamine, and phosphatidylinositol at an average ratio of 1:1:1 (2).

It has been reported that phospholipids are easily oxidized or discolored during heating or storage, and they may become precursors of off-flavors in the finished products (3, 4, 5). However, the literature on the effects of phospholipids on the flavor stability of oil quite often reports conflicting results due to the heterogeneity of the substrates and of the phospholipid fractions studied, as well as the variety of methods employed. It has been reported that phospholipids have an antioxidant activity in the vegetable oils due to the chelating effect or synergistic behavior of the phospholipids with other antioxidants.

The following is a literature review concerning the different functions of phospholipids in soybean oil flavor stability.

### Phospholipids as Chelating Agents

"Lecithin" was the first of a number of materials occurring naturally in edible products to be proposed as an antioxidant in foods (6).

Swift et al. (7) reported that purified phosphatidylethanolamine from cottonseed oil was a good antioxidant and increased the antioxygenic action of alpha-tocopherol. Dutton et al. (8) later suggested that the antioxygenic action of phospholipids was due partly to the metal-scavenging effect of the fat-soluble phosphoric acid. Dutton et al. (9) also found that deodorized soybean oil containing 0.02% added acetone precipitated phospholipids and 0.3 ppm of iron as ferric chloride had a superior flavor stability compared to the sample containing added 0.3 ppm of iron only. They further concluded that the concept of phospholipids functioning as a fat-soluble metal scavenger is strengthened by experiments involving the addition of alkyl acid phosphates such as butyl, octyl and lauryl esters. When these compounds were substituted for phospholipids, similar results were obtained and the metal-scavenging properties of the organic forms of phosphoric acid were apparent. However, the addition of phospholipids at concentrations higher than 0.1% was not recommended because of darkening of the oil.

Evans et al. (4) studied the flavor stability of soybean oil using fractionated commercial lecithin. They found that the addition of 0.1% of crude phospholipids, the alcohol-soluble fraction or the alcohol-insoluble fraction to the soybean oil prior to deodorization improved the oil's oxidative stability and initial flavor. However, all three additives caused significant darkening of the oil and the introduction of undesirable storage flavors when added at levels which improved the oxidative stability.

Privett and Quackenbush (10) found that wheat germ phospholipids reacted with fat peroxides to give complex polymers and prevented the accumulation of peroxides in fresh and oxidized wheat germ oil.

Chipault (6) attributed the antioxidant property of the phosphatidylethanolamine to the phosphoric acid group, and explained the inactivity of phosphatidylcholine by the fact that the latter compound exists mainly in a zwitterion form involving the hydroxyl group of the phosphoric acid radical and the nitrogen of the strongly basic choline. The nitrogen base of phosphatidylethanolamine is much weaker. As a result, the hydroxyl group of phosphoric acid remains mostly free. In his study with distilled fatty esters containing no phenolic antioxidants and metal contamination, he reported that phospholipids may have true antioxidant activity.

Many multivalent metals present in soybean oil are deleterious to its flavor stability. Among these metals, copper and iron are of high significance. Others such as zinc, lead, calcium, magnesium, aluminum, cobalt and chromium are less deleterious to the oil (3, 11). Evans et al. and Cowan et al. reported that phospholipids counteract the effect of metals added to vegetable oils, and undoubtedly act as inactivating agents for trace metals present in the original oils. Cowan et al. (11) were of the opinion that in the

1930s, researchers were not familiar with the metal-inactivating properties of the acidic compounds such as citric and phosphoric acids. When they were talking about a synergistic effect in the combined use of an acid such as citric acid and a phenolic antioxidant, the chelation property of citric acid had not yet been established with certainty. The recognition of the metal-inactivating capacity of phospholipids came around 1950. Cowan et al. (11) reported that all acidic compounds that are effective as antioxidants appear to have inactivating properties for trace metals.

Lunde et al. (12) transferred metal ions of Zn (+2) and Fe (+3) from water into vegetable oils containing phospholipids. They found that ions present in the oil may be exchanged with the same or with other kinds of cations in a water phase in contact with the oils. Using radioactive isotopes of Zn and Fe, they measured the exchange rates and established that the sequestering action in the oil strongly depends on the content of phospholipids and reaches a maximum in the concentration range between three and 60 ppm phospholipids in the oil. The same sequestering effect was demonstrated when purified phosphatidylethanolamine and acetone-refined soybean lecithin were added to edible oils from which the phospholipids had been removed. Free fatty acids do not appear to sequester multivalent ions nor to influence the sequestering action by the phospholipids when added to the oil.

Lunde et al (12) postulated that phosphatidylethanolamine shows a much reduced but still distinct sequestering action and concluded that phospholipids may be the active sequestering agents in the oils. However, they considered the possibility that other compounds, such as water present in the oil, have similar effects and affect sequestering action. Their findings opened the possibility of replacing catalytically-active elements with ions of less catalytic activity associated with the oxidation of edible oils.

Parke (13) concluded that phospholipids are antioxidants for soybean oil at low concentrations of five to ten ppm, but high levels of phospholipids are pro-oxidant. He also indicated that phospholipids may possess the ability to chelate metal ions.

Other investigations involving the contents of heavy metals in edible oils showed that the distribution and concentration of the metals are related to the phospholipid content in the oil. Lunde (14), using autoradiography and spectroscopy techniques, established that trace elements such as arsenic, zinc, bromine, sodium, copper, manganese and iron are distributed in vegetable oils in a way dependent on the phospholipid content in the oil. According to his work, a high phospholipid content seems to give a more homogeneous distribution of organic trace elements.

List et al. (15) reported that the apparent function of phosphoric acid during degumming of soybean oil is to alter the iron compounds to make them more readily removable in the subsequent processing steps.

List et al. (16) found that the phosphorus and iron removal percentages varied among the different processors. They also reported that phosphoric acid has a metal inactivating property. However, fats treated with phosphoric acid often develop melony and cucumber off-flavors. This limits its use in oil processing. List et al. (17) demonstrated that a thorough degumming is necessary to produce high quality oils. They established that an iron concentration of 0.26 ppm in the soybean oil had no significant effect on the flavor scores, but with iron concentration above 0.66 ppm, there was significant difference at the 1% confidence level.

Whether phospholipids chelate metals in crude oils is not clearly proven. The question is complicated by lack of knowledge concerning the structure(s) and the relative pro-oxidant properties native to the crude oil (18).

## Phospholipids as Synergists

Chipault (6) reported that phospholipids are synergists to other antioxidants. Despite the oxidative quality of certain phospholipids, crude commercial phospholipids of vegetable origin were among the first antioxidants to be used for fats and oils (19). Their effectiveness for this purpose is very limited as compared with phenolic oxidation inhibitors, but as natural food constituents, they have the advantage of being unobjectionable on toxicity grounds. Lea (19) also attributed this antioxidant activity mainly to the cephalin fraction which enhances the activity of phenolic antioxidants such as tocopherols. Lea postulated that the main mechanism of this action may be based on the inactivation of trace metals such as copper and iron.

Going (20) and Hudson and Mahgoub (21) found that refined and bleached oils are much more prone to develop peroxides than crude oil. The reason for this difference is unknown. It could be speculated that the phospholipids in crude oil provide an antioxidant function above that of tocopherols. They also found that even partially hydrogenated oil oxidizes more rapidly than the crude oil. List and Erickson (18) attributed the difference between refined-bleached oil and crude oil to crude oil's higher content of tocopherols which serve as natural antioxidants and to the higher content in phospholipids. The antioxidant, synergistic and metal-inactivating properties of soybean phospholipids have also been recognized by Evans (4). Thus the complete removal of phospholipids by degumming and refining may be responsible for decreased oxidative stability of soybean oil. However, the mechanism is open to speculation.

Sims (22) reported that phospholipids are not antioxidants when present alone in the oil, but some phospholipids are capable of reinforcing the action of tocopherols or other antioxidants of the phenolic type. Chipault (6) reported that the antioxidant activity of phospholipids is confined to the cephalin fraction.

Hudson and Mahgoub (21) reported that the two most ubiquitous phospholipids, phosphatidylcholine and phosphatidylethanolamine, although non-antioxidants in their own right, act as powerful synergists in conjunction with representatives of at least two classes of naturally occurring antioxidants, the tocopherols and the flavonols. It seems likely that phosphoric and citric acids work as synergists by chelating traces of pro-oxidant metals. Chelation is undoubtedly enhanced by the presence in most phospholipids of basic nitrogen with its ligand-forming potential. Whether phospholipids work as synergists to have any practical value in oil or fat stabilization remain uncertain. It is considered that phospholipids affect oil flavor adversely and cause foaming problems when oils are heated. However the presence of phospholipids, along with tocopherols, in crude vegetable oil may account, at least in part, for the fact that crude oils normally have superior flavor stability over refined oils.

## Phospholipids as Color and Flavor Precursors

Phospholipids have been shown to easily oxidize or discolor during heating or prolonged storage. It has been reported that phospholipids play a major role in the development of warmed-over flavors in products such as poultry, mutton, beef and pork (5, 23).

Evans et al. (4) reported that the addition of either crude or fractionated phospholipids to soybean oil prior to deodorization and at levels which improved the oxidative stability of the oil caused significant darkening and introduced undesirable storage flavors in the oil.

Goss (24) reported that the practice of phospholipid removal from soybean oil was originated to prevent the formation of sludge during the storage and shipment of crude oil. The primary purpose of phospholipid removal, however, was to produce lecithin to be used as an antispattering ingredient for margarine and, in conjunction with other treatments, to provide resistance to soybean oil reversion flavor.

In the earlier days, improperly refined soybean oil that contained significant amounts of phospholipids produced fishy flavors on heating. This type of deterioration would be expected from the high content of polyunsaturated fatty acids of phospholipids (25). However, oxidized amino compounds may also play a role in flavor reversion, and 0.7% of nitrogen-containing compounds have been isolated from reverted soybean oil (26). One of the possible precursors of nitrogen compounds are the phospholipids. Soybean phospholipids were purified by precipitation from acetone and prepared in granular form. Flavor compounds isolated from these phospholipids at room temperature contained 2.72% nitrogen.

List (27) believed that the removal of nearly all of the phospholipids is essential to produce a finished oil with optimum color and flavor stabilities. Evans et al. (4) reported that the presence of substantial amounts of phospholipids can lead to dark-colored oils and can also serve as precursors of off-flavors.

Lea (19) reported that some phospholipids seem to undergo a certain amount of spontaneous hydrolytic decomposition on standing or heating in addition to enzymatic hydrolysis. Oley (28) observed an extreme example of this in the case of synthetic distearoyl phosphatidic acid which decomposes completely into glycerophosphoric acid and free fatty acids when stored without precaution in the solid state, or when held for less than 24 hr in aqueous ethanol solution. Lea (19) also observed that phospholipids containing several different ester linkages are readily attacked by both acid and alkali. Virtanen and Tammelander (29) found that the pH of maximum stability for lecithin was 5.2.

Autoxidation of phospholipids by the action of atmospheric oxygen occurs on the unsaturated fatty acids (Lea, 19). Phospholipids in the crude state show an extreme readiness to decompose and become brown. The very highly unsaturated fatty acids in the phospholipid molecule are susceptible to oxidation and lead to the formation of strongly flavored decomposition products at a relatively early stage of the oxidation process.

Evans et al. (4) tested the activity of fractions prepared from commercial soybean lecithin in stabilizing refined soybean oil by adding them to the oil prior to deodorization. They found that amino-containing phospholipids caused very severe browning during heating, and even the crude phosphatidylcholine fraction could not be used at concentrations higher than 0.1% for this reason. At this level, it had a small but significant flavor stabilizing action. The high sugar-containing fractions had no effect on color or flavor stability. The presence of even as little as 0.1% phospholipids produced characteristic off-flavors when the fat finally oxidized. Phospholipids themselves, of course, can be protected by phenolic antioxidants such as nordihydroguaiaretic acid, or their oxidation can be accelerated by pro-oxidants such as copper, hemecompounds or ascorbic acid.

Lea (19), on the other hand, reported that several kinds of phospholipids, including phosphatidylethanolamine, phosphatidylserine and the ethanolamine and serine plasmalogens, contain free amino groups which can apparently undergo the same kind of Maillard type browning reactions with sugars and other aldehydes as do the amino groups of amino acid and proteins. Lipid browning reactions of this type have only been definitely recognized to date in dried eggs, but probably occur as well in other systems.

Scholfield et al. (30) maintained that the color of crude soybean lecithin is due partly to carotenoids and partly to brown products of lipid-browning type reactions between phospholipid free amino groups and aldehydes, occuring mainly during the heating stages of preparation. Neither removal of free sugar from the crude lecithin nor hydrogenation prevented darkening. For this reason, Sessa et al. (31) postulated that if phospholipids are not completely removed from the oil during refining, they may serve as one of the precursors of nitrogen compounds. In bitter-tasting soybean phosphatidylcholine, they identified a multitude of oxygenated products expected from the decomposition of linolenate and linoleate hydroperoxides. Therefore, they concluded that any small amounts of these phospholipids left in refined soybean oil would be expected to contribute to off-flavors by oxidation.

Phospholipids have been implicated as one of the causes of oxidation in soybean oil and palm oil (15, 32, 33, 34, 35).

Love and Pearson (23) found that on oxidation of egg phophatidylethanolamine dispersions at 38C for 48 hr, the 22:6, 20:4 and 23:3 polyunsaturated fatty acids had been completely oxidized, whereas some 18:2 polyunsaturated fatty acids still remained. This is in agreement with the findings of Holman and Elwer (36) who concluded that each additional double bond in a fatty acid increases the rate of oxidation by a factor of at least two.

According to works of Hornstein et al. (37), the phospholipids and the total lipids developed rancid odors in cooked meats much more quickly than the neutral lipids, a finding further supporting the conclusion of Wilson et al. (5) that phospholipids play a major role in development of warmed-over flavor in cooked meats.

Darkening of the oil has also been observed in oils other than soybean oil. Burchardt (38) reported that dark color formation was due to phospholipids during the processing of safflower oil. Phosphatidylethanolamine had a color-forming potential of four compared to one for phosphatidylcholine, yet total unsaturation was nearly equal for phosphatidilethanolamine and phosphatidylcholine.

Roubal (39) found that in model systems free of pro-oxidants, the ethanolamine derivative exhibited a high rate of oxidation due to its content in fatty acids rich in unsaturation.

Jewell (40) heated 1,2-dipalmitoyl-sn-glycerol-3-phospho-ethanolamine and tripalmitoglycerol separately, at 150C for one hour in air and found alkanes, ketones and lactones as major products in both cases. However, he observed qualitative and quantitative differences, with alkanals absent in the oxidized phospholipid. He concluded that this apparently was largely due to their reaction with the amino groups, as evidenced by thin layer chromatography. The phospholipid produced more 2-pentanol, 2-methyl-propanol and ethyl-hexadecanoate, smaller amounts of hydrocarbons and methylketones than with triacylglycerol.

Phosphatidylcholine was found to develop rancid flavors and bitter tastes during storage (31, 41). On the other hand, Lundberg (42) and Olcott and Venderveen (43) reported that phosphatidylcholine was less effective than phosphatidylethanolamine for the oxidative stability of oils.

Studies in aged, lean beef and lean pork by Hornstein et al. (37) revealed that phospholipids are among the more unstable constituents. They found that cephalin fractions from both beef and pork produced "fishy" odors, probably attributable to the high arachidonic acid content of these fractions. Lecithin and sphingomyelin fractions also resulted in the "fishy" odor although less pronounced than in the cephalin fraction. The same authors also observed that phospholipids exposed to air produced rancid flavors more quickly than in neutral fat, but phospholipids darkened more rapidly and more severely than in neutral lipids.

## Conclusion

A great number of research works on the functions of phospholipids in vegetable oils and other fat and oil systems has been carried out since 1940. Various methods, substrates, and phospholipid fractions have been used to identify the physical and chemical properties of phospholipids, and to substantiate the necessity for their removal because of a pro-oxidant behavior, or their addition as possible antioxidants in vegetable oils.

Phospholipids have been found to be pro-oxidants when present in oil systems due to their unsaturated fatty acid content and amino-containing moieties. As a result, off-flavors and colored oil products can be produced.

Some works have provided evidence that phospholipids have sequestering properties for the pro-oxidant metals present in the oils. Phospholipids are also considered as synergists to other antioxidants such as tocopherols and flavonols. However, their synergistic effect has not yet been entirely clarified. Some researchers attribute it to the phospholipid's chelating effect.

It finally seems that phospholipids have different physical and chemical properties which can result in low or high quality oils, depending on the phospholipid nature and content in the oil, the presence of other compounds and the oil system in which phospholipids exist.

## References

1. Pryde, H. E., in Handbook Soy Oil Processing and Utilization, edited by D. R. Erickson, E. H. Pryde, O. L. Brekke, T. L. Mounts and R. A. Falb, Am. Soybean Assn. and AOCS, 1980.
2. Brekke, O. L., in Handbook of Soy Oil Processing and Utilization, edited by D. R. Erickson, E. H. Pryde, O. L. Brekke, T. L. Mounts and R. A. Falb, Am. Soybean Assn. and AOCS, 1980.
3. Evans, C. D., A. W. Schwab, H. A. Moser, J. E. Hawley and E. H. Melvin, JAOCS 28:68 (1951).
4. Evans, C. D., P. M. Cooney, C. R. Scholfield and H. J. Hutton, Ibid. 31:295 (1954).
5. Wilson, B. R., A. M. Pearson and D. H. Shorland, J. Agric. Food Chem. 24:7 (1976).
6. Chipault, J. R., in Autoxidation and Antioxidants, edited by W. O. Lundberg, Interscience Publishers, N. Y., NY, (1962).
7. Swift, C. E., W. G. Rose and G. S. Jamieson, Oil and Soap 19:176 (1942).
8. Dutton, H. J., A. W. Schwab, H. A. Moser and J. C. Cowan, JAOCS 25:385 (1948).
9. Dutton, H. J., A. W. Schwab, H. A. Moser and J. C. Cowan, Ibid. 26:441 (1949).
10. Privett, O. S., and F. W. Quackenbush, Ibid. 31:169 (1954).
11. Cowan, J. C., P. M. Cooney and C. D. Evans, JAOCS 39:6 (1962).
12. Lunde, G., L. H. Landmark and J. Gether, JAOCS 54:207 (1977).
13. Parke, D. W., M. S. Thesis, Ohio State University, 1981.
14. Lunde, G., JAOCS 48:517 (1971).
15. List, G. R., C. D. Evans, L. T. Black and T. L. Mounts, JAOCS 55:275 (1978).
16. List, G. R., C. D. Evans, K. Warner, R. E. Beal, W. F. Kwolek, L. T. Black and K. J. Moulton, JAOCS 54:8 (1977).
17. List, G. R., T. L. Mounts and A. J. Heakin, JAOCS 55:280 (1978).
18. List, G. R., and D. R. Erickson, in Soy Oil Processing and Utilization, edited by D. R. Erickson, E. H. Pryde, O. L. Brekke, T. L. Mounts and R. A. Falb. AOCS and Am. Soybean Assn., 1980.
19. Lea, C. H., J. Sci. Food Agric. 8:1 (1957).
20. Going, L. H., JAOCS 45:632 (1968).
21. Hudson, B. J. F. and S. E. O. Mahgoub, J. Sci. Food Agric. 32:208 (1981).

22. Sims, R. P. A., in Autoxidation and Antioxidants, edited by W. O. Lundberg, Interscience Publishers, N.Y., NY, 1962.
23. Love, J. D., and A. M. Pearson, J. Agr. Food Chem. 22:1032 (1974).
24. Goss, W. H., Oil and Soap 23:241 (1946).
25. Frankel, E. N., in Handbook of Soybean Processing and Utilization, edited by D. R. Erickson et al., Am. Soybean Assn. and AOCS, 1980.
26. Chang, S. S., K. M. Brobst, H. Tai and C. E. Ireland, JAOCS 38:671 (1961).
27. List, G. R., in Soy Oil Processing and Utilization, edited by D. R. Erickson, E. H. Pryde, O. L. Brekke, T. L. Mounts and R. A. Falb, Am. Soybean Assn. and AOCS, 1980.
28. Oley, G., Chem. and Ind., pp. 1069, from Lea, C. H., J. Sci. Food Agric. 8:1 (1957).
29. Virtanen, A. I., and A. M. Tammelander, Ann. Acad. Sci. (from Lea, C. H., J. Sci. Food Agric. 8:1 (1957).
30. Scholfield, C. R., and H. J. Dutton, JAOCS 31:258 (1954).
31. Sessa, D. J., K. Warner and J. J. Rackis, J. Agric. Food Chem. 24:16 (1976).
32. List, G. R., T. L. Mounts, K. Warner and A. J. Heakin, JAOCS 55:277 (1978).
33. Guttfinger, T., and A. Letan, Ibid. 55:856 (1978).
34. Wiedermann, L. H., Ibid. 58:159 (1981).
35. Goh, S. H., H. T. Khor and P. T. Gee, Ibid. 59:296 (1982).
36. Holman, R. T., and O. C. Elwer, Ibid. 24:122 (1947).
37. Hornstein, I., P. F. Crowe and M. J. Heimberg, J. Food Sci. 26:581 (1961).
38. Burchardt, H. J., in Handbook of Soy Oil Processing and Utilization, edited by D. R. Erickson, E. H. Pryde, O. L. Brekke, T. L. Mounts and R. A. Falb, Am. Soybean Assn. and AOCS, 1980.
39. Roubal, W. T., JAOCS 44:325 (1967).
40. Jewel, N. E., JAOCS 57:398 (1980).
41. Sessa, D. J., K. Warner and J. J. Rackis, J. Food Sci. 39:69 (1974).
42. Lundberg, W. O., (ed.) Autoxidation and Antioxidants, Vol. 1, Interscience Publishers, N.Y., NY, 1961.
43. Olcott, H. S., and J. Vanderveen, J. Food Sci. 28:313 (1963).

# Subject Index

## D

## Q



## R